E. coli Plasmid Vectors

METHODS IN MOLECULAR BIOLOGY™

John M. Walker, SERIES EDITOR

250. **MAP Kinase Signaling Protocols,** edited by *Rony Seger, 2004*
249. **Cytokine Protocols,** edited by *Marc De Ley, 2004*
248. **Antibody Engineering:** *Methods and Protocols,* edited by *Benny K. C. Lo, 2004*
247. ***Drosophila* Cytogenetics Protocols,** edited by *Daryl S. Henderson, 2004*
246. **Gene Delivery to Mammalian Cells:** *Volume 2: Viral Gene Transfer Techniques,* edited by *William C. Heiser, 2004*
245. **Gene Delivery to Mammalian Cells:** *Volume 1: Nonviral Gene Transfer Techniques,* edited by *William C. Heiser, 2004*
244. **Protein Purification Protocols, Second Edition,** edited by *Paul Cutler, 2004*
243. **Chiral Separations:** *Methods and Protocols,* edited by *Gerald Gübitz and Martin G. Schmid, 2004*
242. **Atomic Force Microscopy:** *Methods and Applications,* edited by *Pier Carlo Braga and Davide Ricci, 2004*
241. **Cell Cycle Checkpoint Control Protocols,** edited by *Howard B. Lieberman, 2004*
240. **Mammalian Artificial Chromosomes:** *Methods and Protocols,* edited by *Vittorio Sgaramella and Sandro Eridani, 2003*
239. **Cell Migration in Inflammation and Immunity:** *Methods and Protocols,* edited by *Daniele D'Ambrosio and Francesco Sinigaglia, 2003*
238. **Biopolymer Methods in Tissue Engineering,** edited by *Anthony P. Hollander and Paul V. Hatton, 2003*
237. **G Protein Signaling:** *Methods and Protocols,* edited by *Alan V. Smrcka, 2003*
236. **Plant Functional Genomics**: *Methods and Protocols,* edited by *Erich Grotewold, 2003*
235. ***E. coli* Plasmid Vectors**: *Methods and Applications,* edited by *Nicola Casali and Andrew Preston, 2003*
234. **p53 Protocols,** edited by *Sumitra Deb and Swati Palit Deb, 2003*
233. **Protein Kinase C Protocols,** edited by *Alexandra C. Newton, 2003*
232. **Protein Misfolding and Disease:** *Principles and Protocols,* edited by *Peter Bross and Niels Gregersen, 2003*
231. **Directed Evolution Library Creation:** *Methods and Protocols,* edited by *Frances H. Arnold and George Georgiou, 2003*
230. **Directed Enzyme Evolution**: *Screening and Selection Methods,* edited by *Frances H. Arnold and George Georgiou, 2003*
229. **Lentivirus Gene Engineering Protocols,** edited by *Maurizio Federico, 2003*
228. **Membrane Protein Protocols**: *Expression, Purification, and Characterization,* edited by *Barry S. Selinsky, 2003*
227. **Membrane Transporters:** *Methods and Protocols,* edited by *Qing Yan, 2003*
226. **PCR Protocols,** *Second Edition,* edited by *John M. S. Bartlett and David Stirling, 2003*
225. **Inflammation Protocols,** edited by *Paul G. Winyard and Derek A. Willoughby, 2003*
224. **Functional Genomics:** *Methods and Protocols,* edited by *Michael J. Brownstein and Arkady B. Khodursky, 2003*
223. **Tumor Suppressor Genes:** *Volume 2: Regulation, Function, and Medicinal Applications,* edited by *Wafik S. El-Deiry, 2003*
222. **Tumor Suppressor Genes:** *Volume 1: Pathways and Isolation Strategies,* edited by *Wafik S. El-Deiry, 2003*
221. **Generation of cDNA Libraries:** *Methods and Protocols,* edited by *Shao-Yao Ying, 2003*
220. **Cancer Cytogenetics:** *Methods and Protocols,* edited by *John Swansbury, 2003*
219. **Cardiac Cell and Gene Transfer:** *Principles, Protocols, and Applications,* edited by *Joseph M. Metzger, 2003*
218. **Cancer Cell Signaling:** *Methods and Protocols,* edited by *David M. Terrian, 2003*
217. **Neurogenetics:** *Methods and Protocols,* edited by *Nicholas T. Potter, 2003*
216. **PCR Detection of Microbial Pathogens:** *Methods and Protocols,* edited by *Konrad Sachse and Joachim Frey, 2003*
215. **Cytokines and Colony Stimulating Factors:** *Methods and Protocols,* edited by *Dieter Körholz and Wieland Kiess, 2003*
214. **Superantigen Protocols,** edited by *Teresa Krakauer, 2003*
213. **Capillary Electrophoresis of Carbohydrates,** edited by *Pierre Thibault and Susumu Honda, 2003*
212. **Single Nucleotide Polymorphisms:** *Methods and Protocols,* edited by *Pui-Yan Kwok, 2003*
211. **Protein Sequencing Protocols, Second Edition,** edited by *Bryan John Smith, 2003*
210. **MHC Protocols,** edited by *Stephen H. Powis and Robert W. Vaughan, 2003*
209. **Transgenic Mouse Methods and Protocols,** edited by *Marten Hofker and Jan van Deursen, 2003*
208. **Peptide Nucleic Acids:** *Methods and Protocols,* edited by *Peter E. Nielsen, 2002*
207. **Recombinant Antibodies for Cancer Therapy:** *Methods and Protocols,* edited by *Martin Welschof and Jürgen Krauss, 2002*
206. **Endothelin Protocols,** edited by *Janet J. Maguire and Anthony P. Davenport, 2002*
205. ***E. coli* Gene Expression Protocols,** edited by *Peter E. Vaillancourt, 2002*
204. **Molecular Cytogenetics:** *Protocols and Applications,* edited by *Yao-Shan Fan, 2002*
203. ***In Situ* Detection of DNA Damage:** *Methods and Protocols,* edited by *Vladimir V. Didenko, 2002*
202. **Thyroid Hormone Receptors:** *Methods and Protocols,* edited by *Aria Baniahmad, 2002*
201. **Combinatorial Library Methods and Protocols,** edited by *Lisa B. English, 2002*
200. **DNA Methylation Protocols,** edited by *Ken I. Mills and Bernie H. Ramsahoye, 2002*
199. **Liposome Methods and Protocols,** edited by *Subhash C. Basu and Manju Basu, 2002*
198. **Neural Stem Cells:** *Methods and Protocols,* edited by *Tanja Zigova, Juan R. Sanchez-Ramos, and Paul R. Sanberg, 2002*
197. **Mitochondrial DNA:** *Methods and Protocols,* edited by *William C. Copeland, 2002*
196. **Oxidants and Antioxidants:** *Ultrastructure and Molecular Biology Protocols,* edited by *Donald Armstrong, 2002*
195. **Quantitative Trait Loci:** *Methods and Protocols,* edited by *Nicola J. Camp and Angela Cox, 2002*
113. **DNA Repair Protocols:** *Eukaryotic Systems,* edited by *Daryl S. Henderson, 1999*
112. **2-D Proteome Analysis Protocols,** edited by *Andrew J. Link, 1999*
111. **Plant Cell Culture Protocols,** edited by *Robert D. Hall, 1999*

METHODS IN MOLECULAR BIOLOGY™

E. coli Plasmid Vectors
Methods and Applications

Edited by

Nicola Casali
University of California at Berkeley, School of Public Health, Berkeley, CA

and

Andrew Preston
University of Cambridge, Department of Clinical Veterinary Medicine, Cambridge, UK

Humana Press ✷ Totowa, New Jersey

© 2003 Humana Press Inc.
999 Riverview Drive, Suite 208
Totowa, New Jersey 07512
humanapress.com

All rights reserved. No part of this book may be reproduced, stored in a retrieval system, or transmitted in any form or by any means, electronic, mechanical, photocopying, microfilming, recording, or otherwise without written permission from the Publisher. Methods in Molecular Biology™ is a trademark of The Humana Press Inc.

All papers, comments, opinions, conclusions, or recommendations are those of the author(s), and do not necessarily reflect the views of the publisher.

This publication is printed on acid-free paper. ∞
ANSI Z39.48-1984 (American Standards Institute) Permanence of Paper for Printed Library Materials.

Cover illustration provided by Nicola Casali.

Cover design by Patricia F. Cleary.

Production Editor: Wendy S. Kopf.

For additional copies, pricing for bulk purchases, and/or information about other Humana titles, contact Humana at the above address or at any of the following numbers: Tel.: 973-256-1699; Fax: 973-256-8341; E-mail: humana@humanapr.com; Website: http://humanapress.com

Photocopy Authorization Policy:
Authorization to photocopy items for internal or personal use, or the internal or personal use of specific clients, is granted by Humana Press, provided that the base fee of US $20.00 per copy, is paid directly to the Copyright Clearance Center at 222 Rosewood Drive, Danvers, MA 01923. For those organizations that have been granted a photocopy license from the CCC, a separate system of payment has been arranged and is acceptable to Humana Press Inc. The fee code for users of the Transactional Reporting Service is: [1-58829-151-0/03 $20.00].

Printed in the United States of America. 10 9 8 7 6 5 4 3 2 1

Library of Congress Cataloging in Publication Data

E. coli plasmid vectors : methods and applications / edited by Nicola Casali and Andrew Preston.
  p. cm. -- (Methods in molecular biology ; v. 235)
  Includes bibliographical references and index.
  ISBN 1-58829-151-0 (alk. paper); (e-ISBN) 1-59259-409-3
  ISSN 1064-3745
   1. Plasmids--Laboratory manuals. I. Casali, Nicola II. Preston, Andrew, Ph.D. III. Methods in molecular biology (Totowa, N.J.) ; v. 235.

QH452.6.E13 2003
572.8'69--dc21

2002191946

Preface

Plasmids are autonomously replicating extrachromosomal DNA molecules that are stably inherited and can be present at many copies per cell. These features, coupled to the ease with which plasmids can be manipulated to carry fragments of foreign DNA, have led to their exploitation as one of the critical workhorses of modern molecular biology.

E. coli *Plasmid Vectors* focuses on the manipulation of plasmids in *Escherichia coli*. The well-characterized genetics of this bacterium have established its role as the universal cloning host, and recombinant DNA manipulation is almost exclusively performed in this organism. Despite the astonishing advances in molecular biology technologies and applications witnessed in the last decade, the ability to clone a DNA fragment of interest into a recombinant plasmid vector, and to maintain and manipulate it in an *E. coli* host, remains the foundation of many genetic analyses.

E. coli *Plasmid Vectors* introduces relevant aspects of plasmid biology and describes the development of plasmid vectors. It also provides advice on choosing the right vector and a suitable host strain. The middle segment covers methods that are required to clone DNA into plasmid vectors, transform *E. coli*, and analyze recombinant clones. Protocols for the construction and screening of libraries are included, as well as specific techniques required for specialized cloning vehicles, such as cosmids, bacterial artificial chromosomes (BACs), λ vectors, and phagemids. The final section gives protocols for a variety of commonly used downstream applications. The value of *E. coli*-derived plasmid vectors in providing the means to study diverse organisms is evident from chapters describing the mutagenesis of foreign genes for reintroduction into the homologous host, the production of recombinant proteins, and the uses of reporter genes.

Commercial kits dominate many of these areas of molecular biology. Where pertinent, chapters include overviews of the methods that underpin these kits, give specific protocols for representative techniques, and include practical advice and tips for troubleshooting problems. In doing so, E. coli *Plasmid Vectors* provides not only a basic guide for those new to the field, but also a valuable resource for more experienced researchers.

Nicola Casali
Andrew Preston

Contents

Preface v
Contributors ix

1. The Function and Organization of Plasmids
   Finbarr Hayes 1
2. Choosing a Cloning Vector
   Andrew Preston 19
3. *Escherichia coli* Host Strains
   Nicola Casali 27
4. Chemical Transformation of *E. coli*
   W. Edward Swords 49
5. Electroporation of *E. coli*
   Claire A. Woodall 55
6. DNA Transfer by Bacterial Conjugation
   Claire A. Woodall 61
7. Cosmid Packaging and Infection of *E. coli*
   Mallory J. A. White and Wade A. Nichols 67
8. Isolation of Plasmids from *E. coli* by Alkaline Lysis
   Sabine Ehrt and Dirk Schnappinger 75
9. Isolation of Plasmids from *E. coli* by Boiling Lysis
   Sabine Ehrt and Dirk Schnappinger 79
10. High-Purity Plasmid Isolation Using Silica Oxide
    Stefan Grimm and Frank Voß-Neudecker 83
11. High-Throughput Plasmid Extraction Using Microtiter Plates
    Michael A. Quail 89
12. Isolation of Cosmid and BAC DNA from *E. coli*
    Daniel Sinnett and Alexandre Montpetit 99
13. Preparation of Single-Stranded DNA from Phagemid Vectors
    W. Edward Swords 103
14. Using Desktop Cloning Software to Plan, Track,
    and Evaluate Cloning Projects
    Robert H. Gross 107
15. Cloning in Plasmid Vectors
    Carey Pashley and Sharon Kendall 121

16	Extraction of DNA from Agarose Gels
	Nicholas Downey 137
17	Cloning PCR Products with T-Vectors
	Wade A. Nichols 141
18	Construction of Genomic Libraries in λ-Vectors
	Yilun Wang, Zheng Cao, Darryl Hood,
	and James G. Townsel 153
19	Rapid Screening of Recombinant Plasmids
	Sangwei Lu 169
20	Restriction Analysis of Recombinant Plasmids
	Joanne Goranson-Siekierke and Jarrod L. Erbe 175
21	Screening Recombinant DNA Libraries
	Wade A. Nichols 183
22	Sequencing Using Fluorescent-Labeled Nucleotides
	Allison F. Gillaspy 195
23	Site-Directed Mutagenesis Using the Megaprimer Method
	Zhidong Xu, Alessia Colosimo, and Dieter C. Gruenert 203
24	Site-Directed Mutagenesis by Inverse PCR
	Clifford N. Dominy and David W. Andrews 209
25	Creating Nested DNA Deletions Using Exonuclease III
	Rosamund Powles and Lafras M. Steyn 225
26	Transposon and Transposome Mutagenesis of Plasmids, Cosmids, and BACs
	Alistair McGregor 233
27	In Vitro Transcription and Translation
	Farahnaz Movahedzadeh, Susana González Rico,
	and Robert A. Cox 247
28	Vectors for the Expression of Recombinant Proteins in *E. coli*
	Sally A. Cantrell 257
29	Expression of Recombinant Proteins from *lac* Promoters
	Charles R. Sweet 277
30	Plasmid-Based Reporter Genes: *Assays for β-Galactosidase and Alkaline Phosphatase Activities*
	Minghsun Liu 289
31	Plasmid-Based Reporter Genes: *Assays for Green Fluorescent Protein*
	Sergei R. Doulatov 297
Index 305	

Contributors

DAVID W. ANDREWS • *Department of Biochemistry, McMaster University, Hamilton, ON, Canada*
SALLY A. CANTRELL • *Division of Infectious Diseases, School of Public Health, University of California, Berkeley, CA*
ZHENG CAO • *Department of Anatomy and Physiology, Meharry Medical College, Nashville, TN*
NICOLA CASALI • *Division of Infectious Diseases, School of Public Health, University of California, Berkeley, CA*
ALESSIA COLOSIMO • *University of Rome and CSS-Mendel Institute, Rome, Italy*
ROBERT A. COX • *National Institute for Medical Research, London, UK*
CLIFFORD N. DOMINY • *SYN-X Pharma Inc, Toronto, ON, Canada*
SERGEI R. DOULATOV • *Department of Microbiology, Immunology and Molecular Genetics, University of California, Los Angeles, CA*
NICHOLAS DOWNEY • *Department of Microbiology, Immunology and Molecular Genetics, David Geffen School of Medicine, University of California, Los Angeles, CA*
SABINE EHRT • *Department of Microbiology, Weill Medical College of Cornell University, New York, NY*
JARROD L. ERBE • *Department of Biology, Wisconsin Lutheran College, Milwaukee, WI*
ALLISON F. GILLASPY • *Department of Microbiology and Immunology, University of Oklahoma Health Sciences Center, Oklahoma City, OK*
JOANNE GORANSON-SIEKIERKE • *Department of Microbiology, University of Colorado Health Sciences Center, Denver, CO*
STEFAN GRIMM • *Max-Planck-Institute for Biochemistry, Martinsried, Germany*
ROBERT H. GROSS • *Department of Biological Sciences and Center for Biological and Biomedical Computing, Dartmouth College, Hanover, NH*
DIETER C. GRUENERT • *Department of Medicine, University of Vermont, Colchester, VT*
FINBARR HAYES • *Department of Biomolecular Sciences, University of Manchester Institute of Science and Technology, Manchester, UK*
DARRYL HOOD • *Department of Pharmacology, Meharry Medical College, Nashville, TN*

SHARON KENDALL • *Department of Pathology and Infectious Diseases, Royal Veterinary College, London, UK*
MINGHSUN LIU • *Department of Medicine, School of Medicine, University of Washington, Seattle, WA*
SANGWEI LU • *Division of Infectious Diseases, School of Public Health, University of California, Berkeley, CA*
ALISTAIR MCGREGOR • *Department of Molecular Genetics, University of Cincinnati, Cincinatti, OH*
ALEXANDRE MONTPETIT • *Service d'hématologie-oncologie, Centre de Cancérologie Charles Bruneau, Centre de Recherche, Hôpital Ste-Justine, Département de pédiatrie, Université de Montréal, Montréal, Canada*
FARAHNAZ MOVAHEDZADEH • *Department of Pathology and Infectious Diseases, Royal Veterinary College, London, UK*
WADE A. NICHOLS • *Department of Biological Sciences, Illinois State University, Normal, IL*
CAREY PASHLEY • *Department of Medical Microbiology, St. Bartholomew's and Royal London School of Medicine and Dentistry, London, UK*
ROSAMUND POWLES • *Division of Medical Microbiology, University of Cape Town, Rondebosch, South Africa*
ANDREW PRESTON • *Department of Clinical Veterinary Medicine, University of Cambridge, Cambridge, UK*
MICHAEL A. QUAIL • *Sanger Institute, The Wellcome Trust Genome Campus, Hinxton, UK*
SUSANA GONZÁLEZ RICO • *Sección de Bacteriología, Instituto de Medicina Tropical, Universidad Central de Venezuela, Caracas, Venezuela*
DIRK SCHNAPPINGER • *Department of Microbiology, Weill Medical College of Cornell University, New York, NY*
DANIEL SINNETT • *Service d'Hématologie-Oncologie, Centre de Cancérologie Charles Bruneau, Centre de Recherche, Hôpital Ste-Justine, Département de pédiatrie, Université de Montréal, Montréal, Canada*
LAFRAS M. STEYN • *Division of Medical Microbiology, University of Cape Town, Rondebosch, South Africa*
CHARLES R. SWEET • *Division of Infectious Disease, University of Massachusetts Medical Center, Worcester, MA*
W. EDWARD SWORDS • *Department of Microbiology and Immunology, Wake Forest University School of Medicine, Winston-Salem, NC*
JAMES G. TOWNSEL • *Department of Anatomy and Physiology, Meharry Medical College, Nashville, TN*

FRANK VOß-NEUDECKER • *Max-Planck-Institute for Biochemistry, Martinsried, Germany*
YILUN WANG • *Department of Biological Sciences, Tennessee State University, Nashville, TN*
MALLORY J. A. WHITE • *Department of Biological Sciences, Illinois State University, Normal, IL*
CLAIRE A. WOODALL • *Department of Clinical Veterinary Medicine, University of Cambridge, Cambridge, UK*
ZHIDONG XU • *Department of Surgery, Comprehensive Cancer Center, University of California, San Francisco, CA*

1

The Function and Organization of Plasmids

Finbarr Hayes

1. Introduction

In 1952, Joshua Lederberg coined the term *plasmid* to describe any bacterial genetic element that exists in an extrachromosomal state for at least part of its replication cycle *(1)*. As this description included bacterial viruses, the definition of what constitutes a plasmid was subsequently refined to describe exclusively or predominantly extrachromosomal genetic elements that replicate autonomously. Plasmids are now known to be present in most species of Eubacteria that have been examined, as well as in Archaea and lower Eukarya *(2)*.

Although most of the genetic material that directs the structure and function of a bacterial cell is contained within the chromosome, plasmids contribute significantly to bacterial genetic diversity and plasticity by encoding functions that might not be specified by the chromosome *(3)* (*see* **Subheading 3**). For example, antibiotic resistance genes are often plasmid-encoded, which allows the bacterium to persist in an antibiotic-containing environment, thereby providing the bacterium with a competitive advantage over antibiotic-sensitive species.

Under laboratory conditions, plasmids are generally not essential for the survival of the host bacterium and they have served as invaluable model systems for the study of processes such as DNA replication, segregation, conjugation, and evolution *(3)*. Moreover, ever since their utility was evinced by the first gene-cloning experiments in the early 1970s, plasmids have been pivotal to modern recombinant DNA technology as gene-cloning and gene-expression vehicles, among other uses *(4,5)*.

2. Basic Plasmid Characteristics
2.1. Size and Copy Number

Naturally occurring plasmids vary greatly in their physical properties, a few examples of which are shown in **Table 1**. They range in size from <2-kilobase pair (kbp) plasmids, which can be considered to be elements simply capable of replication, to

Table 1
Examples of Plasmids with Different Physical Characteristics

Plasmid	Host	Plasmid size (kbp)	Plasmid geometry	Plasmid copy number	Ref.
pUB110	*Bacillus subtilis*	2.3	Circular	20–50	*7*
ColE1	*Escherichia coli*	6.6	Circular	10–30	*9*
lp25	*Borrelia burgdorferi*	24.2	Linear	1–2	*6*
pNOB8	*Sulfolobus* sp.[a]	41.2	Circular	2–40	*10*
F	*Escherichia coli*	99.2	Circular	1–2	*11*
SCP1	*Streptomyces coelicolor*	350.0	Linear	4	*12*
pSymA	*Sinorhizobium meliloti*	1354.2	Circular	2–3	*8*

[a]Archaea.

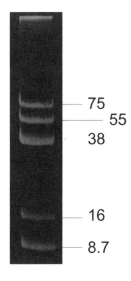

Fig. 1. Plasmid complement of a multiplasmid-containing strain of *Lactococcus lactis* analyzed by agarose gel electrophoresis. The approximate sizes of the plasmids are indicated (kbp).

megaplasmids that are many hundreds of kilobase pairs in size. At the upper end of this scale, the distinction between a megaplasmid and a minichromosome can become obscure. Some bacterial species simultaneously harbor *multiple* different plasmids that can contribute significantly to the overall genome size of the host bacterium (*see* **Fig. 1**) *(6,13)*. As an example, the symbiotic soil bacterium *Sinorhizobium meliloti* has three replicons (3.65, 1.68, and 1.35 megabase pairs [Mbp]) in addition to its chromosome (6.69 Mbp) *(8)*. The smallest megaplasmid, pSymA, can be cured from the host bacterium under laboratory conditions but provides nodulation and nitrogen-fixation functions that are important for the symbiotic interaction of the bacterium and its plant host.

Different plasmids have different *copy numbers* per chromosome equivalent. Some plasmids have a steady-state copy number of one or a few copies, whereas other, mainly small, plasmids are present at tens or even hundreds of copies per chromosome. The plasmid copy number is determined by replication control circuits that are discussed under **Subheading 4**, and in detail by del Solar and Espinosa *(14)*. Therefore, the contribution of plasmid DNA to the host bacterium's genome depends on the number of different plasmids that the bacterium harbors, as well as their size and copy number.

2.2. Geometry

Although most plasmids possess a circular *geometry*, there are now many examples in a variety of bacteria of plasmids that are linear *(15,16)*. As linear plasmids require specialized mechanisms to replicate their ends, which circular plasmids and chromosomes do not, linear plasmids tend to exist in bacteria that also have linear chromosomes *(17)*.

Circular plasmids can have more than one *topology* determined by the opposing actions of DNA gyrases and topoisomerases *(18)*. Plasmid DNA is mostly maintained in a covalently closed circular, supercoiled form (analogous to the behavior of an elastic band that is held fixed at one position while it is twisted at the 180° position). However, if a nick is introduced into one of the strands of the DNA double helix, supercoiling is relieved and the plasmid adopts an open circular form that migrates more slowly in an agarose gel than the covalently closed circular form. If nicks are introduced at opposite positions on both DNA strands, the plasmid is linearized. In addition, the activity of DNA homologous recombination enzymes can convert plasmid monomers to dimers and higher-order species that, because of their larger size, will migrate more slowly during agarose gel electrophoresis than the monomeric forms.

3. Plasmid-Encoded Traits

Many plasmids are phenotypically cryptic and provide no obvious benefit to their bacterial host other than the possible exclusion of plasmids that are incompatible with the resident plasmid (*see* Chapter 2). However, many other plasmids specify traits that allow the host to persist in environments that would otherwise be either lethal or restrictive for growth (*see* **Table 2**).

Antibiotic resistance is often plasmid encoded and can provide the plasmid-bearing host a competitive advantage over antibiotic-sensitive species in an antibiotic-containing environment such as the soil, where many antibiotic-producing micro-organisms reside, or a clinical environment where antibiotics are in frequent use *(35)*. Indeed, plasmid-encoded antibiotic resistance is of enormous impact to human health. The relative ease with which plasmids can be disseminated among bacteria, compared with chromosome-encoded traits, means that antibiotic resistance can spread rapidly and this has contributed to the dramatic clinical failure of many antibiotics in recent years. Furthermore, resistance genes may be located on transposable elements *(36)* within plasmids that can further promote the transmissibility of antibiotic resistance genes. In some instances, plasmids may harbor a number of genes encoding resistance to different antibiotics (multidrug resistance).

Table 2
Examples of Naturally Occurring Plasmids and Relevant Features

Plasmid	Host	Plasmid size (kbp)	Relevant feature	Ref.
pT181	*Staphylococcus aureus*	4.4	Tetracycline resistance	*19*
pRN1	*Sulfolobus islandicus*[a]	5.4	—	*20*
2µ	*Saccharomyces cerevisiae*[b]	6.3	—	*21*
ColE1	*Escherichia coli*	6.6	Colicin production and immunity	*9*
pMB1	*Escherichia coli*	8.5	*Eco*RI restriction–modification system	*22*
pGKL2	*Kluyveromyces lactis*[b]	13.5	Killer plasmid	*23*
pAMβ1	*Enterococcus faecalis*	26.0	Erythromycin resistance	*24*
pSK41	*Staphylococcus aureus*	46.4	Multidrug resistance	*25*
pBM4000	*Bacillus megaterium*	53.0	rRNA operon	*13*
pI258	*Staphylococcus aureus*	28.0	Metal ion resistance	*26*
pSLT	*Salmonella enterica* ssp. *typhimurium*	93.9	Virulence determinants	*27*
pMT1	*Yersinia pestis*	101.0	Virulence determinants	*28*
pADP-1	*Pseudomonas* sp.	108.8	Atrazine (herbicide) catabolism	*29*
pWW0	*Pseudomonas putida*	117.0	Aromatic hydrocarbon degradation	*30*
pBtoxis	*Bacillus thuringiensis* ssp. *israelensis*	137.0	Mosquito larval toxicity	*31*
pX01	*Bacillus anthracis*	181.7	Exotoxin production	*32*
pSOL1	*Clostridium acetobutylicum*	192.0	Solvent production	*33*
pSymB	*Sinorhizobium meliloti*	1683.3	Multiple functions associated with plant symbiosis	*34*

[a] Archaea.
[b] Eukarya (yeast).

Other plasmid-encoded traits also contribute to the persistence of the host bacterium in otherwise inhospitable environments. These include resistance to metal ions such as lead, mercuric, and zinc *(37)*, production of virulence factors that allow the bacterium to colonize hosts and survive host defenses *(38)*, and metabolic functions that allow utilization of different nutrients. The last trait includes the plasmid-mediated biodegradation of a variety of toxic substances such as toluene and other organic hydrocarbons, herbicides, and pesticides *(39)*. The production of plasmid-encoded bacteriocins to which other microorganisms are susceptible can give the plasmid-containing bacterium a competitive edge over other microorganisms in an ecological niche *(39a)*, as can plasmid-located genes for bacteriophage resistance and for the restriction of foreign nucleic acids which enter the cell. Conversely, plasmid-encoded antirestriction systems may protect plasmid DNA from degradation by host restriction enzymes when it first enters a new cell *(39b)*. The profound effects that plasmids can exert on bacterial behavior is sharply illustrated by the recent observation that *Bacillus cereus*, an opportunistic food-borne pathogen; *Bacillus thuringiensis*, a source of commercially useful insecticidal proteins; and *Bacillus anthracis*, the causative agent of anthrax, are mainly discriminated by their plasmids *(40)*.

4. Plasmid Replication

Plasmids, like chromosomes, are replicated during the bacterial cell cycle so that the new cells can each be provided with at least one plasmid copy at cell division *(41)*. To this end, plasmids have developed a number of strategies to initiate DNA replication but have mostly co-opted the host polymerization machinery *(42)* for subsequent stages of DNA synthesis, thereby minimizing the amount of plasmid-encoded information required for their replication. Small plasmids have been identified which consist of a replicon and very little extraneous DNA sequences *(42a)*. These, and other cryptic plasmids, can be viewed as purely selfish genetic elements as they apparently provide no advantage to their host. However, they may exclude related, invading plasmids from the host or may function as the core of lager plasmids which will evolve in the future. Large plasmids often contain multiple replicons dispersed at different locations on the plasmid or express different forms of a replication protein. These phenomena may reflect the different replication requirements of a plasmid that can exist in more than one bacterial host *(43)*.

4.1. Iteron-Containing Replicons

The genetic organization of a stylized plasmid replicon is illustrated in **Fig. 2A**. This replicon consists of a number of elements, including a gene for a plasmid-specific replication initiation protein (Rep), a series of directly repeated sequences (iterons), DnaA boxes, and an adjacent AT-rich region. The relative positions of the operator site, iterons, AT-rich stretch, and DnaA boxes can vary between replicons *(44)*. The numbers of iterons and DnaA boxes and the length of the AT-rich region can also differ.

Rep, which usually negatively autoregulates its own expression, binds to the iterons, which typically are 17–22 bp in length but vary in number and sequence between different replicons *(44)*. The spacing between shorter repeats is greater than that between longer repeats so the distance between equivalent positions within adjacent iterons is always approx 22 bp, corresponding to two turns of the DNA helix. Thus, when Rep proteins bind to the iterons, they are arrayed on the same face of the DNA helix. DnaA is a protein required for initiation of replication of the bacterial chromosome. It also performs a similar function in plasmid replication by binding to the DnaA boxes in the replicon *(45)*. The Rep-DnaA-DNA nucleoprotein complex promotes strand melting at the nearby AT-rich region to which host replication factors subsequently gain access and promote leading and lagging strand synthesis in a manner analogous to initiation of replication at the chromosomal origin, *oriC*.

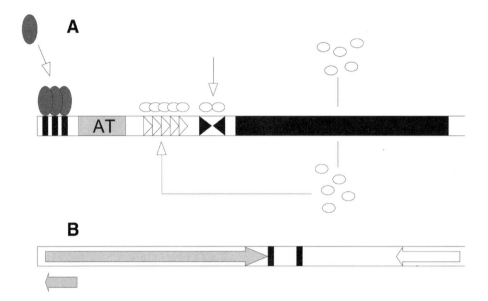

Fig. 2. The genetic organization of plasmid replicons. (**A**) The organization of a generic replicon that contains iterons. The stippled rectangle represents the *rep* gene whose protein product (ovals) binds both the directly repeated iterons (open triangles) and the operator site (filled triangles) upstream of *rep*. The filled boxes represent binding sites for host DnaA protein (shaded spheres). The AT-rich region is also indicated. (**B**) The organization of the ColE1 replicon. The leftward- and rightward-shaded arrows indicate the genes for the RNAI and RNAII transcripts, respectively. The open arrow represents the *rom* gene. The filled and hatched rectangles indicate the origin and primosome assembly sites, respectively.

Plasmid replication is a rigorously controlled process in part because plasmid overreplication would tax the metabolic capacity of the host cell and put the plasmid-bearing cell at a disadvantage compared to a plasmid-free counterpart. Plasmids control their copy number primarily at the stage of replication initiation. The frequency with which initiation of replication of iteron-containing plasmids occurs is modulated in part by sequestration of the origin region in nucleoprotein complexes and intermolecular pairing of complexes on different plasmids, which is referred to as "handcuffing" *(14,44)*.

4.2. ColE1-Type Replicons

The replicon of the ColE1 plasmid of *Escherichia coli* is the basis for many gene-cloning and gene-expression vectors that are commonly used in current molecular biology (*see* Chapters 2 and 28). In contrast to the replication of iteron-containing plasmids, ColE1 replication proceeds without a plasmid-encoded replication initiation protein and instead utilizes an RNA species in initiation and RNA–RNA interactions to achieve copy number control (*see* **Fig. 2B**) *(46)*.

ColE1 uses an extensive RNA primer for leading-strand synthesis. The RNAII preprimer is transcribed from the RNAII promoter by host RNA polymerase. RNAII forms a persistent RNA–DNA hybrid at the plasmid origin of replication. This hybrid is cleaved by RNase H and the resulting free 3'OH group on the cleaved RNAII acts as a primer for continuous leading-strand synthesis, catalyzed by host DNA polymerase I.

ColE1 regulates its copy number with a short RNA countertranscript, RNAI. This species is expressed constitutively from the strong RNAI promoter, is nontranslated, and is fully complementary to part of RNAII. The interaction of RNAI with RNAII results in an RNAII configuration that impairs further elongation of this transcript, thereby reducing the frequency of RNA–DNA duplex formation and initiation of replication. The RNAI–RNAII interaction is counterbalanced by the shorter half-life of RNAI compared to RNAII. The ColE1-encoded Rom protein (also known as Rop) increases the frequency of RNAI–RNAII interactions. The gene for Rom is deleted in many ColE1-based plasmid vectors, resulting in increased copy numbers compared to ColE1 itself. Perturbations of ColE1 plasmid copy number are rapidly mirrored by changes in RNAI concentration, resulting in the enhancement or suppression of replication and the maintenance of ColE1 copy number within a narrow window.

4.3. Rolling-Circle Replication

Many small (<10 kbp) plasmids of Gram-positive Eubacteria replicate by a rolling-circle mechanism, which is distinct from the replication of iteron-containing or ColE1-like plasmids (*see* **Fig. 3**) *(47)*. Rolling-circle plasmids have also been identified in Gram-negative Eubacteria and in Archaea. Some bacteriophage, including M13 of *E. coli*, also replicate in this way.

In rolling-circle replication, binding of a plasmid-encoded replication protein to the leading-strand origin (also known as the double-strand origin) distorts the DNA in this region and exposes a single-stranded region in an extruded cruciform. A nick is introduced at this site by the replication protein and this exposes a 3'OH group from which the leading strand is synthesized by DNA polymerase III. Leading strand initiation differs between rolling circle plasmids, procaryotic chromosomes, and other plasmids, although chain elongation is similar in all systems. As the leading strand is synthesized, the nontemplate strand of the old plasmid is displaced ahead of the replication fork until, eventually, it is removed entirely. The resulting single-stranded intermediate is characteristic of rolling-circle replication and its identification provides evidence that a plasmid replicates by this mechanism *(48)*. The lagging-strand origin (also known as the single-strand origin) is exposed on the displaced single-stranded intermediate and lagging-strand initiation commences at this origin using host replication factors. RNA polymerase synthesizes RNA primers at the lagging strand origin. DNA polymerase I initiates lagging strand synthesis from these RNA primers, after which DNA polymerase III continues elongation.

5. Plasmid Segregation

DNA replication produces precise plasmid copies, but plasmids must also ensure that they are distributed to both daughter cells during bacterial cell division. If the

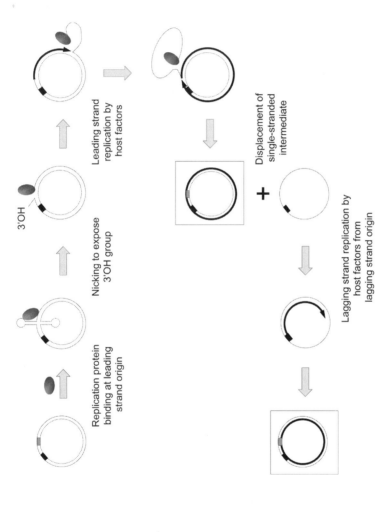

Fig. 3. Replication of rolling-circle plasmids. The two DNA strands of the plasmid are shown as solid and dotted lines. Newly replicated strands are shown as thick lines. Filled and shaded boxes represent the lagging-strand and leading-strand origins, respectively. The plasmid-encoded replication protein is shown as an oval. The replication protein nicks at a specific site in this region exposing a 3'OH group, which host replication factors use to initiate leading-strand synthesis. Synthesis of the leading strand displaces the nontemplate strand from the plasmid and forms the typical single-strand intermediate. The lagging-strand origin on this intermediate serves as an initiation site for RNA-primed synthesis of the complementary strand. The two double-stranded products of rolling-circle replication are boxed.

steady-state copy number of a plasmid is sufficiently high, it is easy to envisage how passive diffusion of these copies might be sufficient to ensure that each daughter cell acquires at least one copy of the plasmid when the cell divides. Plasmid copy number control circuits subsequently modulate the numbers of plasmid copies in the daughter cells to normal levels in preparation for the next round of cell division. Although it is still considered likely that random diffusion is sufficient for the stable inheritance of moderate- or high-copy-number plasmids, recent evidence suggests that these plasmids might not be entirely free to disperse through the cytoplasm but, instead, might be compartmentalized into subcellular regions from which the plasmids are distributed equitably *(49)*. The mechanism for this is unknown.

In contrast to high-copy-number plasmids, plasmids with a copy number of one or a few have evolved specific strategies to guarantee their faithful inheritance, which cannot be achieved by random diffusion.

5.1. Active Partition Systems

Following plasmid replication, active partitioning systems position the plasmids appropriately within the cell such that at cell division, each of the new cells acquires at least one copy of the plasmid (*see* **Fig. 4**). The most well studied active partition system is, arguably, that of the P1 plasmid in *E. coli* *(50,51)*. The plasmid located components of this system are organized in a cassette that consists of an autoregulated operon containing the *parA* and *parB* genes and a downstream *cis*-acting sequence, *parS*. The ParA and ParB proteins and a host protein, integration host factor, form a nucleoprotein complex at *parS* that is presumed to interact with an unknown host partitioning apparatus. This complex guides the tethered P1 plasmid copies to the one-quarter and three-quarter cell-length positions following replication at the midcell. The plasmids remain at these positions as the bacterial cell elongates. When the cell divides at its center the plasmids are again at the midpoint positions of the new cells and the cycles of replication and partition are repeated (*see* **Fig. 4**).

Active partition systems are widely distributed among low-copy-number bacterial plasmids and homologous systems are likely to be implicated in chromosome partition in many bacteria *(52)*.

5.2. Site-Specific Recombination

Many laboratory strains of *E. coli* have been mutated to be deficient in homologous recombination. This reduces the frequency with which genes cloned in multicopy plasmids undergo rearrangements in these strains. In contrast, most wild-type bacteria are recombination proficient and this is critical for bacterial DNA repair and evolution *(53)*. As plasmid copies are identical, homologous recombination in wild-type bacteria can convert plasmid monomers to dimers or higher-order species. The complete dimerization of a plasmid population within a cell will halve the number of plasmids available for partition at cell division and thereby contribute to plasmid segregational instability. Furthermore, because dimers have two replication origins, they may be more favored for replication than plasmids with a single origin, which may further skew intracellular plasmid distribution toward dimeric forms. The formation of trimers and other multimers will have an even more profound effect on plasmid segregation *(54)*.

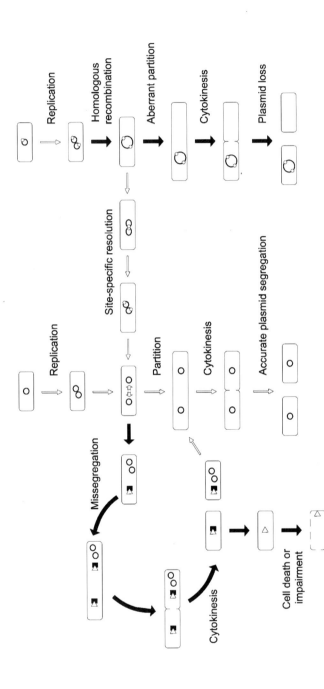

Fig. 4. The contribution of accessory stability mechanisms to plasmid maintenance. Active partitioning systems position replicated plasmids (large circles) appropriately within the elongating bacterial cell (rounded rectangles) such that each daughter cell receives a plasmid copy after cell division (**center**). If homologous recombination dimerizes two plasmid copies, the plasmids cannot be distributed equitably at cell division and this may lead to plasmid loss (**right**). However, site-specific recombination at plasmid recombination sites (small circles) can resolve the plasmid dimer to monomers that can now be partitioned accurately. If a plasmid-free cell arises because of missegregation or a defect in replication, toxin–antitoxin systems can kill or impair the growth of the plasmid-free cell specifically (**left**). The plasmid-encoded toxin (open triangle) is efficiently sequestered by an antitoxin (filled rectangle) in the plasmid-containing cell. In the plasmid-free derivative, the antitoxin is more susceptible to degradation by host enzymes than the toxin, so that the latter is eventually liberated from the former and can poison the host. Open and filled arrows indicate productive and nonproductive steps, respectively, in accurate plasmid segregation. For clarity, the host chromosome is not depicted in this representation.

Both high- and low-copy-number plasmids commonly solve this problem by using site-specific recombination to resolve dimers to monomers (*see* **Fig. 4**). This process involves site-specific recombinases that bind to specific recombination sites on both monomer copies within the plasmid dimer and form a synaptic complex in which the two recombination sites are brought in to close proximity *(55)*. The site-specific recombinases cleave DNA strands within this complex and promote strand exchange between the two sites and results in the monomerization of the plasmid dimer. The XerC and XerD site-specific recombinases encoded by most bacterial chromosomes are involved in the resolution of dimeric forms of many plasmids, including the ColE1 plasmid, and bacterial chromosomes *(56)*.

5.3. Toxin–Antitoxin Systems

An additional mechanism which plasmids use to favor their maintenance in bacterial populations involves the killing or growth impairment of cells that fail to acquire a copy of the plasmid. This has variously been referred to as postsegregational cell killing, plasmid addiction, or toxin–antitoxin systems *(57–60)*. This mechanism involves a plasmid-encoded protein toxin and antitoxin. The antitoxin, which may be either a protein or a nontranslated RNA, neutralizes the toxin by either binding to the toxin protein or by inhibiting its translation. The antitoxin is more susceptible to degradation by host enzymes than the toxin and, thus, replenishment of the antitoxin levels by the presence of plasmid is required to prevent toxin action. When a plasmid-free derivative arises (e.g., as a result of a replication or partitioning defect), the toxin is subsequently liberated to interact with an intracellular target and cause either death or a growth disadvantage of the plasmid-free cell (*see* **Fig. 4**). In the case of the CcdAB toxin–antitoxin system encoded by the F plasmid, the toxin (CcdB) is a DNA gyrase poison. It both entraps a cleavage complex between gyrase and DNA and associates with DNA gyrase to produce a complex that is impaired in supercoiling activity *(61)*. These combined effects are lethal for *E. coli*.

A variety of different toxin–antitoxin systems are widely disseminated on bacterial plasmids, although the intracellular targets for the toxin components of these systems probably differ. Toxin–antitoxin homologs have also been identified on many bacterial chromosomes, where they might function as bacterial programmed cell death systems during periods of nutritional and other stresses *(57)*.

Large, low-copy-number plasmids often utilize partition, recombination, and toxin–antitoxin systems to promote segregational stability. The segregational maintenance of these plasmids is achieved through the activity of all three mechanisms.

6. Plasmid Dissemination in Bacterial Populations

Certain bacterial species can achieve a state of natural competence for the uptake of naked plasmid DNA (transformation) *(62)*, or can acquire DNA that has been packaged into a bacteriophage head and is injected into the host (transduction) *(63)*. However, the conjugative transfer of DNA between donor and recipient cells is probably the most common mechanism by which plasmids are disseminated in bacterial populations *(64,65)*. A wide variety of phenotypes can be conferred by conjugative plasmids, including antibiotic resistance, bacteriocin production and immunity, and catabolic functions.

Conjugative plasmids have been identified in most major groups of Eubacteria, and more recently in Archaea *(66)*. Furthermore, conjugative plasmid transfer is not limited to closely related bacteria but has also been demonstrated between evolutionary–divergent Gram-negative and Gram-positive Eubacteria *(67)*, and from Eubacteria to yeast *(68)*. The T-DNA region of the Ti virulence plasmid of the Gram-negative bacterium *Agrobacterium tumefaciens* is also transferred by a conjugation-like process to susceptible plant hosts, where it integrates in the plant genome and induces the formation of crown gall tumors *(69)*.

Conjugation is mediated by cell-to-cell contact between the donor and recipient. Plasmid DNA is usually transferred through a tube-like structure known as a pilus, which is extruded by the donor and physically connects to the recipient cell. In the Gram-positive bacterium *Enterococcus*, this cell-to-cell contact is promoted by plasmid-encoded aggregation substances that are induced in response to sex pheromones excreted by the recipient cell *(70)*. As a large number of genes may be required for the conjugation process and these genes reside on the conjugative plasmid itself, small plasmids are usually not self-transmissible. Nevertheless, small plasmids that encode relaxase enzymes, which perform the initial nicking reactions at their cognate plasmid origins of transfer (*oriT*), can undergo conjugative mobilization if other conjugation functions are provided *in trans* by a helper plasmid within the cell *(71)*.

Conjugative transfer of the F Plasmid is one of the best-characterized conjugation processes. In this system, the propilin protein encoded by the *traA* gene is processed by host-encoded leader peptidase into the pilin product. The latter is inserted into the inner cell membrane with the aid of a transfer-specific chaperone protein, TraQ *(71a)*. Pilin in the membrane is organized into the extracellular pilus filament through the action of a number of assembly proteins encoded by the F plasmid. Plasmid DNA conjugation involves the transfer of only one strand of the plasmid DNA between the donor and recipient cells. Following transfer, the two single strands act as templates for synthesis of the complementary strands by the DNA replication machinery in both donor and recipient cells. In the case of the F plasmid, a relaxase enzyme, TraI, nicks one DNA strand in the relaxosome complex assembled at *oriT*. TraI is also a helicase which unwinds the two strands after nicking. The nicked strand is transferred through the pilus to the recipient cell where its ends are religated. Following F plasmid transfer, the plasmid-specific TraT and TraS proteins inhibit a second transfer event to the recipient by impeding mating pair stabilization (surface exclusion) and by preventing DNA transfer (entry exclusion), respectively.

7. Plasmid Evolution: Plasmids Are Modular Elements

Whole genome and plasmid-specific sequencing projects have recently begun to provide fascinating glimpses into the genetic organization and evolution of plasmids. These studies have revealed that plasmids, particularly large plasmids, are commonly constructed in a modular fashion by the recombination activities of transposons, insertion sequences, bacteriophages, and smaller plasmids *(72)*. For example, the backbone

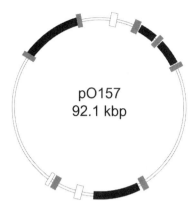

Fig. 5. Simplified representation of the relative distribution of transposable elements (gray boxes) and putative virulence genes (filled arcs) on the pO157 virulence plasmid of *E. coli*. Replicons, one of which is apparently interrupted, are shown as white boxes. For clarity, the locations of partition, conjugation and other genes are not shown. (Adapted from **ref. 73**.)

of the 92-kbp virulence plasmid of *E. coli* O157 bears a striking resemblance to that of the F plasmid. However, this backbone is interrupted by a number of regions containing putative virulence genes (*see* **Fig. 5**) *(73)*. These virulence patches are framed by intact insertion sequences or insertion-sequence remnants, suggesting that an ancestral plasmid related to F was colonized successively by a number of mobile elements conferring different virulence functions. Similarly, the mosaic structure of the 46.4-kbp multidrug-resistance plasmid pSK41 from *Staphylococcus aureus* suggests that this plasmid has acquired its many resistance determinants by the insertion of both transposable elements and smaller plasmids into a conjugative progenitor plasmid *(25)*. A large pathogenicity island bounded by insertion-sequence elements represents one-quarter of the 181.7-kbp virulence plasmid, pXO1, of *B. anthracis* and appears to have been acquired through transposition *(32)*. This plasmid also harbors numerous other insertion sequences indicative of a highly recombinogenic history. These and many other recent examples indicate that large plasmids have evolved by accumulating additional genetic functions through successive, independent recombination events that are frequently mediated by transposable elements. The serial acquisition of virulence, antibiotic resistance and other determinants by plasmids allows the hosts that harbor them to invade and persist in increasingly hostile niches.

Other examples of the modular organization of plasmids include the frequent close association of plasmid replication and maintenance cassettes and the clustering of genes for conjugation functions in specific plasmid regions. Over time, common control circuits have developed in some plasmids that coordinate these core activities of replication, maintenance, and transfer *(74)*. The continued molecular dissection of plasmids, both at the genomic level and in finer detail concerning the molecular function of specific systems, will undoubtedly prove as exciting and informative in the immediate future as the analysis of these versatile elements has proven in the last half century.

Acknowledgments

Work in the author's laboratory is supported by grants from the Biotechnology and Biological Sciences Research Council and by the Wellcome Trust.

References

1. Lederberg, J. (1952) Cell genetics and hereditary symbiosis. *Physiol. Rev.* **32**, 403–430.
2. Summers, D. K. (1996) *The Biology of Plasmids*, Blackwell Science, Oxford.
3. Thomas, C. M. (ed.) (2000) *The Horizontal Gene Pool*. Harwood Academic, Amsterdam.
4. Cohen, S. N. (1993) Bacterial plasmids: their extraordinary contribution to molecular genetics. *Gene* **135**, 67–76.
5. Lederberg, J. (1998) Plasmid (1952–1997). *Plasmid* **39**, 1–9.
6. Casjens, S., Palmer, N., van Vugt, R., et al. (2000) A bacterial genome in flux: the twelve linear and nine circular extrachromosomal DNAs in an infectious isolate of the Lyme disease spirochete *Borrelia burgdorferi*. *Mol. Microbiol.* **35**, 490–516.
7. Muller, R. E., Ano, T., Imanaka, T., et al. (1986) Complete nucleotide sequences of *Bacillus* plasmids pUB110dB, pRBH1 and its copy mutants. *Mol. Gen. Genet.* **202**, 169–171.
8. Barnett, M. J., Fisher, R. F., Jones, T., et al. (2001) Nucleotide sequence and predicted functions of the entire *Sinorhizobium meliloti* pSymA megaplasmid. *Proc. Natl. Acad. Sci. USA* **98**, 9883–9888.
9. Chan, P. T., Ohmori, H., Tomizawa, J., et al. (1985) Nucleotide sequence and gene organization of ColE1 DNA. *J. Biol. Chem.* **260**, 8925–8935.
10. She, Q., Phan, H., Garrett, R. A., et al. (1998) Genetic profile of pNOB8 from *Sulfolobus*: the first conjugative plasmid from an archaeon. *Extremophiles* **2**, 417–425.
11. Frost, L. S., Ippen-Ihler, K., and Skurray, R.A. (1994) Analysis of the sequence and gene products of the transfer region of the F sex factor. *Microbiol. Rev.* **58**, 162–210.
12. Yamasaki, M., Miyashita, K., Cullum, J., et al. (2000) A complex insertion sequence cluster at a point of interaction between the linear plasmid SCP1 and the linear chromosome of *Streptomyces coelicolor* A3(2). *J. Bacteriol.* **182**, 3104–3110.
13. Kunnimalaiyaan, M., Stevenson, D. M., Zhou, Y., et al. (2001) Analysis of the replicon region and identification of an rRNA operon on pBM400 of *Bacillus megaterium* QM B1551. *Mol. Microbiol.* **39**, 1010–1102.
14. del Solar, G. and Espinosa, M. (2000) Plasmid copy number control: an ever-growing story. *Mol. Microbiol.* **37**, 492–500.
15. Kobryn, K. and Chaconas, G. (2001) The circle is broken: telomere resolution in linear replicons. *Curr. Opin. Microbiol.* **4**, 558–564.
16. Meinhardt, F., Schaffrath, R., and Larsen, M. (1997) Microbial linear plasmids. *Appl. Microbiol. Biotechnol.* **47**, 329–336.
17. Volff, J. N. and Altenbuchner, J. (2000) A new beginning with new ends: linearisation of circular chromosomes during bacterial evolution. *FEMS Microbiol. Lett.* **186**, 143–150.
18. Champoux, J. J. (2001) DNA topoisomerases: structure, function, and mechanism. *Annu. Rev. Biochem.* **70**, 369–413.
19. Khan, S. A. and Novick, R. P. (1983) Complete nucleotide sequence of pT181, a tetracycline-resistance plasmid from *Staphylococcus aureus*. *Plasmid* **10**, 251–259.
20. Peng, X., Holz, I., Zillig, W., et al. (2000) Evolution of the family of pRN plasmids and their integrase-mediated insertion into the chromosome of the crenarchaeon *Sulfolobus solfataricus*. *J. Mol. Biol.* **303**, 449–454.

21. Futcher, A. B. (1988) The 2μ circle plasmid of *Saccharomyces cerevisiae*. *Yeast* **4**, 27–40.
22. Betlach, M., Hershfield, V., Chow, L., et al. (1976). A restriction endonuclease analysis of the bacterial plasmid controlling the *Eco*RI restriction and modification of DNA. *Fed. Proc.* **35**, 2037–2043.
23. Schickel, J., Helmig, C., and Meinhardt, F. (1996) *Kluyveromyces lactis* killer system: analysis of cytoplasmic promoters of the linear plasmids. *Nucleic Acids Res.* **24**, 1879–1886.
24. Dunny, G. M. and Clewell, D. B. (1975) Transmissible toxin (hemolysin) plasmid in *Streptococcus faecalis* and its mobilization of a noninfectious drug resistance plasmid. *J. Bacteriol.* **124**, 784–790.
25. Berg, T., Firth, N., Apisiridej, S., et al. (1998) Complete nucleotide sequence of pSK41: evolution of staphylococcal conjugative multiresistance plasmids. *J. Bacteriol.* **180**, 4350–4359.
26. Silver, S. and Phung, L. T. (1996) Bacterial heavy metal resistance: new surprises. *Annu. Rev. Microbiol.* **50**, 753–789.
27. McClelland, M., Sanderson, K. E., Spieth, J., et al. (2001) Complete genome sequence of *Salmonella enterica* serovar *typhimurium* LT2. *Nature* **413**, 852–856.
28. Lindler, L. E., Plano, G. V., Burland, V., et al. (1998) Complete DNA sequence and detailed analysis of the *Yersinia pestis* KIM5 plasmid encoding murine toxin and capsular antigen. *Infect. Immun.* **66**, 5731–5742.
29. Martinez, B., Tomkins, J., Wackett, L. P., et al. (2001) Complete nucleotide sequence and organization of the atrazine catabolic plasmid pADP-1 from *Pseudomonas* sp. strain ADP. *J. Bacteriol.* **183**, 5684–5697.
30. Bertoni, G., Perez-Martin, J., and de Lorenzo, V. (1997) Genetic evidence of separate repressor and activator activities of the XylR regulator of the TOL plasmid, pWW0, of *Pseudomonas putida*. *Mol. Microbiol.* **23**, 1221–1227.
31. Ben-Dov, E., Nissan, G., Pelleg, N., et al. (1999) Refined, circular restriction map of the *Bacillus thuringiensis* subsp. *israelensis* plasmid carrying the mosquito larvicidal genes. *Plasmid* **42**, 186–191.
32. Okinaka, R. T., Cloud, K., Hampton, O., et al. (1999) Sequence and organization of pXO1, the large *Bacillus anthracis* plasmid harboring the anthrax toxin genes. *J. Bacteriol.* **181**, 6509–6515.
33. Nolling, J., Breton, G., Omelchenko, M. V., et al. (2001) Genome sequence and comparative analysis of the solvent-producing bacterium *Clostridium acetobutylicum*. *J. Bacteriol.* **183**, 4823–4838.
34. Finan, T. M., Weidner, S., Wong, K., et al. (2001) The complete sequence of the 1,683-kb pSymB megaplasmid from the N_2-fixing endosymbiont *Sinorhizobium meliloti*. *Proc. Natl Acad. Sci. USA* **98**, 9889–9894.
35. Bush, K. and Miller, G. H. (1998) Bacterial enzymatic resistance: β-lactamases and aminoglycoside-modifying enzymes. *Curr. Opin. Microbiol.* **1**, 509–515.
36. Craig, N. L., Craigie, R., Gellert, M., and Lambowitz, A. M. (eds.) (2002) *Mobile DNA II*, American Society for Microbiology, Washington DC.
37. Silver, S. (1996) Bacterial resistances to toxic metal ions—a review. *Gene* **179**, 9–19.
38. Stephens, C. and Murray, W. (2001) Pathogen evolution: how good bacteria go bad. *Curr. Biol.* **11**, R53–R56.
39. Tan, H. M. (1999) Bacterial catabolic transposons. *Appl. Microbiol. Biotechnol.* **51**, 1–12.
39a. James, R., Kleanthous, C., and Moore, G. R. (1996) The biology of E colicins: paradigms and paradoxes. *Microbology 142*, 1569–1580.

39b. Belogurov, A.A., Delver, E. P., Agafonova, O. V., et al. (2000) Antirestriction protein Ard (Type C) encoded by IncW plasmid pSa has a high similarity to the 'protein transport' domain of TraC1 primase of promiscuous plasmid RP4. *J. Mol. Biol.* **296,** 969–977.
40. Helgason, E., Okstad, O. A., Caugant, D. A., et al. (2000) *Bacillus anthracis, Bacillus cereus,* and *Bacillus thuringiensis*—one species on the basis of genetic evidence. *Appl. Environ. Microbiol.* **66,** 2627–2630.
41. del Solar, G., Giraldo, R., Ruiz-Echevarria, M. J., e al. (1998) Replication and control of circular bacterial plasmids. *Microbiol. Mol. Biol. Rev.* **62,** 434–464.
42. Kelman, Z. and O'Donnell, M. (1995) DNA polymerase III holoenzyme: structure and function of a chromosomal replicating machine. *Annu. Rev. Biochem.* **64,** 171–200.
42a. Burian, J., Guller, L., Macor, M., et al. (1997) Small crytic plasmids of multiplasmid, clinical *Escherichia coli. Plasmid* **37,** 2–14.
43. Caspi, R., Pacek, M., Consiglieri, G., et al. (2001) A broad host range replicon with different requirements for replication initiation in three bacterial species. *EMBO J.* **20,** 3262–3271.
44. Chattoraj, D. K. (2000) Control of plasmid DNA replication by iterons: no longer paradoxical. *Mol. Microbiol.* **37,** 467–476.
45. Messer, W. and Weigel, C. (1997) DnaA initiator—also a transcription factor. *Mol. Microbiol.* **24,** 1–6.
46. Cesareni, G., Helmer-Citterich, M., and Castagnoli, L. (1991) Control of ColE1 plasmid replication by antisense RNA. *Trends Genet.* **7,** 230–235.
47. Khan, S. A. (2000) Plasmid rolling-circle replication: recent developments. *Mol. Microbiol.* **37,** 477–484.
48. te Riele, H., Michel, B., and Ehrlich, S. D. (1986) Are single-stranded circles intermediates in plasmid DNA replication? *EMBO J.* **5,** 631–637.
49. Pogliano, J., Ho, T. Q., Zhong, Z., and Helinski, D. R. (2001) Multicopy plasmids are clustered and localized in *Escherichia coli. Proc. Natl Acad. Sci. USA* **98,** 4486–4491.
50. Bignell, C. and Thomas, C. M. (2001) The bacterial ParA–ParB partitioning proteins. *J. Biotechnol.* **91,** 1–34.
51. Moller-Jensen, J., Jensen, R. B., and Gerdes, K. (2000) Plasmid and chromosome segregation in prokaryotes. *Trends Microbiol.* **8,** 313–320.
52. Gordon, G.S. and Wright, A. (2000) DNA segregation in bacteria. *Annu. Rev. Microbiol.* **54,** 681–708.
53. Arber, W. (2000) Genetic variation: molecular mechanisms and impact on microbial evolution. *FEMS Microbiol. Rev.* **24,** 1–7.
54. Summers, D. K., Beton, C. W., and Withers, H. L. (1993) Multicopy plasmid instability: the dimer catastrophe hypothesis. *Mol. Microbiol.* **8,** 1031–1038.
55. van Duyne, G. D (2001) A structural view of Cre-*loxp* site-specific recombination. *Annu. Rev. Biophys. Biomol. Struct.* **30,** 87–104.
56. Barre, F. X., Soballe, B., Michel, B., et al. (2001) Circles: the replication-recombination-chromosome segregation connection. *Proc. Natl Acad. Sci. USA* **98,** 8189–8195.
57. Engelberg-Kulka, H. and Glaser, G. (1999) Addiction modules and programmed cell death and antideath in bacterial cultures. *Annu. Rev. Microbiol.* **53,** 43–70.
58. Gerdes, K. (2000) Toxin-antitoxin modules may regulate synthesis of macromolecules during nutritional stress. *J. Bacteriol.* **182,** 561–572.
59. Gerdes, K., Gultyaev, A. P., Franch, T., et al. (1997) Antisense RNA-regulated programmed cell death. *Annu. Rev. Genet.* **31,** 1–31.

60. Holcik, M. and Iyer, V. N. (1997) Conditionally lethal genes associated with bacterial plasmids. *Microbiology* **143,** 3403–3416.
61. Couturier, M., Bahassi, el-M., and van Melderen, L. (1998) Bacterial death by DNA gyrase poisoning. *Trends Microbiol.* **6,** 269–275.
62. Tortosa, P. and Dubnau, D. (1999) Competence for transformation: a matter of taste. *Curr. Opin. Microbiol.* **2,** 588–592.
63. Sternberg, N. L. and Maurer, R. (1991) Bacteriophage-mediated generalized transduction in *Escherichia coli* and *Salmonella typhimurium. Methods Enzymol.* **204,** 18–43.
64. Lanka, E. and Wilkins, B. M. (1995) DNA processing reactions in bacterial conjugation. *Annu. Rev. Biochem.* **64,** 141–169.
65. Zechner, E. L., de la Cruz, F., Eisenbrandt, R., et al. (2000) Conjugative-DNA transfer processes, in *The Horizontal Gene Pool* (Thomas, C. M., ed.), Harwood Academic, Amsterdam, pp. 87–174.
66. Charlebois, R. L., She, Q., Sprott, D. P., et al. (1998) Sulfolobus genome: from genomics to biology. *Curr. Opin. Microbiol.* **1,** 584–588.
67. Trieu-Cuot, P., Derlot, E., and Courvalin, P. (1993) Enhanced conjugative transfer of plasmid DNA from *Escherichia coli* to *Staphylococcus aureus* and *Listeria monocytogenes. FEMS Microbiol. Lett.* **109,** 19–23.
68. Bates, S., Cashmore, A. M., and Wilkins, B. M. (1998) IncP plasmids are unusually effective in mediating conjugation of *Escherichia coli* and *Saccharomyces cerevisiae*: involvement of the Tra2 mating system *J. Bacteriol.* **180,** 6538–6543.
69. Christie, P. J. (1997) *Agrobacterium tumefaciens* T-complex transport apparatus: a paradigm for a new family of multifunctional transporters in eubacteria. *J. Bacteriol.* **179,** 3085–3094.
70. Dunny, G. M. and Leonard, B. A. (1997) Cell–cell communication in gram-positive bacteria. *Annu. Rev. Microbiol.* **51,** 527–564.
71. Byrd, D. R. and Matson, S. W. (1997) Nicking by transesterification: the reaction catalysed by a relaxase. *Mol. Microbiol.* **25,** 1011–1022.
71a. Harris, R. L., Sholl, K. A., Conrad, M. N., et al. (1999) Interaction between the F plasmid TraA (F-pilin) and TraQ proteins. *Mol. Microbiol.* **34,** 780–791.
72. Osborn, M., Bron, S., Firth, N., et al. (2000) The evolution of bacterial plasmids, in *The Horizontal Gene Pool* (Thomas, C. M., ed.), Harwood Academic, Amsterdam, pp. 301–361.
73. Burland, V., Shao, Y., Perna, N. T., et al. (1998) The complete DNA sequence and analysis of the large virulence plasmid of *Escherichia coli* O157:H7. *Nucleic Acids Res.* **26,** 4196–4204.
74. Bingle, L. E. and Thomas, C. M. (2001) Regulatory circuits for plasmid survival. *Curr. Opin. Microbiol.* **4,** 194–200.

2

Choosing a Cloning Vector

Andrew Preston

1. Introduction

Since the construction of the first generation of general cloning vectors in the early 1970s, the number of plasmids created has increased to an almost countless number. Thus, a critical decision facing today's investigator is that of which plasmid to use in a particular project? Despite the bewildering choice of commercial and other available vectors, the choice of which cloning vector to use can be decided by applying a small number of criteria: insert size, copy number, incompatibility, selectable marker, cloning sites, and specialized vector functions. Several of these criteria are dependent on each other. This chapter discusses these criteria in the context of choosing a plasmid for use as a cloning vector and **Table 1** displays the features of some commonly used cloning vectors.

2. Criteria for Choosing a Cloning Vector
2.1. Insert Size

For projects in which it is desired that a particular piece of DNA be cloned, one consideration is the size of the insert DNA. Most general cloning plasmids can carry a DNA insert up to around 15 kb in size. Inserts in excess of this place constraints on proper replication of the plasmids (particularly for high-copy-number vectors) and can cause problems with insert stability. Several types of vectors are available for cloning large fragments of DNA. These are most commonly used to construct libraries of clones that are often representative of entire genomes. Library clones are then screened to identify the particular clone that carries the DNA of interest.

2.1.1. Cosmids (1,2)

Cosmids are conventional vectors that contain a small region of bacteriophage λ DNA containing the cohesive end site (cos). This contains all of the *cis*-acting elements for packaging of viral DNA into λ particles. For cloning of DNA in these vec-

**Table 1
Commonly Used Cloning Vectors**

Plasmid	Features	Commercial source
pUC18, pUC19	Small size (2.7 kb) High copy number Multiple cloning site Ampicillin-resistance marker Blue/white selection (*see* Chapter 19)	NEB
pBluescript vectors	As pUC Single-stranded replication origin T7 and SP6 promoters flanking MCS[a]	Stratagene
pACYC vectors	Low copy number (15 copies per cell) p15A origin of replication	NEB
Supercos	Cosmid vector Two cos sites Insert size 30–42 kb Ampicillin-selectable marker T3 and T7 promoters flanking cloning site	Stratagene
EMBL3	λ replacement vector MCS sites: *Sal*I, *Bam*HI and *Eco*RI	Promega
λ ZAP	λ vector In vivo excision into pBluescript phagemid vector Cloning capacity 10 kb Blue/white selection	Stratagene
pBeloBAC11	BAC vector Inserts up to 1 Mb T7 and SP6 promoters flank insertion site Blue/white selection Cos site LoxP site	NEB

[a]MCS = multiple-cloning site.

tors, linear genomic DNA fragments are ligated in vitro to the vector DNA and this is then packaged into bacteriophage particles (*see* Chapter 7). On introduction into *Escherichia coli* host cells, the vector is circularized to form a large plasmid containing the cloned DNA fragment. Cosmids are most commonly used to generate large insert libraries. Because of the constraints of packaging, the vector plus insert should comprise between 28 and 45 kb.

2.1.2. λ Vectors (see Chapter 18)

The bacteriophage λ genome comprises 48,502 bp. On entering the host cell, the phage adopts one of two life cycles: lytic growth or lysogeny. In lytic growth, approx 100 new virions are synthesized and packaged before lysing the host cell, releasing the progeny phage to infect new hosts. In lysogeny, the phage genome undergoes recombination into the host chromosome, where it is replicated and inherited along with the host DNA *(3)*. Which of the two different life cycles is adopted is determined by the multiplicity of infection and the host cell nutritional status. The larger the multiplicity of infection and the poorer the nutritional state of the cell, the more lysogeny is favored *(4)*.

Use of λ as a cloning vector involves only the lytic cycle. This renders the middle third of the λ genome, which encodes functions for gene expression regulation and establishment of lysogeny, redundant for these purposes. It is the ability to replace this portion of the genome with foreign DNA without affecting the lytic life cycle that makes λ useful as a cloning vector. Insertion λ vectors have the nonessential DNA deleted and contain a single site for insertion of DNA. Typically 5–11 kb of foreign DNA can be accommodated in these vectors. Replacement vectors contain specific restriction sites flanking the nonessential genes. Digestion of linear vector DNA at these sites produces two "arms" that are ligated to the foreign DNA. Many commercially available λ vectors are sold as predigested and purified arms. Replacement vectors typically can accommodate between 8 and 24 kb of foreign DNA, depending on the vector.

During the early phase of infection, λ DNA replicates bidirectionally, in circular form from a single origin of replication before shifting to replication via a rolling-circle mode. This produces a concatamer of genomes in a head-to-tail arrangement that is then processed to give individual genomes for packaging. The shift to rolling-circle replication depends on the interplay between host- and phage-encoded recombination functions. As the recombination proficiency of different λ vectors can vary, the investigator is urged to ensure that the *E. coli* strain used for infection is capable of properly replicating the phage. This information is generally supplied with commercially available vectors. A great many features that aid cloning into and screening of recombinant phage have also been incorporated into λ vectors. Often, the use of these features also necessitates the use of particular host strains.

2.1.3. Bacterial Artificial Chromosomes (5,6)

Bacterial artificial chromosomes (BACs) are circular DNA molecules. They contain a replicon that is based on the F factor comprising *oriS* and *repE* encoding an ATP-driven helicase along with *parA*, *parB*, and *parC* to facilitate accurate partitioning (*see* Chapter 1). The F factor is capable of carrying up to one quarter of the *E. coli* chromosome and, thus, BACs are capable of maintaining very large DNA inserts (often up to 350 kb); however, many BAC libraries contain inserts of around 120 kb. Newer versions of BAC vectors contain sites to facilitate recovery of cloned DNA (e.g., *loxP*) *(7)*. A DNA fragment is cloned into BAC vectors in a similar fashion to cloning into general cloning vectors; DNA is ligated to a linearized vector and then introduced into an *E. coli* cloning strain by electroporation.

2.2. Copy Number

Different cloning vectors are maintained at different copy numbers, dependent on the replicon of the plasmid (*see* Chapter 1). In a majority of cases in which a piece of DNA is cloned for maintenance and amplification for subsequent manipulation, the greater the yield of recombinant plasmid from *E. coli* cultures, the better. In this scenario, a high-copy-number vector is desirable such as those whose replication is driven by the ColE1 replicon *(8)*. The original ColE1-based plasmids have a copy number of 15–20. However, a mutant ColE1 replicon, as found in the pUC series of plasmids *(9)*, produces a copy number of 500–700 as a result of a point mutation within the RNAII regulatory molecule (*see* Chapter 1) that renders it more resistant to inhibition by RNAI *(10)*. It should be noted that this mutation is temperature sensitive. Mutant RNAII is resistant to RNAI inhibition at 37°C or 42°C but not at 30°C, at which temperature the copy number of pUC plasmids returns to that of nonmutated ColE1 plasmids.

In some cases, a high-copy-number may cause problems for cloning DNA. For example, the cloned DNA may encode proteins that are toxic to the cell when present at high levels. This is particularly true of membrane proteins. Even if the protein is expressed poorly from the cloned DNA, the presence of many hundreds of copies of the gene on the plasmid may raise the level of protein to toxic levels. In these cases, using a plasmid with a lower copy number may reduce the gene dosage below a level at which toxicity occurs. For example, pBR322 is based on the original ColE1 replicon and thus has a copy number of 15–20 *(11)*. The pACYC series of plasmids are based on the p15A replicon, which has a copy number of 18–22 *(12)*. Low-copy-number plasmids include pSC101 (copy number around 5) *(13)*, whereas BACs are maintained at one copy per cell *(5)*.

2.3. Incompatibility

Incompatibility refers to the fact that different plasmids are sometimes unable to coexist in the same cell. This occurs if the two different plasmids share functions required for replication and/or partitioning into daughter cells. Direct competition for these functions often leads to loss of one of the plasmids from the cell during growth of a culture. Plasmid size can also influence maintenance within a culture, as larger plasmids require longer for replication and, thus, may be outcompeted by faster replicating of smaller plasmids. Thus, ColE1-based plasmids are incompatible with other ColE1-based plasmids but are compatible with R6K- or p15A-based plasmids. Incompatibility only becomes an issue if it requires that two plasmids be maintained-together (e.g., if cloning into an *E. coli* strain that contains a helper plasmid) (*see* Chapter 3).

2.4. Selectable Marker

Introduction of plasmids in to *E. coli* cells is an inefficient process. Thus, a method of selecting those cells that have received a plasmid is required. Furthermore, cells that do not contain a plasmid are at a growth advantage over those that do and, thus, have to replicate both the chromosome and additional plasmid DNA. This is of particular consequence when dealing with high-copy-number or large plasmids. In this case, a selective pressure must be imposed for maintenance of the plasmid. Almost all conventional plasmids use an antibiotic resistance gene as a selectable marker, carried on the backbone of the vector. Thus, the addition of the appropriate antibiotic to the growth medium will kill those cells that do not contain the plasmid and produce a culture in which all cells do contain a plasmid. In many cases, the choice of antibiotic is not restricted. However, some cloning strains of *E. coli* are inherently resistant to some antibiotics and, thus, the same antibiotic cannot be used as a selection for those cells carrying a particular plasmid. The genotype of the desired cloning strain should be checked prior to cloning (*see* Chapter 3). In some situations, downstream applications render some antibiotics as unsuitable choices. For example, the mutation of genes in cloned DNA fragments is often achieved by the disruption of the gene by insertion of an antibiotic-resistance cassette. This both mutates the gene and acts as a marker for the mutation. Often, the mutation is introduced into the organism from which the DNA is derived. In this case, only some antibiotics are suitable for use because of restrictions on introducing particular antibiotic resistances into some bacteria against which the antibiotic is used as a therapeutic or because of the inherent resistance of the original organism to the antibiotic. In these projects, the vector should not confer resistance to the antibiotic to be used in the downstream application.

Some plasmid vectors contain two antibiotic-resistance cassettes. For example, pACYC177 contains both ampicillin- and kanamycin-resistance genes. Many of the cloning sites in this vector lie in these genes and cloning into one of these sites inactivates that particular antibiotic resistance. Profiling the antibiotic resistances of recombinant clones is a way of selecting for those carrying insert DNA fragments. Many newer vectors now carry specialized cloning sites (polylinker, multiple-cloning site; *see* **Subheading 2.5.**) for which cloning of insert DNA does not interfere with inherent vector functions. The most common antibiotic resistances carried on vectors used in *E. coli* are resistance to ampicillin, kanamycin, tetracycline, and chloramphenicol.

2.4.1. Ampicillin

This drug inhibits the bacterial transpeptidase involved in peptidoglycan biosynthesis and thus inhibits cell wall biosynthesis *(14)*. As such, ampicillin inhibits log-phase bacteria but not those in a stationary phase. Resistance to ampicillin is conferred by a β-lactamase, which cleaves the β-lactam ring of ampicillin *(14)*. The β-lactamase most commonly expressed by cloning vectors is that encoded by the *bla* gene *(15)*.

2.4.2. Kanamycin

A member of the aminoglycoside family of antibiotics, kanamycin was first isolated from *Streptomyces kanamyceticus* in Japan in 1957. This polycation is taken into the bacterial cell through outer-membrane pores but crosses the cytoplasmic membrane in an energy-dependent process utilizing the membrane potential. The molecule interacts with three ribosomal proteins and with rRNA in the 30S ribosomal subunit, to prevent the transition of an initiating complex to a chain-elongating complex, and thus inhibits protein synthesis. Resistance to kanamycin is conferred by aminophosphotransferases. Those commonly encoded by vectors are Aph (3')-I from Tn*903* and Aph (3')-II from Tn,5 which transfer phosphate from ATP to the kanamycin to inactivate it *(16)*. It is important to note that these two resistance genes have differing

DNA sequences and, thus, different restriction maps. They will not cross-hybridize under stringent conditions in Southern hybridizations.

2.4.3. Chloramphenicol

First isolated from a soil actinomycete in 1947, chloramphenicol was widely used as a broad-spectrum antibiotic although its clinical use has been curtailed because of drug-induced bone-marrow toxicity and the emergence of bacterial chloramphenicol resistance. Chloramphenicol inhibits the activity of ribosomal peptidyl transferase and thus inhibits protein synthesis *(17)*. Chloramphenicol resistance is conferred by chloramphenicol acetyl transferase (*cat*), which transfers an acetyl group from acetyl CoA to chloramphenicol and inactivates it *(18)*.

2.4.4. Tetracycline

Originally isolated from *Streptomyces aureofaciens* in 1948, there are now many tetracycline derivatives available. They bind to a single site on the 30S ribosomal subunit to block the attachment of aminoacyl tRNA to the acceptor site and thus inhibit protein synthesis *(19)*. Tetracycline resistance is conferred by efflux proteins, TetA (A–E), which catalyze the energy-dependent export of tetracycline from the cell against a concentration gradient *(19)*.

2.5. Cloning Sites

The cloning of DNA into a vector usually involves ligation of the insert DNA fragment to vector DNA that has been cut with a restriction endonuclease. This is facilitated by the insert and vector DNA fragments having compatible cohesive ends. Thus, the vector of choice may be one that has a restriction endonuclease site that is compatible with the insert fragment-generating enzyme. It should be noted, however, that any blunt-end fragment can be ligated to any other blunt-end fragment and that even DNA-fragments generated by restriction enzymes that generate overhangs can be made blunt ended (*see* Chapter 15). In many older vectors, the restriction endonuclease sites were dispersed around the plasmid and were often in one of the vector genes. For example, many of the cloning sites in the pACYC series of vectors are located within one of the antibiotic-resistance genes of these plasmids. Cloning into these sites inactivated the resistance gene and the subsequent sensitivity to the antibiotic was used as a screen for recombinant plasmids containing the insert DNA.

More modern vectors often contain an artificial stretch of DNA that has a high concentration of restriction endonuclease sites that do not occur elsewhere on the plasmid. These multiple-cloning sites (MCSs) or polylinkers give a wide choice of restriction endonucleases for use in the cloning step. They also limit the cloning site to one small region of the vector and thus allow the specific positioning of the insert DNA close to other features of the vector. For example, the MCSs of many vectors such as the pUC series are flanked by sequences complementary to a universal series of primers, the M13 forward and reverse primers. These priming sites are oriented such that extension of the primers annealed to these sites allows sequencing of both ends of an insert DNA in the MCS. In this fashion, one set of universal primers can be used to sequence any insert DNA regardless of which site the DNA was inserted at within the MCS.

Many plasmids contain MCSs that lie within the coding sequence of the α fragment of *lacZ*. This feature (blue/white selection) facilitates the identification of recombinant constructs that carry a cloned fragment by distinguishing them from clones that arise from religation of the cloning vector. This feature is discussed in Chapter 19.

2.6. Specialized Plasmid Functions

Some projects will involve specific downstream applications that will require specialized plasmid functions that are only present on some plasmids. For example, both the pUC and pBluescript series of vectors are high-copy-number, ampicillin-resistance-conferring plasmids that contain MCSs that facilitate the use of a wide range of restriction endonucleases in the cloning step. However, one feature present on pBluescript vectors that is not present on pUC vectors is promoters flanking the MCS that permit transcription of the insert DNA on either strand. The two promoters T7 and SP6 are recognized by bacteriophage RNA polymerases that must be supplied *in trans*. They do not transcribe host genes or other plasmid genes, enabling specific transcription of the insert DNA (*see* Chapter 27).

pBluescript vectors are phagemids. They contain a single-stranded filamentous bacteriophage origin of replication (M13 phage) and, thus, are useful for generating single-stranded DNA (*see* Chapter 13) for applications such as DNA sequencing or site-directed mutagenesis. Single-stranded replication is initiated by infecting with a helper phage encoding the necessary functions. These vectors can also replicate as conventional double-stranded plasmids. The single-stranded origin can exist in two orientations. Those versions in which it is in same orientation as the plasmid origin are denoted as "+," whereas those with the origin in the opposite orientation are denoted as "−."

Many plasmids have been designed to achieve high-level expression of recombinant proteins from the cloned DNA. These expression vectors are discussed in Chapter 28.

3. Summary

When choosing a cloning vector for use in a cloning project, the investigator is faced with an enormous choice. However, the application of a small number of criteria can quickly guide the selection of a suitable vector. Many plasmids contain sufficient features that render them suitable for a wide range of projects. Thus, the investigator needs to be equipped with only a small number of vectors in order to satisfy most needs.

References

1. Hohn, B., Koukolikova-Nicola, Z., Lindenmaier, W., et al. (1988) Cosmids. *Biotechnology* **10,** 113–127.
2. Collins, J. and Hohn, B. (1978) Cosmids: a type of plasmid gene-cloning vector that is packageable in vitro in bacteriophage lambda heads. *Proc. Natl. Acad. Sci. USA* **75,** 4242–4246.
3. Ptashne, M. (1986) A Genetic Switch: Gene Control and Phage λ. Blackwell Scientific, Palo Alto, CA.

4. Herskowitz, I. and Hagen, D. (1980) The lysis–lysogeny decision of phage λ: explicit programming and responsiveness. *Annu. Rev. Genet.* **14,** 399–445.
5. Shizuya, H., Birren, B., Kim, U. J., et al. (1992) Cloning and stable maintenance of 300-kilobase-pair fragments of human DNA in *Escherichia coli* using an F-factor-based vector. *Proc. Natl. Acad. Sci. USA* **89,** 8794–8797.
6. Monaco, A. P. and Larin, Z. (1994) YACs, BACs, PACs and MACs: artificial chromosomes as research tools. *Trends Biotechnol.* **12,** 280–286.
7. Palazzolo, M. J., Hamilton, B. A., Ding, D. L., et al. (1990) Phage lambda cDNA cloning vectors for subtractive hybridization, fusion-protein synthesis and Cre–loxP automatic plasmid subcloning. *Gene* **88,** 25–36.
8. Kahn, M., Kolter, R., Thomas, C., et al. (1979) Plasmid cloning vehicles derived from plasmids ColE1, F, R6K, and RK2. *Methods Enzymol.* **68,** 268–280.
9. Vieira, J. and Messing, J. (1982) The pUC plasmids, an M13mp7-derived system for insertion mutagenesis and sequencing with synthetic universal primers. *Gene* **19,** 259–268.
10. Lin-Chao, S., Chen, W. T., and Wong, T. T. (1992) High copy number of the pUC plasmid results from a Rom/Rop-suppressible point mutation in RNA II. *Mol. Microbiol.* **6,** 3385–3393.
11. Bolivar, F., Rodriguez, R. L., Greene, P. J., et al. (1977) Construction and characterization of new cloning vehicles, II: a multipurpose cloning system. *Gene* **2,** 95–113.
12. Chang, A. C. and Cohen, S. N. (1978). Construction and characterization of amplifiable multicopy DNA cloning vehicles derived from the p15A cryptic miniplasmid. *J. Bacteriol.* **134,** 1141–1156.
13. Stoker, N. G., Fairweather, N. F., and Spratt, B. G. (1982) Versatile low-copy-number plasmid vectors for cloning in *Escherichia coli*. *Gene* **18,** 335–341.
14. Donowitz, G. R. and Mandell, G. L. (1988) Beta-lactam antibiotics (1). *N. Engl. J. Med.* **318,** 419–426.
15. Sutcliffe, J. G. (1978) Nucleotide sequence of the ampicillin resistance gene of *Escherichia coli* plasmid pBR322. *Proc. Natl. Acad. Sci. USA* **75,** 3737–3741.
16. Umezawa, H. (1979) Studies on aminoglycoside antibiotics: Enzymic mechanism of resistance and genetics. *Jpn. J. Antibiot.* **32(Suppl),** S1–S14.
17. Drainas, D., Kalpaxis, D. L., and Coutsogeorgopoulos, C. (1987) Inhibition of ribosomal peptidyltransferase by chloramphenicol: kinetic studies. *Eur. J. Biochem.* **164,** 53–58.
18. Shaw, W. V. (1983) Chloramphenicol acetyltransferase: enzymology and molecular biology. *CRC Crit. Rev. Biochem.* **14,** 1–46.
19. Schnappinger, D. and Hillen, W. (1996) Tetracyclines: antibiotic action, uptake, and resistance mechanisms. *Arch. Microbiol.* **165,** 359–369.

3

Escherichia coli Host Strains

Nicola Casali

1. Introduction

To successfully perform molecular genetic techniques it is essential to have a full understanding of the properties of the various *Escherichia coli* host strains commonly used for the propagation and manipulation of recombinant DNA. *E. coli* is an enteric rod-shaped Gram-negative bacterium with a circular genome of 4.6 Mb *(1)*. It was originally chosen as a model system because of its ability to grow on chemically defined media and its rapid growth rate. In rich media, during the exponential phase of its growth, *E. coli* doubles every 20–30 min; thus, during an overnight incubation period, a single selected organism will double enough times to yield a colony on an agar plate, or 1–2 billion cells per milliliter of liquid media. The ease of its transformability and genetic manipulation has subsequently solidified the role of *E. coli* as the host of choice for the propagation, manipulation, and characterization of recombinant DNA. In the past 60 yr *E. coli* has been the subject of intensive research and more is now known about these bacilli than any other organisms on earth.

A wide variety of *E. coli* mutants have been isolated and characterized. Almost all strains currently used in recombinant DNA experiments are derived from a single strain: *E. coli* K-12, isolated from the feces of a diphtheria patient in 1922 *(2)*. This chapter will discuss characteristics of *E. coli* host strains that are important for recombinant DNA experiments in order to aid in the choice of a suitable host and circumvent possible problems that may be encountered. Common mutations and genotypes that are relevant to recombinant DNA experiments are summarized in **Table 1**. A complete listing of genetically defined genes has been compiled by Berlyn et al. *(3)*.

1.1. Genotype Nomenclature

A genotype indicates the genetic state of the DNA in an organism. It is associated with an observed behavior called the phenotype. Genotypes of *E. coli* strains are described in accordance with a standard nomenclature proposed by Demerec et al. *(4)*. Genes are given three-letter, lowercase, italicized names that are often mnemonics

Table 1
Properties of Common Genotypes of *E. coli* Host Strains

Mutation	Description	Significance
Amy	Expresses amylase	Allows amylose utilization
ara	Mutation in arabinose metabolism	Blocks arabinose utilization
dam	Blocks adenine methylation at GATC sequences	Makes DNA susceptible to cleavage by some restriction enzymes
dcm	Blocks cytosine methylation at CC(A/T)GG sequences	Makes DNA susceptible to cleavage by some restriction enzymes
(DE3)	λ lysogen carrying the gene for T7 RNA polymerase	Used for T7 promoter-based expression systems
deoR	Regulatory gene mutation allowing constitutive expression of genes for deoxyribose synthesis	Allows replication of large plasmids
dnaJ	Inactivation of a specific chaperonin	Stabilizes expression of certain recombinant proteins
dut	dUTPase activity abolished	In combination with *ung*, allows incorporation of uracil into DNA; required for Kunkel mutagenesis
e14⁻	A prophagelike element carrying *mcrA*	See *mcrA*
endA1	Activity of nonspecific endonuclease I abolished	Improves yield and quality of isolated plasmid DNA
F'	Host contains an F' episome with the stated features	Required for infection by M13 vectors
gal	Mutation in galactose metabolism	Blocks galactose utilization
gor	Mutation in glutathione reductase	Facilitates cytoplasmic disulfide bond formation
gyrA	DNA gyrase mutation	Confers resistance to nalidixic acid
hflA	Inactivation of a specific protease	Results in high-frequency lysogenization by lambda
hsdR	Inactivation of *Eco* endonuclease activity	Abolishes *Eco* restriction but not methylation (r⁻ m⁺)
hsdS	Inactivation of *Eco* site-recognition activity	Abolishes *Eco* restriction and methylation (r⁻ m⁻)
Hte	Unknown	Enhances uptake of large plasmids
lacI^q	Constitutive expression of the *lac* repressor	Inhibits transcription from the *lac* promoter
lacY	Lactose permease activity abolished	Blocks lactose uptake; improves IPTG-induced control of *lac* promoters
lacZ	β-Galactosidase activity abolished	Blocks lactose utilization

(continued)

E. coli Hosts

Table 1 *(continued)*

Mutation	Description	Significance
lacZΔM15	Partial deletion of β-galactosidase gene	Allows α-complementation for blue/white selection of recombinant colonies in *lacZ* mutant hosts
leu	Mutation in leucine biosynthesis	Requires leucine for growth on minimal media
lon	Inactivation of Lon protease	Increases yield of some recombinant proteins
Δ(malB)	Mutation in maltose metabolism; deletes most of the region encompassing *malEFG* and *malK lamB malM*	Blocks maltose utilization; eliminates expression of maltose-binding protein (MalE)
mcrA, mcrBC	Mutation in methylcytosine-specific restriction systems	Allows more efficient cloning of DNA containing methylcytosines
metB	Mutation in methionine biosynthesis	Requires methionine for growth on minimal media; promotes high specific activity labeling with ^{35}S-methionine
mrr	Mutation in methyladenosine-specific restriction system	Allows more efficient cloning of DNA containing methyladenines
mtl	Mutation in mannitol metabolism	Blocks mannitol utilization
mutD	Inactivates DNA polymerase III subunit	Increases frequency of spontaneous mutation
mutS	Deficient in mismatch repair	Stabilizes DNA heteroduplexes during site-directed mutagenesis
nupG	Mutation in nucleoside transport	Increases plasmid uptake
ompT	Mutation in outer-membrane protease	Improves yield of some recombinant proteins
φ80	Carries the prophage φ80	Often expresses *lacZ*ΔM15
P1	Carries the prophage P1	Expresses the P1 restriction system
P2	Carries the prophage P2	Inhibits growth of *red⁺ gam⁺* λ vectors
phoA	Mutation in alkaline phosphatase	Blocks phosphate utilization; used for PhoA-based reporter systems
phoR	Regulatory gene mutation	Used for *pho* promoter-based expression systems
pnp	Inactivates polynucleotide phosphorylase	Increases stability of some mRNAs resulting in increased protein expression

(continued)

Table 1 *(continued)*

Mutation	Description	Significance
proAB	Mutations in proline biosynthesis	Requires proline for growth in minimal media
recA	Homologous recombination abolished	Prevents recombination of introduced DNA with host DNA, increasing stability of inserts
recBC	Exonuclease and recombination activity of ExoV abolished	Reduces general recombination; enhances stability of palindromes in λ vectors
recD	Exonuclease activity of ExoV abolished	Enhances stability of palindromes in λ vectors
recE	Recombination deficiency	Reduces recombination between plasmids
recF	Recombination deficiency	Reduces recombination between plasmids
recJ	Recombination deficiency	Reduces recombination between plasmids
relA	Eliminates stringent factor resulting in relaxed phenotype	Allows RNA synthesis in the absence of protein synthesis
rne	Inactivates RNase E	Increases stability of some mRNAs resulting in increased protein expression
rpoH (or *htpR*)	Inactivates a heat-shock sigma factor	Abolishes expression of some proteases; improves yield of certain recombinant proteins at high temperature
rpsL	Mutation in small ribosomal protein S12	Confers resistance to streptomycin
sbcA	Mutation in RecE pathway	Improves growth of *recB* mutant hosts
sbcB	ExoI activity abolished	Allows general recombination in *recBC* mutant strains
sbcC	Mutation in RecF pathway	Enhances stability of long palindromes in λ and plasmid vectors
srl	Mutation in sorbitol metabolism	Blocks sorbitol utilization
sup	Suppressor mutation	Suppresses ochre (UAA) and amber (UAG) mutations (*see* **Table 3**)
thi	Mutation in thiamine biosynthesis	Thiamine required for growth in minimal media
thr	Mutation in threonine biosynthesis	Threonine required for growth in minimal media
Tn*10*	Transposon	Encodes resistance to tetracycline

(continued)

Table 1 *(continued)*

Mutation	Description	Significance
Tn5	Transposon	Encodes resistance to kanamycin
tonA	Mutation in outer-membrane protein	Confers resistance to bacteriophage T1
traD	Mutation in transfer factor	Prevents conjugal transfer of F' episome
trp	Mutation in tryptophan biosynthesis	Tryptophan required for growth in minimal media
trxB	Mutation in thioredoxin reductase	Facilitates cytoplasmic disulfide bond formation
tsp	Mutation in a periplasmic protease	Improves yield of secreted proteins and proteins isolated from cell lysates
tsx	Mutation in outer-membrane protein	Confers resistance to bacteriophage T6
umuC	Mutation in SOS repair pathway	Enhances stability of palindromes
ung	Uracil *N*-glycosylase activity abolished	Prevents removal of uracil incorporated into DNA; *see dut*
uvrC	Mutation in UV repair pathway	Enhances stability of palindromes
xylA	Mutation in xylose metabolism	Blocks xylose utilization

Source: Compiled from **refs.** *3* and *5* and information supplied by Invitrogen, New England Biolabs, Novagen, and Stratagene.

suggesting the function of the gene. If the same function is affected by several genes, the different genes are distinguished with uppercase italic letters, for example *recA*, *recB*, *recC*, and *recD* all affect recombination. By convention, *E. coli* genotypes list only genes that are defective, but the superscript symbols "–" and "+" are occasionally used redundantly for clarity or to emphasize a wild-type locus. Phenotypes are capitalized and the letters are followed by either superscript "+" or "–," or sometimes "r" for resistant or "s" for sensitive. Although convention dictates that phenotypes are not specified in the genotype designation, they are sometimes included, when not easily inferred. For example, *rpsL*(Str^r) indicates that a mutation in the gene for ribosomal protein small subunit S12 confers resistance to streptomycin.

Specific mutations are given allele numbers that are usually italic arabic numerals such as *hsdR17*. If the exact locus is not known, then the capital letter is replaced by a hyphen, as in *arg-3*. An amber mutation (*see* **Subheading 2.1.1.**) is denoted by *am* following the gene designation and a temperature-sensitive mutation that renders the gene inactive at high temperature, is denoted by *ts*. A constitutive mutation is denoted by superscript q; thus *lacI*^{q} indicates constitutive expression of the gene for the *lac* repressor.

Deletions are denoted by Δ. If Δ is followed by the names of deleted genes in parentheses, as in Δ(*lac-pro*), then all of the genes between the named genes are also deleted. An insertion is indicated by "::" preceded by the position of the insertion and followed by the inserted DNA; for example, *trpC22*::Tn*10* denotes an insertion of Tn*10* into *trpC*. Alternatively, the map position of an insertion can be denoted by a three-letter code. The first letter is always *z*, the second and third letters indicate 10-min and 1-min intervals, respectively, and are designated by the letters *a–i*. Thus, *zhg*::Tn*10* indicates an insertion of Tn*10* at 87 min. A fusion is denoted by the symbol φ followed by the fused genes in parentheses. A prime denotes that a fused gene is incomplete and can be used before or after the gene designation to denote deletions in the 5' or 3' regions, respectively. A superscript "+" indicates that the fusion involves an operon rather than a single gene. For example, φ(*ompC'-lacZ*$^+$) indicates a fusion between *ompC*, deleted in the 3' region, and the *lac* operon.

F$^+$ and Hfr (*see* **Subheading 2.2.**) strains are denoted by the relevant symbol at the start of the genotype and strains are assumed to be F$^-$ unless indicated. If the strain is F', then this is indicated at the end of the genotype with the genes carried by the F plasmid listed in square brackets. Plasmids and lysogenic phage, carried by the strain, are listed in parentheses at the end of the genotype and may include relevant genetic information.

2. General Properties of Cloning Hosts

The genotypes and features of a representative selection of popular host strains used for general recombinant DNA cloning procedures are listed in **Table 2**. An extended listing of available strain genotypes can be found in **ref. 5**. Many useful strains are available through the American Type Culture Collection (www.atcc.org) and the *E. coli* Genetic Stock Center at Yale (cgsc.biology.yale.edu), as well as from commercial suppliers such as Stratagene, Promega, Novagen, Invitrogen, and New England Biolabs.

2.1. Disablement

Many laboratory *E. coli* strains carry mutations that reduce their viability in the wild and preclude survival in the intestinal tract (*6*). These often confer auxtrophy, that is, they disable the cell's ability to synthesize a critical metabolite, which, therefore, must be supplied in the medium. Such mutations can also serve as genetic markers and may be useful for correct strain confirmation.

2.1.1. Suppressor Mutations

Some vectors contain nonsense mutations in essential genes as a means of preventing spread to natural bacterial populations. Nonsense mutations are chain-termination codons; they are termed amber (UAG) or ochre (UAA) mutations (*5*). Vectors containing these mutations can only be propagated in strains of *E. coli* that contain the appropriate nonsense suppressors. Amber and ochre suppressors are usually found in tRNA genes, and alter the codon-recognition loop so that a specific amino acid is occasionally inserted at the site of the nonsense mutation. Nonsense suppressors commonly used in cloning strains are given in **Table 3**.

Table 2
Properties of Representative *E. coli* Strains Used for Vector Propagation and Cloning Procedures

Strain[a]	Genotype[b]	Blue-white screening	Cloning methylated DNA	Generation of unmethylated DNA	Reduced recombination	Production of ssDNA	Transformation of large plasmids	Supression of amber mutations	Suppliers[c]
DH10B	Δ(*araABC-leu*)7697 *araD139 deoR endA1 galK galU* Δ(*lac*)X74 *mcrA* Δ(*mcrCB-hsdSMR-mrr*) *nupG recA1 rpsL*(Str^r) (φ80 *lacZ*ΔM15)	•			•				I
DH5α	*deoR endA1 gyrA96 hsdR17* Δ(*lac*)U169 *recA1 relA1 supE44 thi-1* (φ80 *lacZ*ΔM15)	•			•				AI
DM1	*ara dam dcm gal1 gal2 hsdR lac leu mcrB thr tonA tsx zac*::Tn9(Cam^r)		•	•					I
GeneHogs	*araD139* Δ(*ara-leu*)7697 *deoR endA1 galU galK* Δ(*lac*)X74 *mcrA* Δ(*mrr-hsdRMS-mcrBC*) *nupG recA1 rpsL*(Str^r) (φ80 *lacZ*ΔM15) λ$^-$	•			•				I
HB101[d]	*ara-14 galK2 proA2 lacY1 hsdS20 mtl-1 recA13 rpsL20*(Str^r) *supE44 xyl-5*	•	•	•	•		•		BIP
INV110	*ara dam dcm dupE44 endA galK galT* Δ(*lac-proAB*) *lacY leu* Δ(*mcrCB-hsdSMR-mrr*)*102*::Tn*10*(Tet^r) *thi-1 thr tonA tsx* F'[*lacI^q lacZ*ΔM15 *proAB*$^+$ *traD36*]				•	•			I
JM109	*endA1 gyrA96 hsdR17* Δ(*lac-proAB*) *recA1 relA1 supE44 thi-1* F'[*lacI^q lacZ*ΔM15 *proAB*$^+$ *traD36*]	•			•	•		•	AP
JS5	Δ(*araABC-leu*)7697 *araD139 galU galK hsdR2* Δ(*lac*)X74 *mcrA mcrBC recA1 rpsL*(Str^r) *thi* F'[*lacI^q lacZ*ΔM15 *proAB*$^+$ Tn*10*(Tet^r)]					•			B
LE392	*galK2 galT22 hsdR514 lacY1 mcrA metB1 supE44 supF28 trpR55*							•	A

(continued)

Table 2 (continued)

Strain[a]	Genotype[b]	Blue-white screening	Cloning methylated DNA	Generation of unmethylated DNA	Reduced recombination	Production of ssDNA	Transformation of large plasmids	Suppression of amber mutations	Suppliers[c]
NM522	Δ(hsdMS-mcrB)5 Δ(lac-proAB) supE thi-1 F'[lacIq lacZΔM15 proAB+]	•						•	AS
SCS110	ara dam dcm endA galK galT Δ(lac-proAB) lacY leu rpsL(Str^r) supE44 thi-1 thr tonA tsx F'[lacIq lacZΔM15 proAB+ traD36]	•		•					S
STBL4	endA1 gal gyrA96 Δ(lac-proAB) mcrA Δ(mcrCB-hsdSMR-mrr) recA1 relA1 supE44 thi-1 F'[lacIq lacZΔM15 proAB+ Tn10(Tet^r)]		•			•	•	•	I
SURE	endA1 gyrA96 lac mcrA Δ(mcrCB-hsdSMR-mrr)171 recB recJ relA1 sbcC supE44 thi-1 umuC::Tn5(Kan^r) uvrC F'[lacIq lacZΔM15 proAB+]	•			•	•		•	AS
TG1	Δ(hsdMS-mcrB)5 Δ(lac-proAB) supE thi-1 F'[lacIq lacZΔM15 proAB+ traD36]	•				•		•	S
XL10-Gold	endA1 gyrA96 lac Δ(mcrA)183 Δ(mcrCB-hsdSMR-mrr)173 recA1 thi-1 relA1 supE44 Hte F'[lacIq lacZΔM15 proAB+ Tn10(Tet^r) Amy Cam^r]		•		•		•	•	S
XL1-Blue MRF'	endA1 gyrA96 lac Δ(mcrA)183 Δ(mcrCB-hsdSMR-mrr)173 recA1 relA1 supE44 thi-1 F'[lacIq lacZΔM15 proAB+ Tn10(Tet^r) Amy Cam^r]	•			•			•	S

Note: Data compiled from suppliers' catalogs.

[a] All strains are derived from *E. coli* K-12 unless otherwise stated.
[b] Cam is chloramphenicol; Kan is kanamycin; Str is streptomycin; Tet is tetracycline.
[c] A is ATCC; B is Bio-Rad; I is Invitrogen; P is Promega; S is Stratagene.
[d] This strain is a hybrid of *E. coli* K-12 and *E. coli* B.

Table 3
Properties of Common *E. coli* Suppressor Mutations

Mutation	Codons suppressed	Amino acid inserted	tRNA gene supplied
supB	Amber, ochre	Glutamine	*glnU*
supC	Amber, ochre	Tyrosine	*tyrT*
supD	Amber	Serine	*serU*
supE	Amber	Glutamine	*glnV*
supF	Amber	Tyrosine	*tyrT*

2.2. Fertility Status

Some *E. coli* strains carry an F episome or fertility factor, which can be found in several different forms *(7)*. It may be carried as a double-stranded single-copy circular extrachromosomal plasmid, designated F$^+$, or if it harbors additional genes, F'. These extrachromosomal forms can transfer themselves to recipient cells, which are F$^-$, and occasionally cause the mobilization of other plasmids (*see* Chapter 6). In Hfr cells (high-frequency chromosome donation), the F factor is integrated into the bacterial chromosome and can cause chromosomal transfer. Mutations in the locus *tra* inhibit transfer and mobilization.

Strains containing the F factor produce surface pili, which are required for infection by vectors based on filamentous phage. The F factor also permits the production and rescue of single-stranded DNA from M13 vectors when coinfected with a helper phage (*see* Chapter 13).

2.3. Restriction and Modification Systems

Restriction–modification systems play a role in preventing genetic exchange between groups of bacteria by enabling the host to recognize and destroy foreign DNA. An archetypal system consists of a DNA methylase and its cognate restriction endonuclease. The methylase covalently modifies host DNA, by transfer of a methyl group from *S*-adenosylmethionine to a cytosine or adenine residue, within the recognition sequence of its cognate restriction enzyme. Methylation prevents digestion at this site, limiting digestion to incoming foreign DNA *(8)*. The restriction–modification systems present in an *E. coli* host will affect the pattern and extent of recombinant DNA methylation and can significantly affect the success of restriction digestions and bacterial transformations. Many common laboratory strains of *E. coli* that are deficient in one or more restriction–modification systems are available to counteract this problem.

2.3.1. Dam and Dcm Methylation

Derivatives of *E. coli* K-12 normally contain three site-specific DNA methylases: Dam, Dcm and *Eco*K. DNA adenine methylase, encoded by *dam*, methylates adenine residues in the sequence GATC *(9,10)*. This sequence will occur approximately once every 256 bp in a theoretical piece of DNA of random sequence. DNA cytosine methylase, encoded by *dcm*, methylates the internal cytosine residue in the sequence CC(A/T)GG, which occurs on average once every 512 bp *(9,11)*. Almost all commonly used cloning strains are Dam⁺ Dcm⁺. Strains that are $recA^-$ (see **Subheading 2.4.**) are always dam^+, because the combination $recA^-$ dam^- results in a lethal phenotype.

Methylation may interfere with cleavage of DNA cloned and propagated in dam^+ and dcm^+ *E. coli* strains. Not all restriction endonucleases are sensitive to methylation. For example, Dam-modified DNA is not cut by *Bcl*I (TGATCA); however, it is cut by *Bam*HI (GGATCC) *(12)*. The restriction enzyme database, REBASE (rebase.neb.com), contains comprehensive information on the methylation sensitivity of restriction endonucleases *(13)*. Not all DNA isolated from *E. coli* is completely methylated. For example, only about 50% of λ DNA sites are Dam methylated, presumably because λ DNA is rapidly packaged into phage heads. Thus, restriction of such DNA with a Dam-sensitive restriction endonuclease will yield a partial digestion pattern.

The presence of Dam or Dcm methylation can also affect the efficiency of plasmid transformation. For example, Dam-modified DNA cannot be efficiently introduced into a dam^- strain, because replication initiation is inhibited when DNA is hemimethylated. Thus, a transformed plasmid is able to replicate once but not again *(14)*.

Dam⁻ Dcm⁻ strains have the disadvantage that these mutations are mutagenic. This is because in wild-type strains, newly synthesized DNA is hemimethylated and any errors introduced by the polymerase are corrected by mismatch repair systems to the original methylated strand. However, in Dam⁻ Dcm⁻ strains, neither strand is methylated and the mismatch is equally likely to be resolved to the newly synthesized strand as to the correct one *(15)*.

2.3.2. EcoK System

The *E. coli* K-12 *Eco*K methylase modifies the indicated adenine residues of the target sequence $A(^mA)CN_6GTGC$, and its complement $GC(^mA)CN_6GTT$ *(8,16)*. The cognate endonuclease will cleave DNA that is unmodified at this sequence. The *Eco*K system is encoded by the *hsdRMS* locus, where *hsdR* encodes the endonuclease, *hsdM* the methylase, and *hsdS* the site-recognition subunit. *E. coli* strains used for cloning are generally either $hsdR^-$, resulting in a restriction minus phenotype (r_K^- m_K^+), or $hsdS^-$, resulting in a restriction and methylation deficiency (r_K^- m_K^-). Strains derived from *E. coli* B are (r_B^+ m_B^+) and carry the equivalent *Eco*B endonuclease and methylase, which modify the adenosine in the sequence $TGAN_8TGCT$ *(17)*.

Because *Eco*K sites are rare, occurring approximately once every 8 kb, this type of methylation does not generally interfere with restriction digestion. However, transformation of unmodified plasmid DNA into $hsdR^+$ strains results in more than a 1000-fold reduction in efficiency and can lead to underrepresentation of fragments containing *Eco*K sites in libraries. Thus, if transferring DNA between strains with different *Eco*K genotypes, a plasmid should be passed through an $hsdM^+$ strain before introduction into an $hsdR^+$ strain.

2.3.3. McrA, McrBC, and Mrr Restriction

E. coli K-12 also contains several methylation-dependent restriction systems, namely McrA, McrBC, and Mrr. The methylcytosine restricting endonucleases, McrA and McrBC, cleave methylcytosines in the sequences CG and (A/C)G, respectively *(18–21)*. Mrr (methyladenine recognition and restriction) cleaves methyladenines, but the precise recognition sequence is unknown *(22,23)*. None of these three systems cleave Dcm- or Dam-modified DNA and are, thus, generally of little concern when subcloning DNA from *dam⁺ dcm⁺ E. coli*, but using strains mutant in these systems may be desirable if cloning highly methylated DNA from other sources. In addition, when cytosine methylases are used in cloning procedures, such as adding linkers, the recombinant DNA should be transformed into an *mcrA⁻ mcrBC⁻* strain to avoid Mcr restriction *(8)*.

Most of these restriction determinants are clustered in a single "immigration control" locus allowing the removal of *hsdRMS*, *mcrBC*, and *mrr* by a single deletion: Δ(*mcrCB-hsdSMR-mrr*) *(19)*.

2.4. Recombination

Following successful transformation of a plasmid vector into *E. coli*, host recombination systems can catalyze rearrangement of the recombinant molecule. This is a particular problem when the cloned DNA contains direct or inverted repeats and can result in duplications, inversions, or deletions. If the resulting product is smaller than the original molecule, it will replicate faster and quickly dominate the population. Mutations in the host that suppress recombination can help maintain the integrity of cloned DNA. Recombination properties are especially relevant to the choice of hosts for library propagation in order to avoid misrepresentation because of the unequal growth of specific clones. However, recombination-deficient strains are generally unfit and suffer from enhanced sensitivity to DNA-damaging agents, deficiency in repairing double-strand breaks in DNA, slow growth rate, and the rapid accumulation of nonviable cells *(24)*; thus, depending on the application, Rec⁺ strains may still be preferable.

E. coli contains three main recombination pathways encoded by *recBCD*, *recE*, and *recF (25,26)*. All three pathways depend on the product of *recA*, with the notable exception of recombination of certain plasmids and phage promoted by the RecE pathway. Hence, *recA⁻* is the most stringent Rec⁻ condition and mutations in *recA* reduce recombination 10,000-fold compared to wild type, almost completely blocking recombination.

The RecBCD, or exonuclease V (ExoV), pathway is predominant in wild-type *E. coli* K-12. Strains with single mutations in *recB* or *recC*, and *recBC* double mutants are defective in this pathway and have indistinguishable phenotypes exhibiting recombination rates 100- to 1000-fold lower than wild type *(27)*. These strains are unfit and tend to accumulate extragenic suppressor mutations in both *sbcB* (suppressor of RecBC⁻), encoding ExoI, and *sbcC (28–30)*. The secondary mutations enable efficient recombination to be catalyzed by the RecF pathway and restore viability *(25)*. In *recBC⁻* strains, the RecE (ExoVIII) pathway is activated by mutations in *sbcA (31)*. Both *recE* and *sbcA* map to the cryptic lambdoid prophage *rac* that is present in most *E. coli* K-12 strains *(32)*. In contrast, mutation in *recD*, which encodes the nuclease activity of ExoV, results in a healthy Rec⁺ phenotype that does not acquire secondary mutations *(33)*.

Cloned palindromes or interrupted palindromes are highly unstable in wild-type *E. coli*. Both *recBC⁻ (34)* and *recD⁻ (35,36)* strains are good hosts for palindrome stabilization in λ-derived vectors. However, most cloning plasmids are unstable in *recBC⁻* and *recD⁻* strains and are difficult to maintain, even with selection *(33,37)*. The problem is especially severe with high-copy-number ColE1 derivatives; this is probably the result of recombination-initiated rolling-circle replication, which results in long linear multimers that do not segregate properly at cell division *(38)*. Mutation in *recA* or *recF* is able to suppress this effect *(37,39)*. Mutations in *sbcBC* also independently stabilize cloned palindromes and *sbcC⁻* strains are permissive for palindromes in plasmids as well as phage *(35,36,40)*.

2.4.1. Recombination Systems in λ-Infected Hosts

Bacteriophage λ is injected into the *E. coli* host as a linear molecule that rapidly circularizes and, during the early phase of infection, replicates by a bidirectional θ-type mechanism, yielding monomeric circles. Subsequently, replication converts to a rolling-circle σ-type mechanism, generating linear concatemers that are suitable substrates for packaging into phage heads *(41)*.

Rolling-circle replication is inhibited by host RecBCD, which degrades the linear concatameric DNA. Thus, efficient propagation by rolling-circle replication requires a *recBC⁻ sbcB⁻* or *recD⁻* host. Alternatively, the exonucleolytic activity RecBCD can be inhibited by the product of the λ *gam* gene, which may be carried on the λ vector itself or on a separate plasmid *(42,43)*.

Infection of *recBCD⁺* strains with *gam⁻* λ will result in the production of the progeny phage only if a suitable recombination pathway exists to convert monomeric circles, produced by θ-replication, to multimeric circles that are acceptable substrates for packaging. Either λ-encoded Red recombinase or host RecA are able to catalyze this reaction *(42)*. Most λ are *gam⁻ red⁻* and, therefore, require a RecA⁺ host for propagation.

The presence of the octameric sequence GCTGGTGG, termed a χ (chi) site *(44)*, in the *gam⁻* λ genome can overcome inefficient multiplication in a *recBC⁺* background *(45)*. The χ site in the λ recombinant causes increased recombination, by a RecBCD-dependent pathway, requiring RecA, resulting in more efficient conversion from monomeric to multimeric circular forms. It should be noted that cloned sequences containing a χ site will be overrepresented in libraries constructed in *gam⁻ χ⁻* vectors if propagated in a *recBC⁺* host.

2.5. α-Complementation

Many current molecular biology techniques rely on the pioneering studies of the *lac* operon by Jacob and Monod in the 1960s *(46)*. The *lac* operon consists of three genes: *lacZYA*, encoding β-galactosidase, which cleaves lactose to glucose and galactose, a permease, and a transacetylase. The *lac* repressor, encoded by the neighboring *lacI* gene, derepresses transcription of the *lac* operon in the presence of lactose *(47)*.

Cells bearing 5' deletions in *lacZ* produce an inactive C-terminal fragment of β-galactosidase termed the ω-fragment; similarly, cells with a 3' deletion in *lacZ* (*lacZ'*) synthesize an inactive N-terminal α-fragment. However, if both fragments are produced in the same cell then β-galactosidase activity is restored *(48)*. This phenomenon, known as α-complementation, is the basis for the visual selection of clones containing recombinant vectors by "blue-white screening" (*see* Chapter 19). The vector expresses the α-fragment and requires a host that expresses the ω-fragment. Generally, the host is engineered to carry the chromosomal deletion Δ*(lac-proAB)*; this mutation is partially complemented by *lacZ*ΔM15, which consists of the lac operon minus the *lacZ'* segment and is often carried, along with *lacI*^{q} *(49)*, on the lambdoid prophage φ80 or the F' plasmid. The F' episome is also usually *proAB*$^+$ to rescue proline auxotrophy and allow maintenance of the plasmid on proline-deficient minimal media.

To select for recombinant *E. coli*, bacilli are grown on media containing the nonfermentable lactose analog isopropyl-β-D-thiogalactoside (IPTG), which inactivates the *lac* repressor and derepresses ω-fragment synthesis. In the presence of IPTG, the chromogenic lactose analog 5-bromo-4-chloro-3-indoxyl-β-D-galactopyranoside (X-Gal) is cleaved by β-galactosidase to a blue-colored product. Cloning vectors that allow blue-white screening contain a multiple cloning site embedded within the α-fragment. Insertion of a DNA fragment within this region abolishes production of the α-fragment, and colonies grown on IPTG and X-Gal appear white.

3. Hosts for Mutagenesis

The frequency of spontaneous mutation in *E. coli* may be increased by three to four orders of magnitude by mutations in *mutD*, which encodes the 3'g5' exonuclease subunit of the DNA polymerase III holoenzyme *(50,51)*. Thus, random mutagenesis can be achieved by maintaining plasmids in a *mutD*$^-$ strain for a number of generations and subsequently transforming the mutated plasmid into a *mutD*$^+$ "tester" strain. This method provides a useful alternative to chemical mutagenesis.

Site-directed mutagenesis methods frequently involve intermediates that contain wild-type/mutant heteroduplexes. Such heteroduplexes are stabilized in *mutS* mutants, which are deficient in mismatch repair, leading to high mutation efficiencies.

Kunkel mutagenesis requires a specialized *dut*$^-$ *ung*$^-$ host strain, which does not express dUTPase or uracil-*N*-glycosylase, resulting in the occasional substitution of uracil for thymine in newly synthesized DNA *(52)*. In this procedure, single-stranded template DNA is prepared from a *dut*$^-$ *ung*$^-$ host; next, a mutant primer is annealed to the template and the second strand is synthesized. Subsequent transformation of the heteroduplex into an *ung*$^+$ strain will result in digestion of the uracil-containing parental strand, enriching for the mutant strand.

Various hosts that are useful for mutagenesis procedures are listed in **Table 4**.

4. Specialized Strains for Protein Expression

E. coli is a popular host for the overexpression of recombinant proteins (*see* Chapters 28 and 29). There are a number of factors that can influence protein yields and careful strain choice can greatly improve the chance of successful expression. Recent innovations have resulted in the availability of many new host strains, a selection of which are given in **Table 5**.

4.1. Repressors

E. coli expression vectors utilize highly active inducible promoters and the correct host strain must be used to ensure proper tight regulation *(53)*. Many common vectors

Table 4
Properties of E. coli Strains Used as Hosts for Mutagenesis

Strain[a]	Genotype[b]	Application	Supplier[c]
BMH 71-18 mutS	Δ(lac-proAB) mutS::Tn10(Tet^r) supE thi-1 F'[lacI^q lacZΔM15 proAB^+]	Used for site-directed mutagenesis	P
CJ236	dut1 mcrA relA1 spoT1 thi-1 ung1 (pCJ105 F^+ Cam^r)	Used for generation of uracil-substituted DNA for Kunkel mutagenesis	B
MV1190	Δ(lac-proAB) Δ(srl-recA)306::Tn10(Tet^r) supE thi F'[lacI^q lacZΔM15 proAB^+ traD36]	Used for enrichment of mutant DNA	B
XL1-Red	endA1 gyrA96 hsdR17 lac mutD5 mutS mutT relA1 supE44 thi-1 Tn10(Tet^r)	Used for random mutagenesis	S
XL-mutS	endA1 gyrA96 lac ΔmcrA183 Δ(mcrCB-hsdSMR-mrr)173 mutS::Tn10(Tet^r) relA1 supE44 thi-1 F'[lacI^q lacZΔM15 proAB^+ Tn5(Kan^r)]	Used for site-directed mutagenesis	S

Note: Date compiled from suppliers' catalogs.
[a] All strains are derived from *E. coli* K-12.
[b] Cam is chloramphenicol; Kan is kanamycin; Tet is tetracycline.
[c] B is Bio-Rad; P is Promega; S is Stratagene.

use the *lac* promoter, the related *lacUV5* promoter, or the *tac* promoter, which is a synthetic hybrid of the *lac* and *trp* promoters (*see* Chapter 29). These promoters are repressed in the presence of the chromosomal *lacI*q allele; however, high-copy-number plasmids require *lacI* or *lacI*q to be supplied *in trans*, on a compatible plasmid, to prevent leakiness. The *lac* promoters can be regulated by the lactose analog IPTG. Improved control can be achieved by using *lacY* mutants that prevent Lac permease-mediated active transport of IPTG. IPTG thus enters the cell in a concentration-dependent manner and the recombinant protein is uniformly expressed in all cells.

Another popular system is based on the bacteriophage T7 RNA polymerase (RNAP) and puts the recombinant protein under the control of the T7 late promoter. The T7 RNAP is regulated by the IPTG-inducible *lacUV5* promoter and is usually supplied *in trans* from the λ(DE3) lysogen. For the expression of toxic proteins, tighter control can be achieved in hosts that express T7 lysozyme, a natural inhibitor of T7 RNAP; by inhibiting basal levels of RNAP, expression of the target gene is reduced prior to induction. The plasmids pLysS or pLysE express T7 lysozyme at low and high levels, respectively, enabling variable levels of expression control (*see* Chapter 28).

4.2. Stability

Host proteases can interfere with the isolation of intact recombinant proteins; degradation may be avoided by the use of protease-deficient hosts. In *E. coli*, *lon* encodes a major ATP-dependent protease and strains that contain deletions of this gene greatly improve the yield of many recombinant proteins *(54,55)*. An *rpoH* mutation represses Lon expression and also independently decreases the rate of protein degradation *(56)*. Mutations in the gene for the outer-membrane protease OmpT also improve the recovery of intact recombinant proteins, especially if purified from whole-cell lysates *(57)*.

Rapid degradation of mRNA may be the limiting factor in the expression of certain genes, particularly when using T7 RNAP-based systems in which transcription is not coupled to translation. An *rne* mutation, abolishing RNaseE activity, eliminates a major source of RNA degradation increasing the availability of mRNA for translation *(58,59)*.

ABLE C and ABLE K strains express a heterogenous DNA polymerase I and reduce the copy number of ColE1-derived plasmids by 4-fold and 10-fold, respectively. The resulting reduction in the basal expression level of toxic recombinant proteins improves cell viability. The availability of both strains allows the choice of the highest plasmid copy number that is still permissive for growth *(60)*.

4.3. Codon Bias

The frequency with which amino acid codons are utilized varies between organisms and is reflected by the abundance of the cognate tRNA species. This codon bias can have a significant impact on heterologous protein expression, so that genes that contain a high proportion of rare codons are poorly expressed *(61,62)*. A subset of the codons for arginine, isoleucine, glycine, leucine, and proline are rarely used in *E. coli*. The forced high-level expression of genes containing these codons results in a depletion of internal tRNA pools and can lead to translational stalling, frame shifting, premature termination, or amino acid misincorporation *(63)*. Recombinant protein expression can

Table 5
Properties of *E. coli* Strains Commonly Used for Recombinant Protein Expression

Strain	Genotype[a]	Derivation[b]	Key features	Supplier[c]
ABLE C, ABLE K	hsdS lac mcrA mcrBC mcrF mrr (Kan^r) F′[lacI^q lacZΔM15 proAB$^+$ Tn10(Tet^r)]	C strain	Reduces plasmid copy number; useful for expression of toxic proteins	S
AD494[d,e]	Δ(araABC-leu)7697 ΔlacX74 ΔmalF3 ΔphoAPvuII phoR trxB::Kan^r F′[lacI^q lacZΔM15 proAB$^+$]	K-12	Enhances cytoplasmic disulfide bond formation	N
B834[d,e]	gal hsdS$_B$ met ompT	B strain	Protease deficient; used for labeling with ^{35}S-methionine	N
BL21[d-f]	gal hsdS$_B$ ompT	B834	Protease deficient	INS
BL21 Star[d,e]	gal hsdS$_B$ ompT rne131	BL21	Improves stability of mRNA	I
BL21 CodonPlus-RIL[d]	endA gal ompT hsdS$_B$ Dcm$^+$ Hte Tet^r (pACYC-RIL argU ileY leuW Cam^r)	BL21	Expresses rare tRNAs; useful for AT-rich genomes	S
BL21 CodonPlus-RP[d]	endA gal ompT hsdS$_B$ Dcm$^+$ Hte Tet^r (pACYC-RP argU proL Cam^r)	BL21	Expresses rare tRNAs, useful for GC-rich genomes	S
BL21 trxB[d,e]	gal hsdS$_B$ ompT trxB15::Kan^r	BL21	Enhances cytoplasmic disulfide bond formation	N
BLR[d,e]	gal hsdS$_B$ ompT Δ(srl-recA)306::Tn10(Tet^r)	BL21	Stabilizes repetitive sequences and prevents loss of l prophage	N

Strain	Genotype	Parent	Notable features	Source
Origami[d,e]	araD139 Δ(araABC-leu)7697 galE galK gor522::Tn10(Tet[r]) ΔlacX74 ΔphoAPvuII phoR rpsL(Str[r]) trxB::Kan[r] F'[lacI[q] lacZΔM15 pro⁺]	K-12	Greatly enhances cytoplasmic disulfide bond formation	N
Rosetta[d,e]	gal hsdS$_B$ lacY1 ompT (pRARE araW argU glyT ileX leuW proL metT thrT tyrU thrU Cam[r])	Tuner	Expresses rare tRNAs; improves IPTG-mediated expression control	N
TKB1	gal hsdSB ompT (DE3) (pTK Tet[r])[g]	B strain	Generates phosphorylated proteins	S
TKX1	endA1 gyrA96 lac Δ(mcrA)183 Δ(mcrCB-hsdSMR-mrr)173 recA1 relA1 supE44 thi-1 F'[lacI[q] lacZΔM15 proAB⁺ Tn5(Kan[r])] (pTK Tet[r])[g]	K-12	Generates phosphorylated proteins	S
Tuner[d,e]	gal hsdS$_B$ lacY1 ompT	BL21	Improves IPTG-mediated expression control	N

Note: Data compiled from suppliers' catalogs.

[a]Cam is chloramphenicol; Kan is kanamycin; Str is streptomycin; Tet is tetracycline.
[b]All B strain derivatives are naturally lon and dcm.
[c]I is Invitrogen; N is Novagen; S is Stratagene.
[d]Available as a lysogen of l(DE3).
[e]Available as (DE3)pLysS.
[f]Available as (DE3)pLysE.
[g]ColE1-compatible plasmid harboring elk controlled by the trp promoter.

be rescued by using hosts that express tRNAs for rare codons and thus provide "universal" translation *(64,65)*. *E. coli* hosts are available that supply genes for rare tRNAs, particularly *argU* (AGA/AGG), *ileY* (AUA), *glyT* (GGA), *leuW* (CUA), and *proL* (CCC), in combinations optimized for the expression of genes from AT- or GC-rich genomes. The laborious classical method of altering individual codons in the target gene, by site-directed mutagenesis, is obviated by the availability of these useful hosts.

4.4. Solubility and Posttranslational Processing

Overproduction of heterologous proteins in *E. coli* often results in misfolding and segregation into insoluble inclusion bodies. The cytoplasmic chaperones, DnaK-DnaJ and GroES-GroEL, assist proper folding in wild-type *E. coli* and there is evidence that co-overproduction of either complex increases the yield of soluble proteins from recombinant *E. coli (66)*.

The *E. coli* cytoplasm is a reducing environment that strongly disfavors the formation of stable disulfide bonds. Mutations in *trxB* and *gor*, which encode thioredoxin and glutathione reductases, facilitate cytoplasmic disulfide bond formation and increase the efficiency of oxidized recombinant protein accumulation. Thus $gor^- trxB^-$ mutants are useful for the production of proteins whose solubility depends on proper oxidation *(67–69)*.

Wild-type *E. coli* lack the ability to phosphorylate tyrosine residues. However, specialized host strains that carry the *elk* tyrosine kinase gene are able to produce tyrosine-phosphorylated proteins that may be required for affinity screening of expression libraries or for the purification of SH2 domain-containing proteins *(70,71)*.

5. Conclusion

Since the first mutants of *E. coli* K-12 were isolated in the 1940s, laboratory strains have been heavily mutagenized by treatment with X-rays, ultraviolet irradiation, and nitrogen mustard. Thus, they may carry unidentified mutations and it can be useful to try more than one strain background if experiments are unsuccessful.

References

1. Blattner, F. R., Plunkett, G., Bloch, C. A., et al. (1997) The complete genome sequence of *Escherichia coli* K-12. *Science* **277,** 1453–1474.
2. Tatum, E. L. and Lederberg, J. (1947) Gene recombination in the bacterium *Escherichia coli*. *J. Bacteriol.* **53,** 673–684.
3. Berlyn, M. K. B., Low, K. B., Rudd, K. E. et al. (1996) Linkage map of *Escherichia coli* K-12, in Escherichia coli *and* Salmonella: *Cellular and Molecular Biology* (Niedhardt, F. C., ed.), ASM, Washington, DC, pp. 1715–1902.
4. Demerec, M., Adelberg, E. A., Clark, A. J. et al. (1966) A proposal for a uniform nomenclature in bacterial genetics. *Genetics* **54,** 61–76.
5. Brown, T. A. (ed.) (1998) *Molecular Biology LabFax I: Recombinant DNA*. BIOS, Oxford.
6. Curtiss, R., III, Pereira, D. A., Hsu, J. C., et al. (1977) Biological containment: the subordination of *Escherichia coli* K-12, in *Recombinant Molecules: Impact on Science and Society* (Beers, R. F., Jr. and Bassett, E. G., eds.), Raven, New York.

7. Frost, L. S., Ippen-Ihler, K., and Skurray, R. A. (1994) Analysis of the sequence and gene products of the transfer region of the F sex factor. *Microbiol. Rev.* **58**, 162–210.
8. Raleigh, E. A. (1987) Restriction and modification in vivo by *Escherichia coli* K12. *Methods Enzymol.* **152**, 130–141.
9. Marinus, M. G. and Morris, N. R. (1973) Isolation of deoxyribonucleic acid methylase mutants of *Escherichia coli* K-12. *J. Bacteriol.* **114**, 1143–1150.
10. Geier, G. E. and Modrich, P. (1979) Recognition sequence of the dam methylase of *Escherichia coli* K12 and mode of cleavage of DpnI endonuclease. *J. Biol. Chem.* **254**, 1408–1413.
11. May, M. S. and Hattman, S. (1975) Analysis of bacteriophage deoxyribonucleic acid sequences methylated by host- and R-factor-controlled enzymes. *J. Bacteriol.* **123**, 768–770.
12. McClelland, M., Nelson, M., and Raschke, E. (1994) Effect of site-specific modification on restriction endonucleases and DNA modification methyltransferases. *Nucleic Acids Res* **22**, 3640–3659.
13. Roberts, R. J. and Macelis, D. (2001) REBASE: restriction enzymes and methylases. *Nucleic Acids Res.* **29**, 268–269.
14. Russell, D. W. and Zinder, N. D. (1987) Hemimethylation prevents DNA replication in *E. coli*. *Cell* **50**, 1071–1079.
15. Marinus, M. G. (1987) DNA methylation in *Escherichia coli*. *Annu. Rev. Genet.* **21**, 113–131.
16. Bickle, T. A. and Kruger, D. H. (1993) Biology of DNA restriction. *Microbiol. Rev.* **57**, 434–450.
17. Ravetch, J. V., Horiuchi, K., and Zinder, N. D. (1978) Nucleotide sequence of the recognition site for the restriction-modification enzyme of *Escherichia coli* B. *Proc. Natl. Acad. Sci. USA* **75**, 2266–2270.
18. Raleigh, E. A. and Wilson, G. (1986) *Escherichia coli* K-12 restricts DNA containing 5-methylcytosine. *Proc. Natl. Acad. Sci. USA* **83**, 9070–9074.
19. Kelleher, J. E. and Raleigh, E. A. (1991) A novel activity in *Escherichia coli* K-12 that directs restriction of DNA modified at CG dinucleotides. *J. Bacteriol.* **173**, 5220–5223.
20. Sutherland, E., Coe, L., and Raleigh, E. A. (1992) McrBC: a multisubunit GTP-dependent restriction endonuclease. *J. Mol. Biol.* **225**, 327–348.
21. Raleigh, E. A. (1992) Organization and function of the mcrBC genes of *Escherichia coli* K-12. *Mol. Microbiol.* **6**, 1079–1086.
22. Heitman, J. and Model, P. (1987) Site-specific methylases induce the SOS DNA repair response in *Escherichia coli*. *J. Bacteriol.* **169**, 3243–3250.
23. Waite-Rees, P. A., Keating, C. J., Moran, L. S., et al. (1991) Characterization and expression of the *Escherichia coli* Mrr restriction system. *J. Bacteriol.* **173**, 5207–5219.
24. Capaldo, F. N., Ramsey, G., and Barbour, S. D. (1974) Analysis of the growth of recombination-deficient strains of *Escherichia coli* K-12. *J. Bacteriol.* **118**, 242–249.
25. Mahajan, S. K. (1988). Pathways of homologous recombination in *Escherichia coli*, in *Genetic Recombination* (Kucherlapati, R. and Smith, G. R., eds.), ASM, Washington, DC.
26. Camerini-Otero, R. D. and Hsieh, P. (1995) Homologous recombination proteins in prokaryotes and eukaryotes. *Annu. Rev. Genet.* **29**, 509–552.
27. Howard-Flanders, P. and Theriot, L. (1966) Mutants of *Escherichia coli* K-12 defective in DNA repair and in genetic recombination. *Genetics* **53**, 1137–1150.
28. Templin, A., Kushner, S. R., and Clark, A. J. (1972) Genetic analysis of mutations indirectly suppressing *recB* and *recC* mutations. *Genetics* **72**, 105–115.

29. Kushner, S. R., Nagaishi, H., and Clark, A. J. (1972) Indirect suppression of *recB* and *recC* mutations by exonuclease I deficiency. *Proc. Natl. Acad. Sci. USA* **69,** 1366–1370.
30. Lloyd, R. G. and Buckman, C. (1985) Identification and genetic analysis of *sbcC* mutations in commonly used *recBC sbcB* strains of *Escherichia coli* K-12. *J. Bacteriol.* **164,** 836–844.
31. Barbour, S. D., Nagaishi, H., Templin, A., et al. (1970) Biochemical and genetic studies of recombination proficiency in *Escherichia coli* II: Rec$^+$ revertants caused by indirect suppression of *rec*$^-$ mutations. *Proc. Natl. Acad. Sci. USA* **67,** 128–135.
32. Kaiser, K. and Murray, N. E. (1979) Physical characterisation of the "Rac prophage" in *E. coli* K12. *Mol. Gen. Genet.* **175,** 159–174.
33. Biek, D. P. and Cohen, S. N. (1986) Identification and characterization of *recD*, a gene affecting plasmid maintenance and recombination in *Escherichia coli*. *J. Bacteriol.* **167,** 594–603.
34. Leach, D. R. and Stahl, F. W. (1983) Viability of lambda phages carrying a perfect palindrome in the absence of recombination nucleases. *Nature* **305,** 448–451.
35. Wyman, A. R., Wertman, K. F., Barker, D., et al. (1986) Factors which equalize the representation of genome segments in recombinant libraries. *Gene* **49,** 263–271.
36. Wertman, K. F., Wyman, A. R., and Botstein, D. (1986) Host/vector interactions which affect the viability of recombinant phage lambda clones. *Gene* **49,** 253–262.
37. Bassett, C. L. and Kushner, S. R. (1984) Exonucleases I, III, and V are required for stability of ColE1-related plasmids in *Escherichia coli*. *J. Bacteriol.* **157,** 661–664.
38. Cohen, A. and Clark, A. J. (1986) Synthesis of linear plasmid multimers in *Escherichia coli* K-12. *J. Bacteriol.* **167,** 327–335.
39. Silberstein, Z. and Cohen, A. (1987) Synthesis of linear multimers of OriC and pBR322 derivatives in *Escherichia coli* K-12: role of recombination and replication functions. *J. Bacteriol.* **169,** 3131–3137.
40. Chalker, A. F., Leach, D. R., and Lloyd, R. G. (1988) *Escherichia coli sbcC* mutants permit stable propagation of DNA replicons containing a long palindrome. *Gene* **71,** 201–205.
41. Hendrix, R. W., Roberts, J. W., Stahl, F. W., et al. (eds.) (1983) *Lambda II*, CSHL, New York.
42. Enquist, L. W. and Skalka, A. (1973) Replication of bacteriophage lambda DNA dependent on the function of host and viral genes I: Interaction of *red, gam* and *rec*. *J. Mol. Biol.* **75,** 185–212.
43. Crouse, G. F. (1985) Plasmids supplying the Q-*qut*-controlled *gam* function permit growth of lambda *red*$^-$ *gam*$^-$ (Fec$^-$) bacteriophages on *recA*$^-$ hosts. *Gene* **40,** 151–155.
44. Stahl, F. W. (1979) Special sites in generalized recombination. *Annu. Rev. Genet.* **13,** 7–24.
45. Lam, S. T., Stahl, M. M., McMilin, K. D., et al. (1974) Rec-mediated recombinational hot spot activity in bacteriophage lambda II: a mutation which causes hot spot activity. *Genetics* **77,** 425–433.
46. Jacob, F. and Monod, J. (1961) Genetic regulatory mechanisms in the synthesis of proteins. *J. Mol. Biol.* **3,** 318–356.
47. Miller, J. H. (1978). The *lacI* gene: its role in *lac* operon control and its use as a genetic system, in *The Operon* (Miller, J. H. and Reznikoff, W. S., eds.), CSHL, New York, pp. 31–88.
48. Ullmann, A., Jacob, F., and Monod, J. (1967) Characterization by in vitro complementation of a peptide corresponding to an operator-proximal segment of the β-galactosidase structural gene of *Escherichia coli*. *J. Mol. Biol.* **24,** 339–343.

49. Muller-Hill, B., Crapo, L., and Gilbert, W. (1968) Mutants that make more *lac* repressor. *Proc. Natl. Acad. Sci. USA* **59**, 1259–1264.
50. Maki, H. and Kornberg, A. (1985) The polymerase subunit of DNA polymerase III of *Escherichia coli* II: Purification of the alpha subunit, devoid of nuclease activities. *J. Biol. Chem.* **260**, 12,987–12,992.
51. Degnen, G. E. and Cox, E. C. (1974) Conditional mutator gene in *Escherichia coli*: isolation, mapping, and effector studies. *J. Bacteriol.* **117**, 477–487.
52. Kunkel, T. A. (1985) Rapid and efficient site-specific mutagenesis without phenotypic selection. *Proc. Natl. Acad. Sci. USA* **82**, 488–492.
53. Makrides, S. C. (1996) Strategies for achieving high-level expression of genes in *Escherichia coli*. *Microbiol. Rev.* **60**, 512–538.
54. Phillips, T. A., Van Bogelen, R. A., and Neidhardt, F. C. (1984) *lon* gene product of *Escherichia coli* is a heat-shock protein. *J. Bacteriol.* **159**, 283–287.
55. Gottesman, S. (1996) Proteases and their targets in *Escherichia coli*. *Annu. Rev. Genet.* **30**, 465–506.
56. Goff, S. A., Casson, L. P., and Goldberg, A. L. (1984) Heat shock regulatory gene *htpR* influences rates of protein degradation and expression of the *lon* gene in *Escherichia coli*. *Proc. Natl. Acad. Sci. USA* **81**, 6647–6651.
57. Grodberg, J. and Dunn, J. J. (1988) *ompT* encodes the *Escherichia coli* outer membrane protease that cleaves T7 RNA polymerase during purification. *J. Bacteriol.* **170**, 1245–1253.
58. Lopez, P. J., Marchand, I., Joyce, S. A., et al. (1999) The C-terminal half of RNaseE, which organizes the *Escherichia coli* degradosome, participates in mRNA degradation but not rRNA processing in vivo. *Mol. Microbiol.* **33**, 188–199.
59. Grunberg-Manago, M. (1999) Messenger RNA stability and its role in control of gene expression in bacteria and phages. *Annu. Rev. Genet.* **33**, 193–227.
60. Greener, A. (1993) Expand your library by retrieving toxic clones with ABLE strains. *Strategies* **6**, 7–9.
61. Kane, J. F. (1995) Effects of rare codon clusters on high-level expression of heterologous proteins in *Escherichia coli*. *Curr. Opin. Biotechnol.* **6**, 494–500.
62. Zahn, K. (1996) Overexpression of an mRNA dependent on rare codons inhibits protein synthesis and cell growth. *J. Bacteriol.* **178**, 2926–2933.
63. Kurland, C. and Gallant, J. (1996) Errors of heterologous protein expression. *Curr. Opin. Biotechnol.* **7**, 489–493.
64. Brinkmann, U., Mattes, R. E., and Buckel, P. (1989) High-level expression of recombinant genes in *Escherichia coli* is dependent on the availability of the *dnaY* gene product. *Gene* **85**, 109–114.
65. Baca, A. M. and Hol, W. G. (2000) Overcoming codon bias: a method for high-level overexpression of *Plasmodium* and other AT-rich parasite genes in *Escherichia coli*. *Int. J. Parasitol.* **30**, 113–118.
66. Thomas, J. G., Ayling, A., and Baneyx, F. (1997) Molecular chaperones, folding catalysts, and the recovery of active recombinant proteins from *E. coli*: to fold or to refold. *Appl. Biochem. Biotechnol.* **66**, 197–238.
67. Derman, A. I., Prinz, W. A., Belin, D., et al. (1993) Mutations that allow disulfide bond formation in the cytoplasm of *Escherichia coli*. *Science* **262**, 1744–1747.
68. Prinz, W. A., Aslund, F., Holmgren, A., et al. (1997) The role of the thioredoxin and glutaredoxin pathways in reducing protein disulfide bonds in the *Escherichia coli* cytoplasm. *J. Biol. Chem.* **272**, 15,661–15,667.

69. Bessette, P. H., Aslund, F., Beckwith, J., et al. (1999) Efficient folding of proteins with multiple disulfide bonds in the *Escherichia coli* cytoplasm. *Proc. Natl. Acad. Sci. USA* **96,** 13,703–13,708.
70. Lhotak, V., Greer, P., Letwin, K., et al. (1991) Characterization of *elk*, a brain-specific receptor tyrosine kinase. *Mol. Cell Biol.* **11,** 2496–2502.
71. Simcox, M. E., Huvar, A., Simcox, T. G., et al. (1994) TK *E. coli* strains for producing tyrosine-phosphorylated proteins in vivo. *Strategies* **7,** 68–69.

4

Chemical Transformation of *E. coli*

W. Edward Swords

1. Introduction

Transformation is defined as the transfer of genetic information into a recipient bacterium using naked DNA, without any requirement for contact with a donor bacterium. The ability to transform or accept exogenous DNA is generally referred to as competence, although the term has been so widely used in different systems that it is difficult to generate an all-inclusive definition for competence. Natural competence occurs in a defined subset of bacterial species that have the capacity to take up linear, and sometimes circular, DNA, usually dependent on a specific uptake system. As natural competence is restricted to a subset of bacteria, methods for the chemical induction of a competent state in otherwise nontransformable bacteria are an important tool in bacterial genetics. For these species, competence refers to the ability to take up and propagate plasmid DNA, usually with no sequence specificity for uptake.

Although not fully understood, chemical methods for the transformation of *Escherichia coli* probably work by transiently opening gated membrane channels, and they require treatment with polyvalent cations and incubation at low temperature. Transient periods of heat and ionic shock probably result in a rapid influx of extracellular medium into the bacterium, after which a recovery period on rich, nonselective medium is usually necessary to ensure full viability of the transformants.

The introduction of plasmids into *E. coli* is an essential step for molecular cloning experiments, and a number of different procedures have been described for this purpose. These may generally be divided into electroporation and chemical transformation methods. Transformation using electroporation is covered in Chapter 5. Advantages of chemical transformation include ease, relative efficiency, and lack of need for a specialized apparatus such as an electroporator. It is important to note that many of the pitfalls encountered in chemical transformation procedures relate to basic issues of bacteriology. The use of isolated colonies for inocula and careful monitoring of the growth phase of the bacterial cultures are essential in the generation of highly chemically competent bacterial cells. A detailed analysis of factors critical to chemical transformation is provided in the classic article by Hanahan *(1)*. The author intends this chapter as a concise, workable summary of several easy and reproducible methods rather than a full review of chemical transformation methods and how they evolved. References provided for each method give more detailed reviews.

2. Materials
2.1. Preparation of Competent Cells
2.1.1. Classical Calcium Chloride Method
1. Host bacterial strain (*see* **Note 1**).
2. Luria–Bertani (LB) broth: 5 g/L tryptone, 10 g/L yeast extract, 5 g/L NaCl. Sterilize by autoclaving.
3. 0.1 M $CaCl_2$; filter sterilize (*see* **Note 2**).
4. 80% Glycerol, sterile.

2.1.2. Modified Calcium Chloride Method (see **Note 3**)
1. **Items 1** and **2** from **Subheading 2.1.1.**
2. 1 M 2-(*N*-morpholino)ethanesulfonic acid (MES) buffer. Adjust the pH to 6.3 using concentrated KOH, filter sterilize, and store at –20°C.
3. TFB buffer: 10 mM MES (pH 6.3), 45 mM $MnCl_2$, 10 mM $CaCl_2$, 100 mM RbCl (KCl may be substituted), 3 mM hexamine cobalt chloride. Prepare by adding salts to diluted MES buffer (*see* **Note 2**).
4. 1 M Potassium acetate solution. Adjust the pH to 7.0, filter sterilize, and store at –20°C.
5. ESB buffer: 10 mM potassium acetate, 100 mM KCl, 45 mM $MnCl_2$, 10 mM $CaCl_2$, 100 mM RbCl (optional), 10% glycerol. Prepare by adding salts, as solids, to diluted potassium acetate solution and adjust the pH of the complete solution to 6.4 with HCl (*see* **Note 2**).
6. 80% Glycerol, sterile.

2.1.3. PEG Method
1. **Items 1** and **2** from **Subheading 2.1.1.**
2. TSS (transformation and storage) medium: LB broth (**Subheading 2.1.1., item 2**), 10% (w/v) polyethylene glycol (PEG), 5% dimethyl sulfoxide (DMSO), 50 mM $MgCl_2$ (pH 6.5) (*see* **Note 2**).

2.2. Transformation of Competent Cells
1. Competent *E. coli*, prepared as described in **Subheading 3.1.** (*see* **Note 4**).
2. Plasmid DNA.
3. LB broth; *see* **Subheading 2.1.1., item 2**.
4. LB agar: 5 g/L tryptone, 10 g/L yeast extract, 5 g/L NaCl, 15g/L bacteriological agar. Sterilize by autoclaving and add antibiotics as appropriate.

3. Methods
3.1. Preparation of Competent Cells
3.1.1. Classical Calcium Chloride Method **(2,3)**

This method was the first generally applicable method for transformation of *E. coli* with plasmid DNA with typical yields of 1×10^7 transformants per microgram of DNA and is still in wide use. Major factors influencing yield include the growth phase

Chemical Transformation

of the bacteria, which seems more crucial with this method than any of the others listed, and the purity of the water used in making the CaCl$_2$ solution.

1. Inoculate 100 mL LB broth with a suitable *E. coli* host strain. The culture should be incubated at 37°C and shaken vigorously (250–300 rpm in a rotary shaker) to ensure sufficient aeration.
2. Monitor the bacterial growth by measuring the optical density (OD$_{650}$) in a spectrophotometer. Logarithmic-phase cultures of *E. coli* typically have an OD$_{650}$ of 0.5–0.7 (*see* **Note 5**).
3. At a late logarithmic growth phase (OD$_{650}$ 0.6–0.7), transfer the bacterial culture to two 50-mL Falcon tubes and place on ice for 5 min.
4. Harvest bacterial cells by centrifugation at 5000*g* for 10 min, discard the supernatant, and resuspend the bacterial pellet in 10 mL (0.1X original culture volume) of ice-cold 0.1 *M* CaCl$_2$. Place the suspended cells on ice for 10 min.
5. Harvest bacterial cells by centrifugation at 5000*g* for 10 min, discard the supernatant, and resuspend the bacterial pellet in 4 mL (0.04X original culture volume) of ice-cold 0.1 *M* CaCl$_2$.
6. Competent cells may be stored at this stage at 4°C for up to 48 h or, following the addition of sterile 80% glycerol to a final concentration of 10%, at –70°C for up to 1 yr (*see* **Note 6**).

3.1.2. Modified Calcium Chloride Method (1)

This procedure yields highly competent cells (10^7–10^9 transformants per microgram of plasmid DNA) and is a preferred method for the generation of large quantities of competent cells for cryostorage. The main disadvantages are the number of different solutions required and the amount of preparation time required.

1. Inoculate *E. coli* into 100 mL of LB broth and monitor growth as described in **Subheading 3.1.1.**, **steps 1** and **2**.
2. At a late logarithmic phase (OD$_{650}$ 0.6–0.7), transfer the bacterial culture to two 50-mL Falcon tubes and place on ice for 5 min.
3. Harvest bacterial cells by centrifugation at 5000*g* for 10 min, discard the supernatant, and resuspend the bacterial pellet in 10 mL (0.1X original culture volume) of ice-cold TFB (*see* **Note 3**). Place the suspended cells on ice for 10 min.
4. Harvest bacterial cells by centrifugation at 5000*g* for 10 min, discard the supernatant, and resuspend the bacterial pellet in 4 mL (0.04X original culture volume) of ice-cold TFB.
5. Competent cells may be stored at this stage at 4°C for up to 48 h or, following the addition of sterile 10% glycerol, at –70°C for up to 1 yr (*see* **Note 6**).

3.1.3. PEG Method (4)

This is a rapid and simple procedure that offers the option of long-term storage of unused competent cells with no further modifications. The major advantages of this method are the speed and ease of preparation of competent cells. The major disadvantage is the reduced transformation efficiency (10^6 transformants per microgram of plasmid DNA), which makes this method unsuitable for the construction of libraries. Note that the growth phase at which the bacteria are harvested is earlier than the other two methods.

1. Inoculate *E. coli* into 100 mL of LB broth and monitor growth as described in **Subheading 3.1.1**, **steps 1** and **2**.

2. During an early logarithmic phase (OD_{650} 0.3–0.4), transfer the bacterial culture to two 50-mL Falcon tubes and place on ice for 5 min.
3. Harvest bacterial cells by centrifugation at 5000g for 10 min, discard the supernatant, and resuspend the bacterial pellet in 10 mL (0.1X original culture volume) of ice-cold TSS. Place the suspended cells on ice for 10 min (*see* **Note 7**).
4. Harvest bacterial cells by centrifugation at 5000g for 10 min, discard the supernatant, and resuspend the bacterial pellet in 5 mL (0.05X original culture volume) of ice-cold TSS. Competent cells may be stored at this stage at –70°C for up to 1 yr.

3.2. Transformation of Competent Cells

1. Transfer 200 µL of bacterial suspension into a sterile Eppendorf tube. Add approx 1 ng of plasmid DNA and mix gently (*see* **Notes 8–10**). Do not pipet up and down or vortex the tube. For each set of transformations, prepare a negative control that consists of competent cells without DNA, and a positive control using a standard plasmid (*see* **Note 11**).
2. Place the transformation mix on ice for 30 min.
3. Transfer the transformation mix to a 42°C water bath and incubate for exactly 30 s. The temperature and time are crucial in this step. Do not shake the tubes.
4. Place the transformation mix on ice for 2 min.
5. Add 1 mL of prewarmed (37°C) LB broth and transfer the bacteria to a suitably sized tube (such as a 15-mL Falcon tube). Incubate at 37°C for 1 h, shaking vigorously (approx 150 rpm) to ensure good aeration (*see* **Note 12**).
6. Prepare a range of serial dilutions of the bacteria from the transformation mix and spread on selective LB agar plates (*see* **Note 13**).

4. Notes

1. For factors affecting the choice of host strain, *see* Chapter 3.
2. Place solution on ice early in the growth of the bacteria to ensure that it is thoroughly chilled before use.
3. If the cells are to be stored at –70°C, use ESB buffer rather than TFB.
4. Excellent competent *E. coli* bacteria are also available commercially through a number of vendors (e.g., Invitrogen and Gibco). Obviously, the cost per transformation is substantially higher, but some commercial preparations provide efficiencies of >10^8 transformants/µg of DNA.
5. A common cause of reduced transformation efficiency is the failure to harvest the bacterial cultures at the proper phase of growth. The use of buffered or rich media such as terrific broth or SOC may be helpful, as the logarithmic phase is somewhat elongated in these media. The OD at which the bacterial cultures are harvested may need to be optimized for different bacterial strains. The reader is encouraged to perform pilot experiments to determine conditions for optimal transformation efficiency. An excellent description of optimization of transformation efficiency is provided in an article by Huff and colleagues (*5*).
6. Freshly prepared competent cells typically yield the highest transformation efficiencies. However, for reasons of convenience, competent cells may be frozen at –70°C. It is recommended that for applications in which optimal transformation efficiency is necessary, such as library construction, freshly prepared cells be used.
7. In the original procedure, the authors note that TSS may be prepared as a 2X solution and an equal volume added to the bacterial cells following this stage. This further streamlines

the preparation time and does yield competent cells; however, the transformation efficiency is further reduced.
8. There is a linear correlation between plasmid size and transformation efficiency. Plasmids >10 kb in size may require special optimization of transformation conditions.
9. The correlation between DNA concentration and transformation efficiency is roughly linear between picogram and nanogram concentrations. However, at concentrations greater than 1 µg, the transformation efficiency is actually reduced slightly. Therefore, it is generally recommended that approx 1 ng of plasmid DNA be used.
10 Although nicked or relaxed plasmids can be transformed into *E. coli*, efficiency is greatest with supercoiled plasmid. Thus, plasmid DNA that is freshly prepared or carefully stored should be used where possible.
11. Positive and negative controls are important in transformation experiments. The negative control (bacteria with no plasmid DNA) should be plated onto the selective agar plates at the lowest dilution in the range. A positive control (e.g., pUC19) is especially important in library construction and other ligations so as to evaluate the transformation efficiency.
12. Some labs report improved transformation efficiency using rich medium such as SOC or terrific broth during the recovery phase. The author has found that the use of prewarmed broth and tubes large enough to allow for sufficient aeration, along with vigorous shaking, result in comparable yields using LB broth. A crucial factor is an understanding of how much time is needed for expression of the antibiotic resistance determinant. If problems are encountered, optimization by testing different recovery times may be helpful.
13. Transformation efficiencies of 10^6–10^7 transformants/µg of DNA are typical, and efficiencies as high as 10^8 transformants/µg are not unusual. Therefore, plating a wide range of dilutions is recommended.

References

1. Hanahan, D. (1983) Studies on the transformation of *Escherichia coli* with plasmids. *J. Mol. Biol.* **166,** 557–580.
2. Cohen, S. N., Chang, A. C. Y., and Hsu, L. (1972) Nonchromosomal antibiotic resistance in bacteria: genetic transformation of *Escherichia coli* by R-factor DNA. *Proc. Natl. Acad. Sci. USA* **69,** 2110–2114.
3. Dagert, M. and Erlich, S. D. (1979) Prolonged incubation in calcium chloride improves the competence of *Escherichia coli* cells. *Gene* **6,** 23–28.
4. Chung, C. T., Niemela, S. L., and Miller, R. H. (1989). One-step preparation of competent *Escherichia coli*: transformation and storage of bacterial cells in the same solution. *Proc. Natl. Acad. Sci. USA* **86,** 2172–2175.
5. Huff, J. P., Grant, B. J., Penning, C. A., et al. (1990) Optimization of routine transformation of *Escherichia coli* with plasmid DNA. *Biotechniques* **9,** 570–576.

5

Electroporation of *E. coli*

Claire A. Woodall

1. Introduction

Electroporation, originally developed as a method to introduce DNA into eukaryotic cells *(1)*, has subsequently been extensively used for bacterial transformation *(2,3)*. This procedure is an effective method for the transfer of DNA to a wide range of Gram-negative bacteria, such as *Escherichia coli*, and reports indicate that 10^9 electrotransformants per microgram of DNA can be achieved in this species *(4,5)*. Electroporation is probably the most efficient and reliable method for the transformation of *E. coli* strains using plasmid DNA.

When bacteria are subject to an electrical pulse, it is thought that the bacterial membrane is polarized, thus forming reversible transient pores through which DNA travels into the cell (for a detailed explanation of the mechanism of electroporation, *see* **ref. 6**). Pore formation starts as a membrane dimple, which then forms a transient hydrophobic pore; some of these become more stable hydrophilic pores *(7,8)*. The exact changes in membrane structure that occur during electroporation are not known. In most cases, electroporation causes the membrane to rupture resulting in between 50% and 70% cell death. Therefore, to obtain an optimal number of transformants, it is often necessary to determine the number of cells killed, which is dependent on the cell type, field strength (explained below), and length of electric pulse.

To achieve a high number of transformants, it is important to consider variables such as electrical field strength, pulse length, and buffer choice. It may be worth determining what type (or shape) of pulse the electroporation apparatus can produce, although there is little difference between the transformation efficiencies obtained with different electroporation pulses *(9)*. The most common type is the exponential waveform pulse that is the decay pattern from the discharged capacitor. Most manufacturers supply detailed information regarding the optimal conditions (including the pulse length, buffer, and voltage) for the electroporation of a wide variety of cells. The manufacturers' information booklets also contain up-to-date literature on methods for the electroporation of cells, and it is advisable to refer to these.

Electroporators usually have three variable parameters: capacitance (measured in farads [F]), field strength (voltage, measured in volts [V]), and resistance (measured in ohms [Ω]), each of which can be altered to optimize the number of transformants recovered. The capacitor discharge produces a controlled exponential pulse, which, combined with the resistance, determines the pulse length or time constant (measured in milliseconds). Thus, to change the time constant of the pulse, either the capacitance can be adjusted (an additional capacitance extender can also be attached to the main equipment) or the resistance can be altered. It is important to consider the buffer in which the cells have been resuspended. The resistance of a buffer is ionic strength dependent; a high-ionic-strength buffer results in time constants 10 times longer than a low-ionic-strength buffer (such as phosphate-buffered saline [PBS]). To obtain efficient transformation, it is not only important to vary the length of the pulse but also the field strength. The field strength is the voltage (V) applied to the sample, which is inversely proportional to the size of the electroporation cuvet gap though which the pulse travels (cuvet gap sizes range from 0.1 to 0.4 cm). Thus, when 100 V are applied to a 0.1-cm cuvet, the resulting field strength is 1000 V/cm. The same voltage applied to a 0.4-cm cuvet results in a field strength of 250 V/cm. For the electroporation of *E. coli* cells, a higher field strength (12.5–15 kV/cm) is required than for eukaryotic cells because bacterial cells are smaller *(4)*.

Several other factors can also influence the transformation efficiency. Most *E. coli* strains are relatively easy to prepare for transformation by electroporation by washing extensively in ice-cold sterile distilled water or buffer *(10)*. However, the efficiency of the transformation can be reduced if electrocompetent cells are not kept on ice at all times *(6)*. The amount of DNA added to the electrocompetent cells can also affect transformation efficiencies, as a certain concentration of DNA may result in the greatest number of transformants. Also, conformation of the DNA may alter the transformation efficiency; for example, a lower transformation efficiency may be obtained with linear plasmid DNA than with supercoiled plasmid DNA.

If resources are plentiful and to save valuable time, electrocompetent cells can be purchased from commercial suppliers. For example, Invitrogen ElectroMAX™DH5α-E™ *E. coli* cells are preprepared for electrotransformation, and electroporation conditions for use with these cells have already been optimized (*see* **Note 1**) *(11)*.

2. Materials

1. *E. coli* strain (*see* **Note 2**).
2. Luria–Bertani (LB) broth medium: 10 g/L of tryptone, 5 g/L of yeast extract, 10 g/L of sodium chloride. Adjust to pH 7.0 by addition of 5 N NaOH; autoclave.
3. LB agar: add 15 g/L Bacto agar to LB broth prior to autoclaving. LB agar plates should contain an appropriate antibiotic for transformant selection. Add the antibiotic once the agar has cooled to 50°C prior to pouring plates.
4. SOC broth medium: 20 g/L tryptone, 5 g/L yeast extract, 0.5 g/L NaCl. Adjust to pH 7.0 by addition of 5 N NaOH, sterilize by autoclaving, and then add 20 mL of sterile 1 M glucose (final concentration of 20 mM).
5. Appropriate antibiotics for selection of plasmid.
6. Sterile ice-cold distilled water.

7. Sterile ice-cold 10% glycerol.
8. Dry-ice bath. Add dry-ice pellets to a 70% ethanol solution.
9. Electroporation apparatus (e.g., Gene Pulser™, Bio-Rad).
10. Electroporation cuvet (0.1-cm cuvet gap), prechilled (*see* **Note 3**).

3. Methods
3.1. Preparation of E. coli *Electrocompetent Cells*

1. Streak a suitable *E. coli* strain onto an LB agar plate for single colonies and incubate at 37°C overnight.
2. Inoculate 50 mL of LB medium with a single colony of freshly grown *E. coli* and incubate overnight at 37°C with vigorous shaking.
3. Add the 50 mL of overnight *E. coli* culture to 500 mL of fresh LB medium (*see* **Note 4**) and incubate at 37°C with vigorous shaking until the OD_{600nm} is approx 0.4 (*see* **Note 5**).
4. Pellet the cells by centrifugation at 6000g for 10 min at 4°C and discard the supernatant (*see* **Note 6**). Resuspend the cells in 500 mL of ice-cold sterile distilled water. It is important to keep the cells on ice from this point forward in order for a high transformation efficiency to be achieved.
5. Repeat **step 4** until the cells have been washed in water at least three times (*see* **Note 7**).
6. Resuspend the cells in a final volume of 1 mL ice-cold sterile 10% glycerol (*see* **Note 8**).
7. If cells are to be used immediately, place the cells on ice. To store electrocompetent cells, aliquot into 100-µL amounts and immediately snap-freeze using a dry-ice ethanol bath and store at –80°C.

3.2. Electroporation of Electrocompetent Cells

1. Remove a 100-µL aliquot of cells from –80°C storage and thaw on wet ice (*see* **Note 9**) or use freshly prepared ice-cold electrocompetent cells.
2. Aliquot 50 µL of cells into an ice-cold electroporation cuvet (0.1-cm gap). Test an aliquot of cells to check that the sample does not arc (*see* **Note 10**), using the electroporation conditions in **step 5**. If arcing does occur, wash cells in 1 mL of ice-cold sterile distilled water and resuspend in 1 mL ice-cold sterile 10% glycerol as described in **Subheading 3.1.**, until arcing does not occur.
3. To each of the 50-µL cell aliquots, add prechilled plasmid DNA (between 5 pg and 100 ng) in a low volume (< 5 µL). Mix by gentle tapping and incubate on ice for 10 to 30 min (*see* **Note 11**).
4. Insert the cuvet containing the DNA/cell mix into the electroporation apparatus (wiping dry the sides of the cuvet).
5. Set the electroporation conditions on a Bio-Rad Gene Pulser to 2.5 kV, 25 µF, and 200 Ω. To deliver an electric pulse, press the pulse button until a beep sounds and a time constant appears in the apparatus window (*see* **Note 12**). If there is a popping sound, the sample has arced, probably because the plasmid DNA has too high a salt concentration (*see* **Note 13**).
6. Immediately after cells have been pulsed, add 950 µL of room temperature SOC medium and gently resuspend the cells. Transfer the cells to a 15-mL tube and incubate at 37°C with vigorous shaking for 1 h.
7. Spread aliquots of the cells onto LB agar plates containing an antibiotic appropriate for the selection of transformants. Several different dilutions of the cell suspension should be spread onto the plate to obtain single colonies (*see* **Note 14**).

8. Invert the plates and incubate at 37°C overnight. Determine the transformation efficiency (*see* **Note 15**).

4. Notes

1. To electroportate ElectroMAX™DH5α-E *E. coli* cells using the Bio-Rad Gene Pulser unit, the following conditions are used to yield approx 1.0×10^{10} transformants per microgram pUC plasmid DNA: 1.8 kV, 25 µF, 200 Ω, 0.1-cm cuvet, and 40 µL cells.
2. For factors affecting the choice of host strain, *see* Chapter 3.
3. Electroporation cuvets must be cold prior to use; for convenience, they may be stored at 4°C.
4. For growth of the 500-mL culture of *E. coli*, it is important to use at least a 2-L flask so that the culture is sufficiently aerated.
5. Optimal transformation efficiencies are achieved when *E. coli* cells are at an optical density, OD_{600nm}, of between 0.4 and 0.6. To obtain this biomass, the cell suspension should be checked every 30 min using a spectrophotometer. It usually takes between 2 and 3 h after inoculation of the culture to reach this point.
6. Just before the cells are ready for harvesting, chill the rotor and then maintain the cells at 4°C from this point onward. An increase in cell temperature will result in lower transformation efficiencies.
7. The cells must be washed extensively in sterile distilled water to remove the growth medium, which may not be suitable for electroporation; for example, it may have a high salt concentration, which will result in sample arcing (explained in **Note 10**).
8. The final 1 mL of cells resuspended in 10% glycerol should result in a cell density of approx 10^{11} cells/mL. However, for optimal transformation frequencies, dilute the cells in ice-cold 10% glycerol to $(2–5) \times 10^{10}$ cells/mL.
9. When the cells are taken from storage at –80°C, place them directly onto ice to thaw, and, once thawed, do not handle the cells vigorously, as this will reduce the transformation efficiency.
10. Arcing, indicated by a popping sound from the electroporation unit, will reduce the transformation frequency because of an uneven electrical current passing though the sample. A high ionic strength (10 m*M* salt or 20 m*M* Mg^{2+}) can often result in arcing of the sample.
11. It is often found that mixing the DNA and cells together and incubating on ice prior to electroporation will increase the transformation efficiency.
12. Time constants will change depending on the apparatus variables, resuspension buffer, and DNA added (*see* **Subheading 1.**). It is worth noting the time constants, as they may correlate with either a high or a low transformation frequency.
13. Plasmid DNA should be free of contaminants such as phenol, ethanol, salts, proteins, and detergents. This is to ensure maximum transformation efficiency and to prevent arcing of the sample. If arcing occurs at this stage, the DNA can be precipitated with ethanol to remove the excess ions (*see* Chapter 8).
14. If a high number of transformants is expected, the original aliquot of cells can be diluted, 1 in 10 serial dilutions of 100 µL should be spread onto plates. If a low number of transformants is expected, it is better to spread aliquots of 200 µL onto five plates, rather than the whole 1 mL on one agar plate, to avoid growth inhibition because of dead cells.
15. The transformation efficiency (cfu/µg) can be calculated using the following formula:

$$\text{Efficiency} = (\text{cfu on plate/ng DNA}) (1 \times 10^3 \text{ ng/µg}) \times \text{Dilution factor}$$

For example, if 10 ng plasmid DNA results in 500 colonies when 100 µL of the undiluted reaction is plated, then

$$(500 \text{ cfu}/10 \text{ ng}) (1 \times 10^3 \text{ ng/µg}) (1 \text{ mL}/0.1 \text{ mL}) = 5 \times 10^4 \text{ µg/cfu}$$

References

1. Neumann, E., Schaefer-Ridder M., Wang Y. et al. (1982) Gene transfer into mouse lyoma cells by electroporation in high electric fields. *EMBO J.* **1,** 841–845.
2. Chassy, B. M. and Flickinger, J. L. (1987) Transformation of *Lacobacillus casei* by electroporation. *FEMS Microbiol. Lett.* **44,** 173–177.
3. Fiedler, S. and Wirth R. (1988) Transformation of bacteria with plasmid DNA by electroporation. *Anal. Biochem.* **170,** 38–44.
4. Dower, W. J., Miller, J. F., and Ragsdale, C. W. (1988) High efficiency transformation of *E. coli* by high voltage electroporation. *Nucleic Acids Res.* **16,** 6127–6145.
5. Wirth, R., Friesenegger, A., and Fiedler S. (1989) Transformation of various species of gram-negative bacteria belonging to 11 different genera by electroporation. *Mol. Gen. Genet.* **216,** 175–177.
6. Sambrook, J. and Russell, D. W. (eds.) (2001) *Molecular Cloning: A Laboratory Manual*, Cold Spring Harbor Laboratory, Cold Spring Harbor, NY.
7. Weaver, J. C. (1993) Electroporation: a general phenomenon for manipulating cells and tissues. *J. Cell. Biochem.* **51,** 426–435.
8. Weaver, J. C. (1995) Electroporation theory: concepts and mechanisms. *Methods Mol. Biol.* **48,** 3–28.
9. Elvin, S. and Bingham, A. H. A. (1991) Electroporation-induced transformation of *Escherichia coli*: evalutaion of a square waveform pulse. *Lett. Appl. Microbiol.* **12,** 39–42.
10. Miller, E. M. and Nickoloff, J. A. (1995) *Escherichia coli* electrotransformation. *Methods Mol. Biol.* **47,** 105–113.
11. Donahue, R. A. and Bloom, F. R. (1998) Electromax DH5 alpha-E Cells: A new addition to the DH5 alpha family. *Focus* **20(3),** 75–76. Available at http://www.invitrogen.com/content/Focus/03075.pdf. Accessed March 26, 2003.

6

DNA Transfer by Bacterial Conjugation

Claire A. Woodall

1. Introduction

Bacterial conjugation is defined as contact-dependent transmission of genetic information from a donor bacterium to a recipient cell *(1)*. Transfer of DNA by conjugation is often termed *lateral* or *horizontal gene transfer*, as opposed to vertical transfer by which genetic information is transferred from mother to daughter cells. When genetic information is transferred by conjugation from the donor strain to the recipient strain, this population of recipient bacteria are called transconjugants. The genetic information is usually transferred to the recipient bacterium on a plasmid; however, conjugative transposons are also known. The mechanism and proteins involved in conjugative DNA transfer vary depending on the type of plasmid or transposon present in the donor strain. For example, RP4 and F (fertility) plasmids isolated from *Escherichia coli* are self-transmissible plasmids and thus contain genes that encode for all of the necessary proteins involved in the mobilization and transfer of the plasmid from one bacterium to another *(2)*. Other plasmids that are conjugative, but non-self-transmissible, such as RSF1010 isolated from *E. coli*, can only be mobilized when the necessary functions are supplied *in trans (3)*. The cointegration of a conjugative and nonconjugative circular plasmid can result in the transfer of both plasmids into a recipient strain. For example, in a high-frequency recombinant (H*fr*) *E. coli* K-12 strain, where the F plasmid has integrated into the bacterial chromosome, the whole chromosome can be conjugated to a recipient strain *(4,5)*. Finally, conjugative transposons can be transferred from one bacterium to another. Examples include Tn*916* isolated from *Enterococcus faecalis* DS16 and Tn*1545* isolated from *Streptococcus (6–11)*.

A detailed understanding of the regulation and biochemistry of bacterial conjugation is still emerging (reviewed in **ref. 12**). However, the basic mechanism and genetics of bacterial conjugation have been established. Early experiments by Lederberg and Tatum in the 1940s showed that DNA from one bacterium could be transferred to another bacterium *(1)*. The recipient bacterial strain subsequently develops phenotypic characteristics of the donor strain. A liquid mating system for bacterial conjugation using the *E. coli* F plasmid is highly efficient and has been investigated in the most detail. The F transfer region (33.3 kb) contains 36 open-reading frames that encode for several proteins involved in bacterial conjugation *(4–14)* (GenBank accession number U01159). When a donor bacterium, containing the F plasmid, and a fertility minus (F–) recipient bacterium come into close contact, a conjugative pilus on the donor bacterium makes contact with the recipient. F factor DNA is transferred though this pilus into the recipient bacterium. The conjugative pilus consists of several pilin subunits comprising three proteins encoded by the genes *traA*, *traQ*, and *traX* (reviewed in **ref. *15***). Once the conjugative pilus has connected the two bacteria, an uncharacterized "mating signal" initiates the formation of a nucleoprotein complex, at the *ori*T (origin of transfer) on the F factor. The nucleoprotein complex consists of F-plasmid-encoded proteins TraI and TraY *(16,17)*. TraY binds to the *ori*T site, which initiates the binding of TraI and activates the nucleoprotein complex *(18)*. The *ori*T on the F plasmid DNA is "nicked" by the transesterase function of TraI and duplex DNA is unwound by the helicase mechanism of TraI *(19,20)*. Single-stranded DNA (ssDNA) is then transferred in a 5' to 3' direction into the recipient cell. In the donor strain, replacement strand synthesis occurs, and complementary strand replacement occurs in the recipient strain. Finally, the transfer of DNA is terminated when the DNA strand is recircularized at *ori*T *(20)*.

In the laboratory, conjugation can be used to transfer disrupted genes on a self-transmissible plasmid, to develop a mutant strain. For example marker-exchange mutagenesis is a technique that can be used to help elucidate the function of a gene, based on the characteristics of the mutant strain compared to the wild-type strain. A gene of interest from a recipient *E. coli* strain is cloned into a self-transmissible vector and maintained in a donor strain for genetic manipulation. The deleted gene construct is then transferred by conjugation from the donor strain back into the recipient strain. In addition, classic laboratory experiments have exploited the conjugation of H*fr E. coli* K-12 strains to map genes and for complementation studies *(21)*. Genetic mapping of the *E. coli* chromosome using interrupted pairing studies with F prime factors can identify the time of entry and order of genes on the bacterial genome *(22)*.

The transfer of genetic information by conjugation in the environment is important for bacterial diversity *(23)*. Conjugative plasmids can mediate the lateral transfer of antibiotic resistance or virulence determinants between bacteria, allowing bacteria to adapt to otherwise hostile environments. For example, a total of 234 lateral transfer events were reported to have occurred since *Salmonella enterica* and *E. coli* MG1655 diverged 100 million years ago, increasing the potential of *E. coli* to inhabit previously unobtainable ecological niches (for reviews, *see* **refs. *24–26***). Conjugative plasmids can also induce biofilm development *(27)*. A biofilm niche is often found in an aquatic environment or at sites of bacterial infection. At these sites, conjugative plasmids may also play an important role in extensive lateral gene transfer, where virulence genes are mobilized between bacterial species. Conjugation can also occur from bacteria to plant and eukaryotic cells; examples are the conjugative transfer of the Ti plasmid from *Agrobacterium tumifaciens* to plants *(28)* and F and RP4 plasmids from

E. coli to the yeast, *Sacromyces cerevisiae* *(29)*. Recently, the RP4 plasmid (also known as RK2 and RP1) was conjugated from *E. coli* into Chinese hamster ovary (CHO K1) cells *(30)*. Thus, conjugation is an invaluable mechanism for the mobilization of DNA among the three kingdoms: bacteria, plants, and animals.

2. Materials

1. Luria–Bertani (LB) broth medium: 10 g/L tryptone, 5 g/L yeast extract, 10 g/L sodium chloride. Adjust to pH 7.0 by addition of 5 N NaOH and autoclave.
2. Antibiotics for selection of transconjugants.
3. LB agar: Add 15 g/L Bacto agar to LB broth prior to autoclaving. LB agar plates should contain an antibiotic for transformant selection. Add appropriate antibiotics, once the agar has cooled to around 50°C, prior to pouring plates.
4. Filter mating system and holder, aseptic closed system for biological samples (Millipore, Nalgene, Whatmann, Sartorius, PALL Gelman Laboratories).
5. 0.45-µm Pore-size Filter-mating disks (smaller pore sizes can also be used). Polycarbonate or nitrocelluose membrane disks are recommended (Millipore, Nalgene, Whatmann, Sartorius, PALL Gelman Laboratories, Nucleopore), also polyvinylidene fluoride (PVDF) Durapore membranes (Millipore).

3. Method

1. Dilute overnight cultures of the donor and recipient strains 1 in 50 in fresh LB broth. Incubate at 37°C with vigorous shaking until an OD_{600} of 0.6 – 0.8 is reached.
2. Mix different ratios of the donor and recipient strains in a sterile universal. For example, donor/recipient ratios of 1 : 1, 1 : 2, and 1 : 10 might be set up in the first instance (*see* **Note 1**). Dilute strains in LB broth if required.
3. Pour the cell mix into the upper chamber of a filter-mating unit. Use a hydro-powered suction pump to collect cells on the membrane filter; the liquid medium will flow directly into the lower chamber. Any remaining cells in the upper chamber can be washed onto the membrane filter with a small volume of LB broth (*see* **Note 2**).
4. Immediately place the membrane filter, cell side upward, onto a nonselective LB agar plate and incubate at 37°C from 30 min to 16 h (*see* **Note 3**).
5. Carefully remove the membrane filter from the agar plate and place in a sterile universal bottle. Aseptically add 5 mL of LB broth. Shake the universal vigorously to wash the cells off of the membrane filter into the liquid medium.
6. Spread plate serial dilutions (*see* **Note 4**) of cells onto selective agar plates and incubate overnight at 37°C to obtain transconjugates (*see* **Note 5**).
7. Determine the conjugation frequency (*see* **Note 6**).

4. Notes

1. Different ratios of donor to recipient strains are used to optimize the conjugation procedure. It may also be necessary to increase the cell biomass of the bacterial cultures. Optimization is particularly important if the recipient strain is not *E. coli.*
2. A variation of the filter-mating assay can be carried out using liquid cultures. In this method, the mixed cell suspension of donor and recipient cells is added to a sterile universal and incubated at 37°C from 30 min to 16 h (*see* **Note 3**). After incubation, the conjugated cell mix should be vortexed vigorously to disrupt the mating pairs. The cell mix is spread plate as described in **step 6**.

3. For optimal recovery of conjugated cells, the incubation period may vary. It may be worth setting up several plates and harvesting the bacteria at time periods from 30 min to 16 h.
4. Serial dilutions of the conjugated cells are done best as a 10-fold dilution series (i.e., 1 mL cell mix in 9 mL LB broth). Spread plate a portion of the diluted mix (e.g., 200 µL).
5. Antibiotic resistance selection must not only select for bacteria carrying the conjugating DNA but also select against the donor bacteria. All donor bacteria will carry the antibiotic resistance conferred by the DNA to be conjugated. Thus, the donor and recipient bacteria must have different antibiotic sensitivities or be distinguishable in another way (e.g., different abilities to grow on the medium on which the transconjugants are selected).
6. The conjugation frequency can be calculated as the number of transconjugants divided by the total number of bacteria isolated on nonselective medium.

Acknowledgments

I gratefully acknowledge Dr. Andrew Grant for proofreading this chapter.

References

1. Lederberg, J. and Tatum, E. L. (1946) Gene recombination in *Escherichia coli*. *Nature* **158**, 558.
2. Firth, N., Ippen-Ihler, K., and Skurray, R. A. (1996) Structure and function of the F factor and mechanism of conjugation, in *Escherichia coli* and *Salmonella*: Celluar and Molecular Biology (Neidhardt, F. C., Curtiss, R., III, Ingraham, J. L., et al., eds.), ASM, Washington, DC, pp. 2377–2401.
3. Haase, J. Lurz, R. Grahn, A. M., et al. (1995) Bacterial conjugation mediated by plasmid RP4: RSF1010 mobilization, donor-specific phage propagation, and pilus production require the same Tra2 core components of a proposed DNA transport complex. *J Bacteriol.* **177**, 4779–4791.
4. Gross, J. D. and Caro, L. G. (1966) DNA transfer in bacterial conjugation. *J. Mol. Biol.* **16**, 269–284.
5. Lloyd, R. G. and Buckman, C. (1995) Conjugational recombination in *Escherichia coli*: genetic analysis of recombinant formation in Hfr × F– crosses. *Genetics* **139**, 1123–1148.
6. Franke, A. E. and Clewell, D. B. (1981) Evidence for a chromosome-borne resistance transposon (Tn916) in *Streptococcus faecalis* that is capable of "conjugal" transfer in the absence of a conjugative plasmid. *J. Bacteriol.* **145**, 494–502.
7. Courvalin, P. and Carlier, C. (1986) Transposable multiple antibiotic resistance in *Streptococcus pneumoniae*. *Mol. Gen. Genet.* **205**, 291–297.
8. Courvalin, P. and Carlier, C. (1987) Tn*1545*: a conjugative shuttle transposon. *Mol. Gen. Genet.* **206**, 259–264.
9. Caillaud, F., Carlier, C., and Courvalin, P. (1987) Physical analysis of the conjugative shuttle transposon Tn*1545*. *Plasmid* **17**, 58–60.
10. Clewell, D. B., Flannagan, S.E., and Jaworski, D.D. (1995) Unconstrained bacterial promiscuity: The Tn*916*–Tn*1545* family of conjugative transposons. *Trends Microbiol.* **3**, 229–236.
11. Hinerfeld, D. and Churchward, G. (2001) Xis protein of the conjugative transposon Tn*916* plays dual opposing roles in transposon excision. *Mol. Microbiol.* **41**, 1459–1467.
12. Frost, L. S., Ippen-Ihler, K., and Skurray, R. A. (1994) Analysis of the sequence and gene products of the transfer region of the F sex factor. *Microbiol. Rev.* **58**, 162–210.
13. Penfold, S. S., Simon, J., and Frost, J. S. (1996) Regulation of the expression of the *traM* gene of the F sex factor of *Escherichia coli*. *Mol. Microbiol.* **20**, 549–558.

14. Anthony, K. G., Klimke, W. A., Manchak, J., et al. (1999) Comparison of proteins involved in pilus synthesis and mating pair stabilization from the related plasmids F and R100-1: insights into the mechanism of conjugation. *J. Bacteriol.* **181,** 5149–5159.
15. Silverman, P. M. (1997) Towards a structural biology of bacterial conjugation. *Mol. Microbiol.* **23,** 423–429.
16. Nelson, W., Howard, M., Sherman, J., et al. (1995) The *traY* gene product and integration host factor stimulate *Escherichia coli* DNA helicase I-catalyzed nicking at the F plasmid *ori*T. *J. Biol. Chem.* **270,** 28,374–28,380.
17. Howard, M., Nelson, W., and Matson, S. (1995) Stepwise assembly of a relaxosome at the F plasmid origin of transfer. *J. Biol. Chem.* **270,** 28,381–28,386.
18. Lahue, E. E. and Matson S. W. (1990) Purified *Escherichia coli* F-factor TraY protein binds *ori*T. *J. Bacteriol.* **172,** 1385–1391.
19. Tsai, M. M, Fu, Y. H., and Deonier, R. C. (1990) Intrinsic bends and integration host factor binding at F plasmid *ori*T. *J. Bacteriol.* **172,** 4603–4609.
20. Matson, S. W., Sampson, J. K., and Byrd, D. R. N. (2001) F plasmid conjugative DNA transfer. *J. Biol. Chem.* **276,** 2372–2379.
21. Dosch, D. C, Helmer, G. L., Sutton, S.H., et al. (1991) Genetic analysis of potassium transport loci in *E. coli*: evidence for three constitutive systems mediating uptake of potassium. *J. Bacteriol.* **173,** 687–696.
22. Holloway, B. and Low, B. (1996) F-prime and R-prime factors, in *Escherichia coli* and *Salmonella: Cellular and Molecular Biology* (Neidhardt, F. C., Curtiss, R., III, Ingraham, J. L., et al., eds.), ASM, Washington DC, pp. 2413–2419.
23. Davison, J. (1999) Genetic exchange between bacteria in the environment. *Plasmid* **42,** 73–91.
24. Lawerence, J. G. and Ochman, H. (1998) Molecular archaeology of the *Escherichia coli* genome. *Proc. Natl. Acad. Sci. USA* **95,** 9413–9417.
25. Ochman, H. and Moran, N. A. (2001) Genes lost and genes found: evolution of bacterial pathogenesis and symbiosis. *Science* **292,** 1096–1099.
26. Ochman, H., Lawrence, J. G., and Groisman, E. A. (2000) Lateral gene transfer and the nature of bacterial innovation. *Nature* **405,** 299–304.
27. Ghigo, J. M. (2001) Natural conjugative plasmids induce bacterial biofilm development. *Nature* **412,** 442–445.
28. Stachel, S. E. and Zambryski, P. C. (1986) *Agrobacterium tumefaciens* and the susceptible plant cell: a novel adaptation of extracellular recognition and DNA conjugation. *Cell* **47,** 155–157.
29. Heinemann, J. A. and Sprague, G. F. (1989) Bacterial conjugative plasmids mobilize DNA transfer between bacteria and yeast. *Nature* **340,** 205–209.
30. Waters, V. L. (2001) Conjugation between bacterial and mammalian cells. *Nat. Genet.* **29,** 375–376.

7

Cosmid Packaging and Infection of *E. coli*

Mallory J. A. White and Wade A. Nichols

1. Introduction

Cosmids are cloning vectors that were developed to enable large fragments of DNA to be cloned and maintained *(1–3)*. Cosmid vectors allow the cloning of fragments up to 45 kilobases (kb) and are commonly used in genomic library construction. The distinguishing feature of a cosmid is the presence of bacteriophage λ *cos* sites, which enable the vector, and cloned genomic DNA, to be packaged into bacteriophage heads *(1,4)*. Cosmids also contain several other components, including a drug-resistance marker that is necessary for the selection of cosmid-containing bacteria, a plasmid origin of replication, usually ColE1, which regulates cosmid replication and copy number within the *Escherichia coli* host *(1,5)*, and restriction sites that facilitate cloning of the desired DNA fragments and insert verification by restriction digestion (*see* **Fig. 1**).

1.1. Overview of Cosmid Library Construction

The protocol to construct cosmid libraries involves basic molecular techniques employed in constructing other forms of genomic libraries *(1–3,6)* (*see* **Fig. 2**).

1. The cosmid DNA vector is digested with an appropriate restriction endonuclease, often *Bam*HI, and dephosphorylated to prevent vector religation.
2. Genomic DNA fragments are generated by partial digestion of chromosomal DNA with *Mbo*I or *Sau*3AI to produce fragments 30–50 kb in length. The genomic fragments are visualized by agarose gel electrophoresis to ensure that the appropriate fragment size has been generated.
3. The genomic insert DNA is ligated to the linearized vector DNA, producing concatemeric hybrid structures.
4. Ligated DNA is mixed with λ bacteriophage packaging extract. Molecules with two *cos* sites, approx 37–50 kb apart, will be packaged into phage heads.
5. *E. coli* strains are infected with the packaged phage and the recombinant cosmid is propagated as a large plasmid in the host cells.
6. The cosmid library can then be screened appropriately for clones of interest.

From: *Methods in Molecular Biology, Vol. 235:* E. coli *Plasmid Vectors*
Edited by: N. Casali and A. Preston © Humana Press Inc., Totowa, NJ

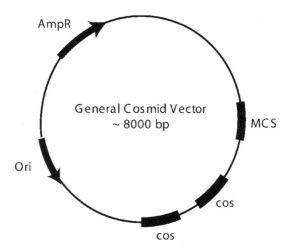

Fig. 1. Representation of a general cosmid vector.

Detailed protocols for cloning genomic DNA and ligation are provided in Chapters 15 and 18. The methods described in this chapter outline the following: (1) the preparation of concatemeric DNA, suitable for packaging, by ligation of prepared cosmid vector and genomic DNA, (2) packaging of the cosmids into λ phage heads in vitro, and (3) infection of the bacterial host cells for titering of the packaged cosmid library. A method for screening cosmid libraries by DNA hybridization is given in Chapter 21.

1.2. In Vitro Packaging

Packaging extracts consist of a lysate of phage-infected *E. coli* cells and contain empty phage heads, unattached phage tails, and phage-encoded proteins required for DNA packaging. The phage proteins Nu1 and A are essential for the function of the bacteriophage terminase to package cosmid DNA at the *cos* sites. The B, C, D, E, and Nu3 proteins are necessary for the synthesis and assembly of the phage head. The G, H, I, J, K, L, M, T, U, V, and Z proteins are required for the assembly of the phage tail *(2,4,7)*.

In the presence of high concentrations of phage head precursors and packaging proteins, exogenous concatameric DNA molecules are cut by the λ terminase at the *cos* sites, allowing the DNA of interest to be packaged into the phage heads. The "full" phage heads are then matured in vitro *(8)*.

1.3. Infection of Bacterial Host Cells

Packaged λ heads are mixed with indicator bacteria and absorb at the λ-receptor protein, encoded by the bacterial *lamB* gene, located in the outer membrane of the host bacteria *(9)*. The linear cosmid recombinant DNA is injected into the bacterial cell via a specialized phage structure, known as a connector, which acts as an orifice for DNA entry. The *E. coli* gene *ptsM* encodes an inner membrane protein that plays a necessary role in the DNA injection process *(9)*. The linear recombinant DNA is circularized at

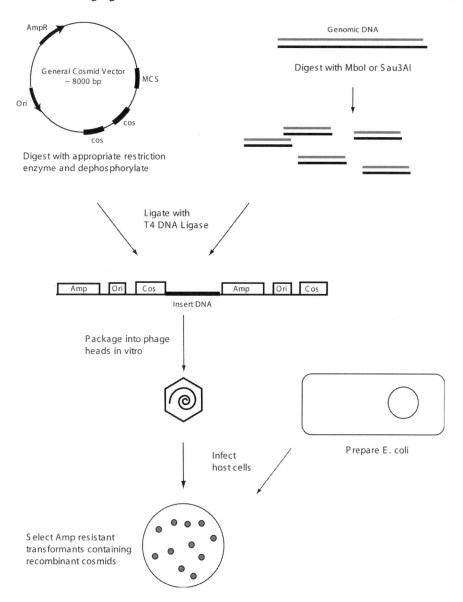

Fig. 2. Schematic outline for cosmid library construction.

the cohesive sticky ends of the *cos* sites by the host ligase enzyme. The resulting circular molecule contains a complete copy of the cosmid vector and will replicate as a plasmid conferring drug resistance upon the bacterial host. Thus, bacteria containing recombinant clones can be selected on appropriate antibiotic-containing media.

It is of utmost importance when constructing the cosmid library that an optimal host strain be utilized for recombinant selection. Commonly used host strains for the growth of λ cosmid libraries are *E. coli* K-12 derivatives. Most often, the genotypes of the host strains are $recA^-$, allowing for the reduction of recombination and the increase of the stability of the cosmid DNA. Examples of some useful *E. coli* host strains that may be employed in cosmid library infection procedures are XL1-Blue MR, VCS257, JM101, and NM554 (*see* Chapter 3). If an inappropriate host strain is utilized for the experimental cosmid library, recombinant clones may not be selected and colony-forming units (CFUs) will not be obtained.

Prior to infection, the host strain is grown in media supplemented with Mg^{2+} ions to facilitate phage absorption *(2)*. Maltose is added to the supplemented media to induce LamB expression at the outer membrane *(2)*. This improves the plating efficiency of the packaging mixtures *(3)*.

2. Materials

2.1. Ligation Reaction

1. Prepared (restriction digested and phosphatase treated) vector DNA (e.g., SuperCos I [Stratagene]). Store at –20°C.
2. Prepared (restriction digested and phosphatase treated) genomic DNA. Store at 4°C.
3. T4 DNA ligase.
4. 10X Ligase buffer: 50 m*M* Tris-HCl (pH 7.5), 7 m*M* $MgCl_2$, 1 m*M* dithiothreitol (DTT), 1 m*M* ATP.

2.2. In Vitro Packaging

1. Packaging extract (Stratagene) (*see* **Note 1**).
2. Substrate DNA, from **Subheading 3.1.**
3. λ Control DNA (*see* **Note 2**).
4. SM buffer: 0.1 *M* NaCl, 0.008 *M* $MgSO_4$, 0.01% gelatin. Sterilize by autoclaving.
5. Chloroform.

2.3. Infection of Bacterial Host Cells

1. Suitable *E. coli* host (*see* **Note 3**).
2. Luria–Bertani (LB) Broth: 1% Bacto-tryptone, 0.5% yeast extract, 0.5% NaCl. Sterilize by autoclaving.
3. LB agar: LB broth, 1.5% agar. Sterilize by autoclaving.
4. 10% Maltose; filter sterilized (do not autoclave).
5. LB supplemented with MgSO4 and maltose: LB broth, 10 m*M* $MgSO_4$; sterilize by autoclaving. Once the media has cooled, add maltose to a final concentration of 0.2%.
6. 10 m*M* $MgSO_4$.
7. Appropriate antibiotics for the selection of the cosmid.

3. Methods

3.1. Ligation Reaction (1,3,6,10–13)

1. Set up the following ligation reaction in a microcentrifuge tube:

    1.5–3.0 µg prepared genomic DNA (32–45 kb in length) (*see* **Note 4**).
    1.0–3.0 µg prepared vector DNA.

2 μl 10X ligase buffer (*see* **Note 5**).
1 μL T4 DNA ligase (1–2 Weiss U).
Sterile water to a final volume of 20 μL.

Also, prepare a negative-control ligation reaction containing all of the above components except genomic DNA fragments.

2. Incubate the ligation reactions at 4°C overnight.

3.2. In Vitro Packaging (1,3,6,12)

1. Remove the appropriate number of prepared packaging extracts from –80°C (or liquid nitrogen) storage and place on dry ice.
2. Thaw the packaging extracts quickly between your fingers until the contents of the tube are just beginning to thaw (*see* **Note 1**).
3. Immediately add 1–4 μL of experimental or negative-control ligation reactions to the packaging extract (*see* **Note 6**).
4. Gently mix the contents of the tube (*see* **Note 7**).
5. Spin the tube for 3–5 s in a microcentrifuge to ensure that all of the contents are at the bottom of the tube.
6. Add 0.5–1.0 mL of SM buffer to the packaging reaction.
7. Add 20 μL of chloroform and mix the contents gently.
8. Briefly centrifuge the reaction tubes in a microcentrifuge.
9. Remove the upper aqueous phase for titering. Alternatively, the packaging reactions may be stored at 4°C for 1 mo (*see* **Note 8**).

3.3. Infection of Bacterial Host Cells (1,3,10–12,14)

3.3.1. Preparation of Bacterial Host Strains

1. The day before the infection procedure, streak the appropriate *E. coli* host strain from a glycerol freezer stock onto appropriate media and incubate at 37°C overnight (*see* **Note 9**).
2. Inoculate a single bacterial colony into 15 mL of LB media, supplemented with $MgSO_4$ and maltose, in a 50-mL conical tube.
3. Incubate the culture, with shaking at approx 200 rpm, until late log-phase growth (optical density $[OD_{600}] = 1.0$), either at 37°C for 4–6 h, or overnight at 30°C (*see* **Note 10**).
4. Centrifuge the culture at 500*g* for 10–30 min at room temperature or 4°C to pellet the bacterial cells.
5. Gently resuspend the cells in 7.5 mL of sterile 10 m*M* $MgSO_4$.
6. Dilute the bacterial cells to an OD_{600} of approx 0.5 with sterile 10 m*M* $MgSO_4$ (*see* **Note 11**).

3.3.2. Infection for Colony-Forming Units

1. Prepare a 1 : 10 and 1 : 50 dilution of the cosmid packaging reactions in SM buffer.
2. Mix 25 μL of each dilution with 25 μL of prepared bacterial host strain in a 1.5-mL microcentrifuge tube.
3. Incubate the tubes at room temperature for 30 min.
4. Add 200 μL of LB broth and incubate at 37°C for 1 h. Gently shake the tube every 15 min to mix the cells.
5. Centrifuge the samples for 30–60 s in a microcentrifuge to pellet the cells.
6. Gently resuspend the cell pellet in 50 μL of fresh LB broth.
7. Plate the cells on LB agar plates containing the appropriate antibiotic selection.
8. Incubate the plates at 37°C overnight.

9. After overnight incubation, count the number of bacterial colonies present (*see* **Note 12**).
10. The library may be stored as a pooled library in a single tube or individual clones may be picked and isolated in multiwell plates. When using multiwell plates, transfer colonies using sterile toothpicks and suspend the bacteria in the appropriate antibiotic-containing broth supplemented with 15% glycerol. Store at –80°C.

4. Notes

1. One of the most important things to consider when constructing a cosmid library is the efficiency of the packaging extracts. It is extremely important that the packaging extracts are not allowed to thaw before use. Prepared packaging extracts are very sensitive to minor variations in temperature. It is highly recommended that the packaging extracts be stored at the bottom of a –80°C freezer and that they are not transferred from one freezer to another *(11)*. Another important step in the packaging procedure is the correct thawing of the packaging extracts. The extracts should be very carefully warmed between fingers but not allowed to thaw completely before adding the experimental DNA. Complete thawing of the extract prior to adding the experimental DNA will increase the viscosity of the extract and will decrease the efficiency of packaging *(3,6)*.
2. Most commercial packaging extracts include a λ DNA positive control that is used to determine the efficiency of packaging. This often involves infection of a λ-sensitive bacterial strain and calculating the efficiency of packaging by counting the resulting number of plaque-forming units (pfus) (*see* Chapter 18). Commercial packaging extracts yield 2×10^8 to 2×10^9 pfus/µg of concatamerized phage DNA *(6,8)*.
3. Choosing an appropriate host strain is an important step in the titering of cosmid libraries (*see* Chapter 3). Some manufacturers do not make it clear that some host strains are only suitable for use with λ DNA controls and are not suitable for the construction of experimental cosmid libraries. XL1-Blue MR is a commonly used host strain that allows the cosmid to replicate as a plasmid and results in the formation of CFUs. VCS257, a derivative of wild-type *E. coli* K-12, is a host strain that is suitable for use with wild-type λ controls.
4. For optimal ligation reactions, use insert DNA at a concentration of greater than 0.1 µg/µL.
5. Add 2 µL of 10 m*M* ATP if it is not present in the ligase buffer.
6. The volume of ligated DNA to be utilized in the packaging reaction is of extreme importance. Do not package more than 4 µL of the ligation reaction. Packaging volumes greater than 4 µL will result in dilution of the proteins present in the packaging extract and thus reduce the efficiency of packaging *(6)*.
7. To mix the packaging reaction, it is recommended that pipetting be avoided and that the reaction be mixed by gently tapping the tube. Gentle pipetting is permissible, but air bubbles must not be introduced into the reaction *(6)*.
8. The packaging reactions are stored in chloroform to prevent bacterial contamination. The chloroform will not affect the phage particles. Prior to removing aliquots of the prepared packaging supernatant, be sure to centrifuge the packaging reactions to ensure that all of the chloroform is at the bottom of the tube. Carryover of chloroform will inhibit the growth of a bacterial lawn and/or plaques during subsequent steps *(6)*.
9. When streaking the bacterial stocks prior to infection, do not add antibiotics to the media *(6)*. Some antibiotics bind to the bacterial cell wall and, in doing so, inhibit phage infection.
10. Growth of the culture at 30°C overnight prevents overgrowth of the bacterial culture and reduces the number of nonviable bacterial cells. Phage can adhere to nonviable cells and this may cause the phage titer to be reduced *(6)*.

11. Diluted bacterial host cells must be used immediately for infection. Do not let the cells stand for any extended period of time *(6,11)*.
12. This is an important step in the protocol, as it evaluates the CFUs/mL of library clones. This information is required when determining the appropriate volume to plate for screening procedures (*see* Chapter 21) and also helps to determine the coverage of the genome represented in the library. The expected number of transformants obtained should range from 1×10^5 to 8×10^5 per microgram of chromosomal DNA, or 5×10^4 to 5×10^5 colonies per ligation reaction *(3,15)*.

References

1. Sambrook, J., Fritsch, E. F., and Maniatis, T. (1989) *Molecular Cloning: A Laboratory Manual*, 2nd ed., Cold Spring Harbor Laboratory, Cold Spring Harbor, NY.
2. Ausbel, F. M., Brent, R., Kingston, R. E., et al. (eds.) (1987) *Current Protocols in Molecular Biology*, Greene Publishing Associates/Wiley–Interscience, New York.
3. DiLella, A. G. and Woo, S. L. C. (1987) Cloning large segments of genomic DNA using cosmid vectors. *Methods Enzymol.* **152,** 199–212.
4. Feiss, M. A. and Becker, A. (1983) DNA packaging and cutting, in *Lambda II* (Hendrix, R. W., Roberts, J. W., Stahl, F. W., et al., eds.), Cold Spring Harbor Laboratory, Cold Spring Harbor, NY, pp. 305–325.
5. Qiagen (1999) *QIAprep Miniprep Handbook*, Qiagen, Valencia, CA, pp. 24–39.
6. Stratagen (2001) *Gigapack III Gold Packaging Extract, Gigapack III Plus Packaging Extract, and Gigapack III Xl Packaging Extract* Stratagene, LA Jolla, CA, pp. 1–13.
7. Furth, M. E. and Wickner, S. H. (1983) Lambda DNA replication, in *Lambda II* (Hendrix, R. W., Roberts, J. W., Stahl, F. W., et al., eds.) Cold Spring Harbor Laboratory, Cold Spring Harbor, NY, pp. 145–167.
8. Hohn, B. (1979) Packaging recombinant DNA molecules into bacteriophage particles in vitro. *Proc. Natl. Acad. Sci. USA* **74,** 3259–3263.
9. Katsura, I. (1983) Tail assembly and injection, in *Lambda II* (Hendrix, R. W., Roberts, J. W., Stahl, F. W., et al., eds.) Cold Spring Harbor Laboratory, Cold Spring Harbor, NY, pp. 326–343.
10. Ish-Horowicz D. and Burke, J. F. (1981) Rapid and efficient cosmid cloning. *Nucleic Acids Res.* **9,** 2989–2998.
11. Stratagene (1998) *SuperCos 1 Cosmid Vector Kit*, Stratagene, La Jolla, CA, pp. 1–32.
12. Kelkar, H. S., Griffith, J., Case, M. E., et al. (2001) The *Neurospora crassa* genome: cosmid libraries sorted by chromosome. *Genetics* **157,** 979–990.
13. Alting-Mess, M. A. and Short, J. M. (1993) Polycos vectors: a system for packaging filamentous phage and phagemid vectors using lambda phage packaging extracts. *Gene* **137,** 93–100.
14. McDaniel, L. D., Zhang, B., Kubiczek, E., et al. (1997) Construction and screening of a cosmid library generated from a somatic cell hybrid bearing human chromosome 15. *Genomics* **40,** 63–72.
15. Evans, G., Lewis, K., and Rothenberg, B. E. (1989) High efficiency vectors for cosmid microcloning and genomic analysis. *Gene* **79,** 9–20.

8

Isolation of Plasmids from *E. coli* by Alkaline Lysis

Sabine Ehrt and Dirk Schnappinger

1. Introduction

Purification of plasmid DNA from *Escherichia coli* using alkaline lysis *(1,2)* is based on the differential denaturation of chromosomal and plasmid DNA in order to separate the two. Bacteria are lysed with a solution containing sodium dodecyl sulfate (SDS) and sodium hydroxide. During this step, chromosomal as well as plasmid DNA are denatured. Subsequent neutralization with potassium acetate allows only the covalently closed plasmid DNA to reanneal and to stay solubilized. Most of the chromosomal DNA and proteins precipitate in a complex formed with potassium and SDS, which is removed by centrifugation. The plasmid DNA is concentrated from the supernatant by ethanol precipitation. Using this procedure, 2–5 µg of DNA can be obtained from a 1.5-mL culture of *E. coli* containing a pBR322-derived plasmid, and three- to five-fold higher yields can be expected from pUC-derived plasmids *(3)*.

As with any plasmid isolation procedure, success in using the alkaline lysis protocol is mainly dependent on the strain of *E. coli* used. Strains that have a high endonuclease A activity, such as HB101 or the JM100 series, yield DNA that often necessitates further purification with a phenol extraction and/or additional precipitation *(3,4)*. However, the alkaline lysis procedure seems to be the most consistent plasmid purification protocol regardless of the strain and it is also better suited for isolation of high-molecular-weight (>10 kb) or low-copy-number plasmids than is the boiling lysis method (*see* Chapter 9). Plasmid DNA isolated by alkaline lysis is suitable for most analyses and cloning procedures without further purification. However, if the isolated plasmid DNA is to be sequenced, an additional purification step, such as phenol extraction, is recommended.

2. Materials
2.1. Growth of *E. coli*

1. Luria–Bertani (LB) medium: 5 g/L yeast extract, 5 g/L NaCl, 10 g/L tryptone.
2. Appropriate antibiotics.

2.2. Plasmid Isolation

1. STE (sucrose/Tris/EDTA) solution: 8% (w/v) sucrose, 50 mM Tris-HCl (pH 8.0), 50 mM EDTA (pH 8.0). Autoclave and store at 4°C.
2. GTE (glucose/tris/EDTA) solution: 50 mM glucose, 25 mM Tris-HCl (pH 8.0), 10 mM EDTA (pH 8.0). Autoclave and store at 4°C.
3. Alkaline–SDS solution: 0.2 N NaOH (*see* **Note 1**), 1% (w/v) SDS. Prepare fresh before use.
4. High-salt solution: 60 mL of 5 M potassium acetate, 11.5 mL glacial acetic acid, 28.5 mL double-distilled H$_2$O. The resulting solution is 3 M acetate and 5 M potassium and has a pH of about 4.8. Store at room temperature, do not autoclave.
5. TE buffer: 10 mM Tris-HCl (pH 8.0), 0.1 mM EDTA (pH 8.0). Autoclave and store at room temperature.
6. 20 mg/mL Lysozyme in 10 mM Tris-HCl (pH 8.0).
7. 10 mg/mL RNase A, DNase-free.
8. 100% Ethanol, 70% ethanol.
9. Phenol : chloroform (1 : 1).

3. Methods
3.1. Growth of E. coli

1. Inoculate 3 mL of sterile LB medium containing the appropriate antibiotic with a single bacterial colony.
2. Grow with shaking at 37°C overnight.

3.2. Plasmid Isolation

1. Centrifuge 1.5 mL of culture for 20 s at maximum speed in a microcentrifuge to pellet the bacteria. Remove the supernatant as completely as possible (*see* **Note 2**).
2. *Optional wash step* (*see* **Note 3**): Resuspend the cell pellet in 0.5 mL STE solution. Centrifuge again and remove the supernatant.
3. Resuspend the bacterial pellet in 100 µL GTE Solution, containing 2 mg/mL of lysozyme (*see* **Note 4**), by vigorous vortexing.
4. Add 200 µL of freshly prepared alkaline–SDS solution and mix by inverting rapidly five times; do not vortex. Place the tube on ice for 5 min.
5. Add 150 µL of high-salt solution and vortex for 2 s to mix. Place the tube on ice for 5 min.
6. Centrifuge at maximum speed for 5 min in a microcentrifuge. Transfer the supernatant to a fresh tube.
7. *Optional extraction with phenol : chloroform* (*see* **Note 5**): Add an equal volume of phenol : chloroform (1 : 1) and mix by vortexing for 5 s. Centrifuge for 1 min in a microcentrifuge at maximum speed to achieve phase separation. Transfer the top aqueous phase to a clean tube.
8. Precipitate DNA by adding 2 volumes of 100% ethanol. Mix by vortexing and let stand for 2 min.
9. Centrifuge at maximum speed for 15 min in a microcentrifuge at 4°C.
10. Carefully remove the supernatant. Add 1 mL of 70% ethanol and centrifuge at maximum speed for 2 min in a microcentrifuge at 4°C. Remove the supernatant as completely as possible and let the DNA pellet air-dry for 10 min (*see* **Note 2**).
11. Dissolve the DNA pellet in 50 µL of TE buffer containing 20 µg/mL of RNase A (*see* **Note 6**).

3.3. Plasmid Analysis

3.3.1. Gel Analysis

Analyze the DNA by agarose gel electrophoresis as described in Chapter 20. It is recommended that undigested supercoiled plasmid DNA be analyzed to verify the integrity of the DNA and to assess the content of chromosomal DNA (*see* **Notes 7** and **8**). In addition, analyze the DNA by cleavage with restriction enzymes. Use 1 µL of the plasmid preparation in the case of high-copy-number plasmids (such as pUC derivatives) and 3 µL or more in the case of medium- or low-copy-number plasmids. If the DNA is resistant to cleavage with restriction enzymes, extract the isolated plasmid DNA with phenol : chloroform and precipitate again with ethanol (*see* **Subheading 3.2.**). The concentration of the plasmid DNA can be roughly estimated by comparing it to a plasmid of known concentration. Run 0.1 µg of the plasmid standard and several amounts (e.g., 1 µL, 2 µL, and 4µL) of the new plasmid prep. After ethidium bromide staining, estimate the concentration of the new plasmid miniprep by comparing the band intensity with those of the plasmid of known concentration.

3.3.2. Spectrophotometric Determination of Plasmid Concentration

The DNA concentration of the plasmid miniprep can be quantitated spectrophotometrically. However, keep in mind that contaminating chromosomal DNA and RNA will contribute to the results from this measurement. To quantitate the nucleic acid concentration, dilute the plasmid DNA 1 : 100 or 1 : 50 (depending on the plasmid copy number) in TE buffer and measure the absorbance (optical density) at 260 nm (A_{260}) and 280 nm (A_{280}) (*see* **Note 9**). Use TE buffer as the blank. This measurement permits the direct calculation of the nucleic acid concentration using the formula

$$[DNA] (\mu g/mL) = A_{260} \times \text{Dilution factor} \times 50$$

where 50 is the extinction coefficient of DNA. The ratio A_{260}/A_{280} provides a reasonable estimate of the purity of the preparation. A pure sample of DNA has an A_{260}/A_{280} ratio of 1.8 ± 0.05.

4. Notes

1. The original protocol asks for 0.2 *N* NaOH. However, if the isolated plasmid DNA is to be used in sequencing reactions, reducing the NaOH concentration to 0.1 *N* is recommended. This reduces the amount of nicked and denatured DNA (*see* **Note 7**) without a significant impact on DNA yield.
2. Insufficient removal of the growth medium and inadequate washing of DNA pellets after ethanol precipitation are the most common reasons for miniprep DNA being resistant to cleavage with restriction enzymes.
3. Some bacterial strains shed cell wall components into the medium that can inhibit the action of restriction enzymes. This can be overcome by washing the cell pellet in STE solution before lysing the bacteria.
4. Addition of lysozyme is not necessary for successful plasmid isolation; however, adding lysozyme to the GTE solution generally increases the DNA yield. However, it may also increase the amount of contaminating chromosomal DNA (*see* **Note 8**).

5. Extraction with phenol : chloroform is optional and will, in many cases, not be necessary. Phenol : chloroform extraction is recommended when working with $endA^+$ strains if problems with cleavage by restriction enzymes arise or if the isolated DNA is to be used in more sensitive applications, such as DNA sequencing. The phenol : chloroform mixture should be equilibrated to room temperature before use.
6. Alternatively, RNase can be added to the GTE solution at a concentration of 100 µg/mL. The addition of RNase A is optional, but recommended, because contaminating RNA may interfere with the detection of DNA fragments upon cleavage with restriction enzymes. It is important to use DNase-free RNase, which is commercially available. Alternatively, DNases can be inactivated by boiling the RNase solution for 10 min.
7. Prolonged exposure of superhelical DNA to heat or alkali results in irreversible denaturation *(5)*. The resulting cyclic coiled DNA migrates through agarose gels at about twice the rate of superhelical DNA. It cannot be cleaved with restriction enzymes, impairs sequencing of the DNA, and stains poorly with ethidium bromide. Traces of this form of DNA can often be seen in plasmid prepared by alkaline or boiling lysis of bacteria.
8. Chromosomal DNA can be identified as a high-molecular-weight band in an undigested plasmid preparation. Upon cleavage with restriction enzymes, this band will disappear and may result in a smear of DNA. In this case, omit the lysozyme from the GTE solution, and, after addition of the SDS–alkaline solution, mix by inverting rather than by vortexing.
9. Do not dilute the nucleic acid below the minimal concentration that allows the reliable quantification of DNA. Optical density measurements of less than 0.05 are generally unreliable; ideally, samples should be diluted so that they yield a minimum reading of 0.1.

References

1. Birnboim, H. C. and Doly, J. (1979) A rapid alkaline extraction procedure for screening recombinant plasmid DNA. *Nucleic Acids Res.* **7,** 1513–1523.
2. Birnboim, H. C. (1983) A rapid alkaline extraction method for the isolation of plasmid DNA. *Methods Enzymol.* **100,** 243–255.
3. Ausubel, F. M., Brent, R., Kingston, R. E., et al. (1997) *Current Protocols in Molecular Biology*, Wiley, New York.
4. Sambrook, J., Fritsch, E. F., and Maniatis, T. (1989) *Molecular Cloning: A Laboratory Manual*, Cold Spring Harbor Laboratory, Cold Spring Harbor, NY.
5. Vinograd, J. and Lebowitz, J. (1966) Physical and topological properties of circular DNA, *J. Gen. Physiol.* **49,** 103–125.

9

Isolation of Plasmids from *E. coli* by Boiling Lysis

Sabine Ehrt and Dirk Schnappinger

1. Introduction

The boiling lysis procedure *(1)* is quick to perform and, therefore, especially suitable for screening large numbers of small-volume *Escherichia coli* cultures. It is described with different adaptations in a variety of protocol books *(2,3)*. The quality of the isolated plasmid DNA is lower than that from an alkaline lysis miniprep, but it is sufficient for restriction analysis.

The bacteria are lysed by treatment with lysozyme, Triton, and heat. The chromosomal DNA remains attached to the bacterial membrane and is removed by centrifugation. The plasmid DNA remains in the supernatant from which it is then precipitated with isopropanol.

The boiling lysis procedure is not recommended when isolating plasmids from *E. coli endA*$^+$ strains that express endonuclease A, such as HB101 and the JM100 series. Contamination of plasmid preps by endonuclease A, which is not completely inactivated by boiling, results in plasmid degradation during subsequent incubation in the presence of Mg^{2+} (e.g., during digestion with restriction enzymes). This problem can be overcome by including an extraction with phenol : chloroform to remove contaminating endonuclease from the plasmid prep. Alternatively, the alkaline lysis method can be used (*see* Chapter 8).

2. Materials
2.1. Growth of E. coli

1. Luria–Bertani (LB) medium: 5 g/L yeast extract, 5 g/L NaCl, 10 g/L tryptone. Autoclaved.
2. Appropriate antibiotics.

2.2. Plasmid Isolation

1. STE solution: 8% (w/v) sucrose, 50 m*M* Tris-HCl (pH 8.0), 50 m*M* EDTA (pH 8.0). Autoclave and store at 4°C.

2. STET solution: 8% (w/v) sucrose, 50 mM Tris-HCl (pH 8.0), 50 mM EDTA (pH 8.0), 5% (w/v) Triton X-100. Filter-sterilize and store at 4°C.
3. 10 mg/mL Lysozyme in 10 mM Tris-HCl (pH 8.0) (*see* **Note 1**).
4. Phenol : chloroform (1 : 1).
5. 100% Isopropanol.
6. 70% Ethanol.
7. TE buffer: 10 mM Tris-HCl (pH 8.0), 0.1 mM EDTA (pH 8.0). Autoclave and store at room temperature.
8. 10 mg/mL RNase A, DNase-free.

3. Methods
3.1. Growth of E. coli

1. Inoculate 3 mL of sterile LB medium containing the appropriate antibiotic with a single bacterial colony.
2. Grow with shaking at 37°C overnight.

3.2. Plasmid Isolation

1. Centrifuge 1.5 mL of culture for 20 s in a microcentrifuge at maximum speed to pellet the bacteria. Remove the supernatant as completely as possible (*see* **Note 2**).
2. *Optional wash step* (*see* **Note 3**): Resuspend the cell pellet in 0.5 mL STE solution. Centrifuge again and remove the supernatant.
3. Resuspend the bacterial pellet in 350 µL of STET by vortexing.
4. Add 25 µL of lysozyme solution and mix by vortexing for 3 s.
5. Place in a boiling water bath for 1 min.
6. Centrifuge for 10 min in a microcentrifuge at maximum speed.
7. Remove the viscous pellet with a sterile toothpick (*see* **Note 4**).
8. *Optional extraction with phenol : chloroform* (*see* **Note 5**): Add an equal volume of phenol : chloroform (1 : 1) and mix by vortexing for 5 s. Centrifuge for 1 min in a microcentrifuge at maximum speed to achieve phase separation. Transfer the top aqueous phase to a clean tube.
9. Mix the supernatant with 450 µL of isopropanol by vortexing and incubate at room temperature for 5 min.
10. Centrifuge for 10 min at 4°C in a microcentrifuge at maximum speed.
11. Carefully remove the supernatant. Add 1 mL of 70 % ethanol and centrifuge for 2 min at 4°C in a microcentrifuge at maximum speed. Remove the supernatant as completely as possible and let the pellet air-dry for 10 min (*see* **Note 2**).
12. Dissolve the pellet in 50 µL of TE buffer containing 20 µg/mL RNase A (*see* **Note 6**).

3.3. Plasmid Analysis
3.3.1. Gel Analysis

Analyze the DNA by agarose gel electrophoresis as described in Chapter 20. It is recommended that undigested supercoiled plasmid DNA be analyzed to verify the integrity of the DNA and to assess the content of chromosomal DNA (*see* **Notes 7** and **8**). In addition, analyze the DNA by cleavage with restriction enzymes. Use 1 µL of the plasmid preparation in the case of high-copy-number plasmids (such as pUC derivatives) and 3 µL or more in the case of medium- or low-copy-number plasmids. If the DNA is resistant to cleavage with restriction enzymes, extract the isolated plasmid DNA with phenol : chloroform and precipitate again with ethanol (*see* **Subheading 3.2.**). The concentration of the plasmid DNA can be roughly estimated by comparing it to a plasmid of known concentration. Run 0.1 µg of the plasmid standard and several amounts (e.g., 1 µL, 2 µL, and 4 µL) of the new plasmid prep. After ethidium bromide staining, estimate the concentration of the new plasmid miniprep by comparing the band intensity with those of the plasmid of known concentration.

3.3.2. Spectrophotometric Determination of Plasmid Concentration

The DNA concentration of the plasmid miniprep can be quantitated spectrophotometrically. However, keep in mind that contaminating chromosomal DNA and RNA will contribute to the results from this measurement. To quantitate the nucleic acid concentration, dilute the plasmid DNA 1 : 100 or 1 : 50 (depending on the plasmid copy number) in TE buffer and measure the absorbance (optical density) at 260 nm (A_{260}) and 280 nm (A_{280}) (*see* **Note 9**). Use TE buffer as the blank. This measurement permits the direct calculation of the nucleic acid concentration using the formula

$$[\text{DNA}] (\mu g/mL) = A_{260} \times \text{dilution factor} \times 50$$

where 50 is the extinction coefficient of DNA. The ratio A_{260}/A_{280} provides a reasonable estimate of the purity of the preparation. A pure sample of DNA has an A_{260}/A_{280} ratio of 1.8 ± 0.05.

4. Notes

1. Many protocols ask for a freshly prepared lysozyme solution, which will work best. However, the lysozyme stock solution can be stored in aliquots at –20°C without a major loss of activity. Avoid repeated freeze–thawing. Lysozyme will not work efficiently if the pH of the solution is less than 8.0.
2. Insufficient removal of the growth medium and inadequate washing of DNA pellet after ethanol precipitation are the most common reasons for miniprep DNA being resistant to cleavage with restriction enzymes.
3. Some bacterial strains shed cell wall components into the medium that can inhibit the action of restriction enzymes. This can be overcome by washing the cell pellet in STE solution before lysing the bacteria.
4. The bacterial pellet should be large and gummy; it contains the cell debris and the chromosomal DNA. A small, dense pellet indicates unsuccessful lysis and will result in a low DNA yield.
5. The extraction with phenol : chloroform is optional. However, if the miniprep DNA is contaminated with DNase and problems with cleavage by restriction enzymes arise or if the DNA is to be used in more sensitive applications such as sequencing reactions, a phenol : chloroform extraction is recommended. The phenol : chloroform mixture should be equilibrated to room temperature before use.
6. The addition of RNase A is optional, but recommended because contaminating RNA may interfere with the detection of DNA fragments upon cleavage with restriction enzymes. It is important to use DNase-free RNase, which is commercially available. Alternatively, DNase can be inactivated by boiling the RNase solution for 10 min.
7. Prolonged exposure of superhelical DNA to heat or alkali results in irreversible denaturation. The resulting cyclic coiled DNA migrates through agarose gels at about twice the rate of superhelical DNA. It cannot be cleaved with restriction enzymes, impairs sequenc-

ing of the DNA, and stains poorly with ethidium bromide. Traces of this form of DNA can often be seen in plasmid prepared by alkaline or boiling lysis of bacteria.
8. Chromosomal DNA can be identified as a high-molecular-weight band in an undigested plasmid preparation. Upon cleavage with restriction enzymes this band will disappear and may result in a smear of DNA. In this case, reduce the concentration of lysozyme solution to 5 mg/mL.
9. Do not dilute the nucleic acid below the minimal concentration that allows the reliable quantification of DNA. Optical density measurements of less than 0.05 are generally unreliable; ideally, samples should be diluted so that they yield a minimum reading of 0.1.

References

1. Holmes, D. S. and Quigley, M. (1981) A rapid boiling method for the preparation of bacterial plasmids. *Anal. Biochem.* **114,** 193–197.
2. Sambrook, J., Fritsch, E. F., and Maniatis, T. (1989) *Molecular Cloning: A Laboratory Manual*, Cold Spring Harbor Laboratory, Cold Spring Harbor, NY.
3. Ausubel, F. M., Brent, R., Kingston, R. E., et al. (1997) *Current Protocols in Molecular Biology*, Wiley, New York.

10

High-Purity Plasmid Isolation Using Silica Oxide

Stefan Grimm and Frank Voß-Neudecker

1. Introduction

The isolation of plasmid DNA from bacteria is a crucial technique in molecular biology and is an essential step in many procedures such as cloning, DNA sequencing, transfection, and gene therapy. These manipulations require the isolation of high-purity plasmid DNA. Commercial anion-exchange columns, even though widely used, are relatively costly. Less expensive methods that use silica oxide as the DNA-binding matrix have the disadvantage that bacterial lipopolysaccharide (LPS), or endotoxin, is copurified and can inhibit downstream applications. In particular, the LPS content of DNA used for transfection heavily influences the efficiency *(1–3)*. Furthermore, these methods usually require the use of hazardous chemicals (e.g., guanidine hydrochloride), which act as chaotropic substances, changing the structure of water to facilitate the binding of the DNA to silica oxide *(4)*.

The method of plasmid isolation described in this protocol relies on silica oxide as the DNA-binding matrix but obviates the disadvantages of this material *(5)*. First, we have found that the addition of chaotropic reagents is not required for the efficient isolation of DNA using silica oxide (*see* **Fig. 1**). The omission of chaotropic reagents from the purification procedure significantly decreases LPS contamination of the DNA. In addition, we have introduced a precipitation step that relies on the similar amphipathic structures of sodium dodecyl sulfate (SDS) and LPS. LPS interacts with SDS that is in suspension in isopropanol and the resulting LPS/SDS complex is removed by centrifugation *(5)*. Together, these improvements lead to a 900-fold reduction of the LPS content of purified DNA relative to published methods (*see* **Fig. 2**). The purity of DNA preparations isolated by this method is comparable with plasmid DNA isolated using CsCl gradients or anion-exchange columns. We routinely obtain approximately 10 µg of plasmid DNA from 900 µL of bacterial culture and the DNA has an A_{260}/A_{280} of 1.8 or higher (*see* Chapter 8), 80% of which is in the supercoiled form. This protocol therefore facilitates an inexpensive isolation of pure plasmid DNA suitable even for demanding applications.

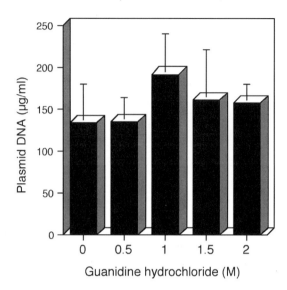

Fig. 1. Binding of plasmid DNA to silica oxide in the presence of different concentrations of a chaotropic reagent. Plasmid DNA was isolated after alkaline lysis of bacteria. Different concentrations of the chaotropic substance, guanidium hydrochloride, were added as indicated, and the amount of DNA that could bind to 7.5 mg silica oxide was determined spectrophotometrically. Presented here are the means and standard deviations of three independent experiments for each condition. Thus, guanidium hydrochloride is not required for binding of plasmid DNA to silica oxide.

2. Materials

1. Silica oxide (Sigma): Dissolve in 250 mL of water, at 50 mg/mL, for 30 min. Remove the fines by suction and reconstitute the original volume. Add 150 µL of 37% HCl and autoclave.
2. P1: 50 mM Tris-HCl, 10 mM EDTA (pH 8.0), 100 µg/mL RNaseA. Store at 4°C.
3. P2: 200 mM NaOH, 1% SDS.
4. P3: 3 M potassium acetate (pH 5.5), chill to 4°C before use.
5. P4: 2.5% SDS in isopropanol (the SDS will not dissolve).
6. Acetone (see **Note 1**).
7. Sterile H$_2$O.

3. Methods

1. Harvest 1.5–2 mL of overnight cultures of *Escherichia coli* clones of interest in Eppendorf tubes by centrifugation at 1000g for 5 min and completely remove the supernatant (see **Notes 2–4**).
2. Resuspend the pellet in 200 µL of P1 by vortexing (see **Note 5**).
3. Add 200 µL of P2 and invert the tube four to eight times to mix the contents (see **Note 6**). Incubate for 5–6 min at room temperature (see **Note 7**).
4. Add 200 µL of ice-cold P3 and invert the tube four to eight times (see **Note 6**). Incubate at –20°C for 5 min.

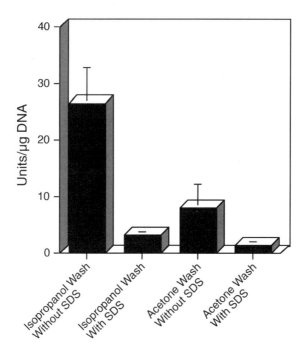

Fig. 2. Levels of LPS in purified plasmid DNA (endotoxin units per microgram of plasmid DNA) following acetone or isopropanol washes and SDS precipitation. The plasmid DNA was isolated as described in the Methods section. Half of the preparations were washed with isopropanol, the other half with acetone. LPS precipitation with SDS in isopropanol was implemented or omitted as indicated. LPS levels were determined by a colorimetric test (BioWhittaker, Walkersville, MD). The means and standard deviations of three different preparations for each condition are shown.

5. Centrifuge at 20,000g (or maximum speed in a microfuge) for 10 min to pellet the cell debris. Carefully remove the supernatant containing the plasmid DNA (*see* **Note 8**).
6. Add 120 µL of P4 (*see* **Notes 9** and **10**) and mix by vortexing. Incubate at –20°C for 20 min (*see* **Note 11**).
7. Centrifuge at 20,000g at 4°C to remove the precipitate and transfer the supernatant into a new Eppendorf tube (*see* **Note 8**).
8. Add 150 µL of prepared silica oxide solution (*see* **Note 10**). Vortex the tube to resuspend the fines and incubate for 20 min at room temperature.
9. Centrifuge at 10,000g for 5 min and discard the supernatant (*see* **Note 12**).
10. Resuspend the pellet in 400 µL of acetone, by vortexing. Centrifuge at 10,000g and remove the supernatant.
11. Repeat the acetone wash step.
12. Dry the pellet using a Speed-vac and then resuspend in 70–90 µL of sterile H_2O by vortexing for approx 1 min (*see* **Notes 13** and **14**).
13. Centrifuge at 20,000g for 5 min and recover the supernatant containing high-purity plasmid DNA.

Table 1
Recommended Volumes of Solution

Format	Culture volume	P1	P2	P3	P4	Silica oxide	Acetone	H$_2$O
96-Well	900 μL	170 μL	170 μL	170 μL	100 μL	150 μL	400 μL	75 μL
Miniprep	1.5–2.0 mL	200 μL	200 μL	200 μL	120 μL	150 μL	400 μL	75–90 μL
Midiprep	10–25 mL	1.5 mL	1.5 mL	1.5 mL	0.9 mL	1.2 mL	4 mL	150–200 μL
Maxiprep	50–100 mL	6 mL	6 mL	6 mL	3.6 mL	4.8 mL	15 mL	500–750 μL
Megaprep	250–500 mL	70 mL	70 mL	70 mL	42 mL	45 mL	100 mL	5–6 mL

4. Notes

1. To facilitate pipetting, acetone should be stored at –20°C.
2. This method can be scaled up for larger cultures; recommended volumes of solutions to use in the different formats are given in **Table 1**. For maxipreps or megapreps, it is advisable to extend **step 6** to 30–45 min. A version of this protocol for a 96-well format is described in **ref. 5**.
3. All centrifugation steps should be performed in a refrigerated centrifuge in order to produce more compact pellets.
4. The bacterial pellets can be stored at –20°C after complete removal of the growth media.
5. Complete resuspension of the pellet in P1 is essential for complete subsequent lysis of the bacteria.
6. Do not vortex the lysate at this stage.
7. Do not extend the lysis time, as genomic DNA will be released from the bacteria.
8. At this stage, the supernatant can be stored at –20°C.
9. For addition of P4, avoid clogging of pipet tips by cutting off their ends to produce a wider opening.
10. Mix P4 and silica oxide stock solutions before use, as SDS and silica oxide sink to the bottom of the solution during storage.
11. Whether the DNA solution freezes or stays liquid during the –20°C incubation has no effect on the yield of plasmid DNA.
12. Do not increase the centrifugation speed of the silica oxide beyond 10,000g, as this may damage the DNA.
13. It is important to dry the silica pellet thoroughly to obtain a good yield of plasmid DNA. Once dry, the plasmid DNA can be stored at –20°C.
14. Elution of the DNA from the silica is facilitated by incubation of the resuspended silica pellet in water pre-warmed to 50°C.

References

1. Wicks, I. P., Howell, M. L., Hancock, T., et al. (1995) Bacterial lipopolysaccharide copurifies with plasmid DNA: implications for animal models and human gene therapy. *Hum. Gene Ther.* **6,** 317–323.
2. Weber, M., Moller, K., Welzeck, M., et al. (1995) Effects of lipopolysaccharide on transfection efficiency in eukaryotic cells. *Biotechniques* **19,** 930–940.
3. Cotten, M., Baker, A., Saltik, M., et al. (1994) Lipopolysaccharide is a frequent contaminant of plasmid DNA preparations and can be toxic to primary human cells in the presence of adenovirus. *Gene Ther.* **1,** 239–246.
4. Hansen, N.J., Kristensen, P., Lykke, J., et al. (1995) A fast, economical and efficient method for DNA purification by use of a homemade bead column. *Biochem. Mol. Biol. Int.* **35,** 461–465.
5. Neudecker, F. and Grimm, S. (2000) High-throughput method for isolating plasmid DNA with reduced lipopolysaccharide content. *Biotechniques* **28,** 107–109.

11

High-Throughput Plasmid Extraction Using Microtiter Plates

Michael A. Quail

1. Introduction

Plasmid extraction is typically performed to produce template DNA for a desired molecular biological reaction, or set of reactions, such as restriction endonuclease digestion (*see* Chapter 20), DNA sequencing (*see* Chapter 22), in vitro mutagenesis (*see* Chapters 23–26), transformation (*see* Chapters 4 and 5), transfection, or probe generation. The principal methodologies of extraction, involving either alkaline lysis or boiling are well established *(1)* and are described in detail in Chapters 8 and 9. These methods are relatively crude but are both inexpensive and rapid, and they yield DNA that is suitable for many molecular biology protocols. DNA of higher purity can be prepared using silica resin-based anion exchange as described in Chapter 10 or by separation on cesium chloride gradients (for details, *see* **ref. 1**, Chapter 1, Protocol 10). Plasmid minipreps are a ubiquitous part of molecular biology. They are simple and easy to perform but can be very tedious when large numbers of samples are involved. In recent years, endeavors such as the human genome project have led to the development of methods for the manageable high-throughput extraction of plasmids *(2–15)*.

High-throughput in molecular biology terms is often defined as anything that is performed in multiwell plates. These are commonly 12 × 8 arrays of 96 wells in a plastic tray. However, the definition is not that easy, as high-throughput covers such a wide spectrum. Although one person may wish to miniprep 96 samples simultaneously, another may be contemplating several thousand, or million, plasmid extractions, and although the basics are the same in each case, the approaches taken will be very different. The key to handling large numbers of samples is organization. The miniprep must now be thought of as a process, and the laboratory's capability to handle the desired throughput at each stage carefully examined.

Most experiments will involve culture in 96-well, deep-well boxes, harvesting of cells, alkaline lysis, removal of the cellular debris, and concentration of the extracted DNA. Thus, this is a multistage process, involving manipulation of samples in microtiter plates and special equipment for manipulation and pipetting into microtiter plates is required. Exactly what will be required for each stage must be clearly thought out.

The first item to consider is the extraction method that is to be employed. The simplest of these is the alkaline lysis miniprep, for which a 96-well microtiter plate protocol is given in **Subheading 3.1.** Here, harvested cells are resuspended and then lysed with sodium hydroxide and sodium dodecyl sulfate (SDS), cellular debris is precipitated with ice-cold potassium acetate and this is removed by passing the sample through a filter plate, before DNA in each well is concentrated by precipitation with alcohol. This is a relatively crude method, yielding DNA of adequate quality for restriction digests and sequencing. For some applications, DNA of higher quality will be required, this is typically prepared using 96-well silica resin-based anion exchange. Plasmid extraction kits in this format are available from a variety of suppliers *(16–21)* and typically involve alkaline lysis and removal of cellular debris, followed by binding of the DNA to the resin, which is immobilized within a filter plate. The DNA is then washed to remove contaminants and eluted from the resin using a low-ionic-strength buffer. Where robotic systems are involved, the use of magnetic capture systems is attractive because it obviates the need for centrifugation and filtration steps. Here, cells are lysed either by alkaline lysis, boiling, or use of detergents and lysozyme to produce a plasmid-containing solution. By using magnetic beads that have been modified to bind DNA (e.g., Sera-Mag particles) *(22)* the plasmid DNA can then be precipitated at will by applying a magnetic field and a series of wash steps performed before pure DNA is eluted.

Having decided on an appropriate method, one must determine the equipment that is to be required. First, the microtiter plates are needed. Plates are commonly available in both 96- and 384-well formats. It is now possible to perform minipreps in a 384-well format; however, as the same principles apply to this as to the 96-well extraction, this chapter will concentrate on use of 96-well microtiter plates. Transferring protocols between the two formats is mostly a matter of scale, although slight modifications to equipment requirements may be necessary.

Microtiter plates are also available in a variety of designs; for instance, deep-well plates that allow for larger culture volumes, V-shaped wells (often thin walled) for polymerase chain reaction (PCR), storage plates in which the well walls stick up from the base (these have gaps between wells and so the risk of cross-well spillage and contamination is lowered, because of this these plates are often used for glycerol-archive plates), flat-bottom plates where the well contents can be visualized spectrophotometrically, and standard round-well U-bottom plates that have high volume capacity and a U-shaped base allowing more complete sample recovery. One can also buy filter plates. These are microtiter plates in which each well is a miniature column that can be a filter membrane that is useful for removing cellular debris after alkaline lysis or a silica-based resin on which the DNA is bound, washed, and, subsequently, eluted.

The liquid-transfer requirements of the process must be considered. Many different pipetting aids are available (*see* **Notes 1** and **2**) including single-channel pipets, with a reagent reservoir, that are used for repeat dispensing, multichannel pipets, multichannel pipets capable of repeat dispensing, 96-well plate fillers, and robotic dispensers.

The laboratory will require equipment that is capable of holding microtiter plates, including shaking incubators for culturing, centrifuge(s) for harvesting bacteria and DNA precipitation, and a vacuum manifold for pulling samples through filter plates (single- and multiple-plate manifolds to which a vacuum line may be attached are available from various suppliers (e.g., **refs. *16*, *19*, and *20***). If large numbers of samples are involved thought needs to be given to storage and organization of plates and a barcoding system should be considered.

There are several different levels of "high-throughput" and the equipment, labor, and organization required for each will vary. One or two sets of 96-well extractions can be performed manually by a single person in a couple of hours with the aid of a multichannel pipet and a microtiter plate centrifuge. If a single worker is organized and has sufficient pipetting aids, centrifuges, and so forth, a throughput of 20–40 plates (96-well) per day is possible using a standard alkaline lysis protocol (such as described in **Subheading 3.1.**). If resin-based purification is required, then throughput is limited to around four plates per day. Off-the-shelf robotic workstations for the minipreparation of plasmid DNA are available from a number of manufacturers *(20,23,24)*. These are typically loaded with harvested cells in deep-well plates and perform the pipetting and filtration stages of an alkaline lysis or resin-based protocol at a rate of 96 samples every 30–60 min. Following the robot performed steps, the samples have to be manually centrifuged to precipitate and concentrate the DNA, but several plates can be loaded simultaneously on the robotic platform and left to process unattended. Higher throughputs than these involve either large amounts of labor or the construction of custom robotic devices (e.g., **refs. *5* and *9***).

The methods detailed in **Subheading 3.** are those in use at the time of writing at the Wellcome Trust Sanger Institute. These are manual methods for microtiter-based preparation of plasmid (*see* **Subheading 3.1.**), large insert clones such as bacterial artificial chromosomes (BACs), PACs, and cosmids (*see* **Subheading 3.2.**) and single-stranded M13 DNA (*see* **Subheading 3.3.**). Each is easily performed at relatively low cost and with a minimum of specialist equipment.

2. Materials
2.1. 96-Well Miniprep of Plasmid DNA

1. Circlegrow (Anachem, UK).
2. Ampicillin, or other antibiotic as appropriate.
3. Deep-well 96-well plates (Beckman).
4. Plate sealer (Costar, Corning).
5. 50% (w/v) glycerol, autoclaved.
6. Glycerol storage plate: Falcon 3077 Microtest U-bottom (Becton Dickinson).
7. GTE buffer: 2.5 mM Tris-HCl (pH 7.4), 10 mM EDTA, 0.92% (w/v) glucose.
8. Serocluster 96-well plate (Costar, Corning).
9. Lysis solution: 0.2 M NaOH, 1% (w/v) SDS. Store at room temperature. For optimal performance, use within 24 h.
10. 3 M Potassium acetate: 294.4 g of potassium acetate, 15 mL of glacial acetic acid, and water to a final volume of 1 L. Store at 4°C.
11. RNase A solution: 20 mg/mL RNase A in 10 mM Tris-HCl (pH 7.4), 15 mM NaCl.

12. Plate sealer pad (3M UPC).
13. Multiscreen filter plate (Millipore).
14. Falcon 3910 96-well plate (Becton Dickinson).
15. Isopropanol.
16. 70% Ethanol, equilibrated to 4°C.

2.2. 96-Well Miniprep of PAC, BAC, or Cosmid DNA

1. 2X TY media: 16 g tryptone, 10 g yeast extract, 5 g NaCl. Adjust pH to 7.4 with NaOH and sterilize by autoclaving.
2. Chloramphenicol, kanamycin, or ampicillin as appropriate.
3. **Items 3–13**, **15**, and **16** in **Subheading 2.1.**

2.3. 96-Well Single-Stranded DNA Prep from M13 Bacteriophage

1. 2X TY media; see **Subheading 2.2.**, **item 1**.
2. Deep-well 96-well plates (Beckman).
3. Overnight culture of an M13 host strain (e.g., TG1).
4. Plate sealer (Costar, Corning).
5. 20% PEG 8000, 2.5 M NaCl.
6. Triton–TE extraction buffer: 0.5% Triton X-100, 10 mM Tris-HCl (pH 8.0), 1 mM EDTA.
7. Metal foil tape (Warth International, Sussex, UK).
8. Serocluster 96-well plate (Costar, Corning).
9. 3 M Potassium acetate; see **Subheading 2.2.**, **item 10**.
10. 96% Ethanol.
11. 70% Ethanol, equilibrated to 4°C.

3. Methods
3.1. Plasmid Preparation in 96-Well Format

1. Fill each well of a 96-well-deep well plate with 1 mL of Circlegrow containing the appropriate antibiotic (typically ampicillin at a final concentration of 100 µg/mL) (*see* **Notes 1, 3,** and **4**).
2. Pick colonies into the media with a toothpick (*see* **Note 5**). Seal the box with a Costar plate sealer (*see* **Note 6**) and pierce each well with a needle to allow aeration during growth.
3. Incubate at 37°C with shaking at 320 rpm for 22 h.
4. Make a glycerol archive, if required, by transfer of 150 µL of culture into the cognate wells of a storage plate containing 100 µL of 50% glycerol in each well. Mix and transfer immediately to storage at –70°C (*see* **Note 7**).
5. Centrifuge the deep-well cultures from **step 3** at 3000g for 5 min to pellet the cells (*see* **Note 8**).
6. Remove the sealant strip and pour off the supernatant into a suitable disinfectant. Turn the plate upside down on a tissue and leave to drain for 1 min.
7. To each well, add 250 µL of GTE buffer (*see* **Note 2**) and vortex the cells for 2 min to resuspend.
8. Centrifuge the plates at 3000g for 5 min to pellet the cells.
9. Discard the supernatant and drain, as in **step 6**.
10. To each well, add 250 µL of GTE buffer and vortex the cells for 2 min to resuspend.
11. Add 4 µL of RNase A solution to each well of a Costar serocluster plate.

12. Transfer 60 µL of the resuspended cells to the plate containing RNase A. To each well, add 60 µL of lysis solution. Seal the plate with a 3M Scotch plate sealer and mix by inversion 10 times.
13. Incubate at room temperature for 10 min.
14. To each well, add 60 µL of cold (4°C) 3 M potassium acetate. Seal the plate with a new 3M Scotch plate sealer and mix by inversion 10 times.
15. Incubate at room temperature for 10 min.
16. Remove the plate sealer and incubate the plate in an oven at 90°C for exactly 30 min.
17. Cool the plate by placing it on ice for 5 min.
18. Tape a Millipore filter plate to the top of a Falcon 3910 plate, ensuring that the filters and the receiving wells are precisely aligned.
19. Transfer the full volume of the serocluster plate to the filter plate and centrifuge for 2 min at 2000g and 20°C (*see* **Note 9**).
20. Remove and discard the filter plate and add 110 µL of isopropanol to each well of filtrate.
21. Seal with a new 3M Scotch plate sealer and mix by inversion twice.
22. Centrifuge for 30 min at 2500g and 20°C.
23. Discard the supernatant and add 200 µL of ice-cold 70% ethanol (*see* **Note 10**).
24. Centrifuge for 5 min at 2500g. Discard the supernatant as before and then leave the plates open to air until liquid has evaporated and the pellet appears dry.
25. When the plates are completely dry, add 50 µL of sterile water and allow the DNA to dissolve overnight at 4°C (*see* **Note 11**).

3.2. 96-Well Miniprep of PAC, BAC, or Cosmid DNA

1. Fill each well of a 96-well deep-well plate (*see* **Notes 1, 3,** and **4**) with 1.25 mL of 2X TY containing the appropriate antibiotic (typically 25 µg/mL of kanamycin for PACs and cosmids or 12.5 µg/mL of chloramphenicol for BACs).
2. Pick colonies into the media with a toothpick (*see* **Note 5**). Seal the box with a Costar plate sealer (*see* **Note 6**) and pierce each well with a needle to allow aeration during growth.
3. Incubate at 37°C, 320 rpm for 22 h.
4. Make a glycerol archive if required (**Subheading 3.1.**, **step 4**).
5. Centrifuge at 3000g for 3 min to pellet the cells (*see* **Note 8**).
6. To each well, add 60 µL of GTE with RNase A (add 1.2 mL of RNase A stock (final concentration 50 mg/mL) to 100 mL of GTE) (*see* **Note 2**).
7. Vortex for approx 4 min, or until the cells have completely resuspended.
8. To each well, add 60 µL of lysis solution.
9. Vortex for 15 s to mix.
10. Incubate at room temperature for 3 min.
11. To each well, add 60 µL of 3 M potassium acetate.
12. Incubate at room temperature for 15 min.
13. Add 100 µL of isopropanol to each well of a Costar Serocluster 96-well plate.
14. Place the serocluster plate under a vacuum manifold and then place a Millipore filter plate, on top of this ensuring that the filters and the receiving wells are precisely aligned.
15. Transfer the full volume from the 96-well lysis plate in **step 12** to the Millipore Multiscreen filter plate and apply a vacuum until all of the fluid has been sucked through (*see* **Note 9**).
16. Discard the filter plate and seal the Serocluster plate with a 3M Scotch plate sealer and mix by inversion twice.
17. Store the plate at 4°C overnight.

18. Centrifuge for 1 h at 3000g and 20°C.
19. Discard the supernatant and add 200 μL of ice-cold 70% ethanol (*see* **Note 9**) to each well.
20. Centrifuge for 5 min. Discard the supernatant, blot the plate on a tissue to remove as much ethanol as possible, and then leave the plates to air-dry.
21. When the plates are completely dry, add 40 μL of sterile water and allow the DNA to dissolve overnight at 4°C (*see* **Note 12**).

3.3. 96-Well Single-Stranded DNA Prep from M13 Bacteriophage

1. Fill each well of a 96-well deep-well plate (*see* **Notes 1, 3,** and **4**) with 1.25 mL of 2X TY that has been seeded with a 1% (v/v) inoculum of an overnight culture of an appropriate M13 host strain.
2. Using a sterile cocktail stick, or similar, pick a single plaque into each well (*see* **Note 5**). Seal the plate with a Costar plate sealer (*see* **Note 6**) and pierce each well with a needle to allow aeration during growth.
3. Incubate at 37°C, 360 rpm for 5.5 h (*see* **Note 13**).
4. Centrifuge at 3000g for 20 min to pellet the cells (*see* **Note 8**).
5. To each well of a fresh 96-well deep-well plate, add 200 μL of 20% PEG 8000/2.5 M NaCl (*see* **Note 2**).
6. Remove 600 μL of supernatant from each well of the plate centrifuged in **step 4** and transfer to the cognate wells of the plate containing PEG/NaCl. Mix by pipetting.
7. Incubate at room temperature for exactly 20 min and then centrifuge for 20 min at 3000g and 20°C to pellet the phage.
8. Decant the supernatants by inverting the plate on a piece of tissue and leave to drain for 1 min.
9. Place a piece of towel into the centrifuge buckets, place the plates upside down into the buckets, and spin at 10g for 2 min to remove residual PEG.
10. Leave the resulting pellet to air-dry for 15–30 min.
11. To each well, add 20 μL of Triton–TE extraction buffer (*see* **Note 14**) and cover with foil tape.
12. Spin briefly to collect samples in the bottom of the wells and then immediately vortex for 2 min to resuspend the phage (*see* **Note 15**). Spin briefly to collect samples in bottom of the wells and then repeat.
13. Place the plate in a water bath at 80°C for exactly 10 min to lyse the phage and then briefly centrifuge.
14. Remove the foil tape and add 40 μL of water to each well and mix by pipetting. Briefly centrifuge to collect samples in the bottom of the wells.
15. Transfer the contents of each well to those of an untreated Corning microtiter plate, containing 10 μL of 3 M potassium acetate in each well, and mix by pipetting.
16. Add 160 μL of 96% ethanol to each well and mix by pipetting.
17. Incubate at −70°C for 20 min, or at −20°C overnight.
18. Centrifuge for 1 h at 3000g and 4°C. As soon as the centrifuge stops, pour off the ethanol supernatant.
19. Invert the plate on a tissue to remove residual traces of ethanol.
20. Add 200 μL of ice-cold 70% ethanol (*see* **Note 10**).
21. Centrifuge for 15 min at 3000g and 4°C.
22. Pour off the supernatant and invert the plate on a tissue to remove residual traces of ethanol.
23. Dry under vacuum and then resuspend the contents of each well in 50 μL of sterile water.

4. Notes

1. Deep-well plates can be filled manually using a reservoir-based repeat pipettor such as an Eppendorf multipipet or by using a 1-mL-capacity multichannel pipet. For filling large numbers of boxes, a 96-well dispensing unit such as a Q-fill *(25)* is recommended.
2. Pipetting is best done either with multichannel pipets, multiwell dispensers, or robotic devices. Manually or electronically operated multichannel pipets, typically 12 channel, are available from a number of suppliers (e.g., **refs.** *16*, and *26–28*). These are available in several series that span the 1- to 1200-µL range and most take standard disposable pipet tips. The 96-well or 384-well dispensers are also useful (e.g., labsystems multidrop *[26]*, and Tomtec Quadra *[29]*). As these units do not have disposable tips, they are normally used for steps that involve dispensing into clean plates.
3. Deep-well plates, sometimes referred to as growth boxes, have a capacity of approx 2 mL per well. To prevent splashing and possible cross-contamination during incubation, add no more than 1.25 mL of media to each well.
4. Media-plate-filling dispensers should be cleaned regularly, as they can be a source of contamination.
5. Picking colonies or plaques can be done either manually or robotically using proprietary (e.g., Q-pick *[25]*) or custom devices. Manual picking is performed by stabbing the colony/plaque with a toothpick and then placing the toothpick in a well of a growth plate. Leave the toothpick in the well to keep track of which wells have been inoculated. When all of the wells have been inoculated, the toothpicks are carefully removed, in small bunches, so as not to contaminate other wells.
6. Before sealing a plate with a sealant strip, make sure that the top surface of the plate is dry. If not, wipe briefly with a tissue. Place the sealant on the plate and press it on by hand to ensure that the plate is adequately sealed and so prevent spillage from one well to the next.
7. Glycerol is toxic to *E. coli* at room temperature. Therefore, for long-term storage, care must be taken to keep glycerol plates cold at all times. Glycerol archives are an excellent resource enabling the generation of extra cultures and DNA if required.
8. Centrifugation speeds will vary between different centrifuges as a function of the rotor radius. The exact speed required can be ascertained by consulting the user manual for each model.
9. Fluid transfer through a Millipore filter plate into the collection plate can be aided either by centrifugation as in **Subheading 3.1.** or by vacuum as in **Subheading 3.2.**
10. Store 96% and 70% ethanol in the refrigerator so that they are prechilled prior to use.
11. The typical yield from the 96-well plasmid preparation described in **Subheading 3.3.** is approx 5 µg.
12. As BACs and PACs are maintained at a single-copy per cell *(30)*, the DNA yield is much lower than that obtained for plasmids. The average yield of BAC or PAC DNA from **Subheading 3.2.** is approx 1–2 µg and is of a quality suitable for both sequencing reaction or restriction digestion.
13. M13 propagation in **Subheading 3.3.** is expedited by inoculation into media that has been seeded (preinoculated) with host cells. Alternatively, plaques can be picked into nonseeded growth media and incubated longer, typically around 16 h.
14. The Triton extraction method was developed by Elaine Mardis at the Washington University Genome Sequencing Centre *(31)*. Current protocols in use at this center are posted on their website *(32)*.

15. Resuspension of PEG-precipitated phage can be difficult. Vortex vigorously while moving the plate about on the head of the vortexing unit. Check that all wells have resuspended before proceeding.

References

1. Sambrook, J. and Russell, D. W. (2001) *Molecular Cloning: A Laboratory Manual*, 3rd ed., Cold Spring Harbor Laboratory, Cold Spring Harbor, NY.
2. Sanger protocols. www.sanger.ac.uk/Teams/Team51/. Accessed February 13, 2003.
3. Gibson, T. J. and Sulston, J. E. (1987) Preparation of large numbers of plasmid DNA samples in microtiter plates by the alkaline lysis method. *Gene Anal. Tech.* **4**, 41–44.
4. Bankier, T. (1993) M13 phage growth and DNA purification using 96 well microtiter trays. *Methods Mol. Biol.* **23**, 41–45.
5. Elkin, C. J., Richardson, P. M., Fourcade, H. M., et al. (2001) High-throughput plasmid purification for capillary sequencing. *Genome Res.* **11**, 1269–1274.
6. Engelstein, M., Aldredge, T. J., Madan, D., et al. (1998) An efficient, automatable template preparation for high throughput sequencing. *Microb. Comp. Genomics* **3**, 237–241.
7. Garner, H. R., Armstrong, B., and Kramarsky, D. A. (1992) High-throughput DNA preparation system. *Genet. Anal. Tech. Appl.* **9**, 134–139.
8. Hilbert, H., Lauber, J., Lubenow, H., et al. (2000) Automated sample-preparation technologies in genome sequencing projects. *DNA Sequence* **11**, 193–197.
9. Itoh, M., Kitsunai, T., Akiyama, J., et al. (1999) Automated filtration-based high-throughput plasmid preparation system. *Genome Res.* **9**, 463–470.
10. Itoh, M., Carninci, P., Nagaoka, S., et al. (1997) Simple and rapid preparation of plasmid template by a filtration method using microtiter filter plates. *Nucleic Acids Res.* **25**, 1315–1316.
11. Konecki, D. S. and Phillips, J. J. (1998) TurboPrep II: an inexpensive, high-throughput plasmid template preparation protocol. *Biotechniques* **24**, 286–293.
12. Marra, M. A., Kucaba, T. A., Hillier, L. W., et al. (1999) High-throughput plasmid DNA purification for 3 cents per sample. *Nucleic Acids Res.* **27**, e37; www.nar.oupjournals.org/methods.
13. Neudecker, F. and Grimm, S. (2000) High-throughput method for isolating plasmid DNA with reduced lipopolysaccharide content. *Biotechniques* **28**, 107–109.
14. Skowronski, E. W., Armstrong, N., Andersen, G., et al. (2000) Magnetic, microplate-format plasmid isolation protocol for high-yield, sequencing-grade DNA. *Biotechniques* **29**, 786–792.
15. Kelley, J. M., Field, C. E., Craven, M. B., et al. (1999) High throughput direct end sequencing of BAC clones. *Nucleic Acids Res.* **27**, 1539–1546.
16. Eppendorf AG, www.eppendorf.com. Accessed February 13, 2003.
17. Invitrogen life-technologies, www.lifetechnologies.com. Accessed February 13, 2003.
18. Macherey-Nagel GmbH, Germany, www.mn-net.com. Accessed February 13, 2003.
19. Promega Corporation, www.promega.com/nap/. Accessed February 13, 2003.
20. Qiagen GmbH, www.qiagen.com. Accessed February 13, 2003.
21. Stratagene, www.stratagene.com. Accessed February 13, 2003.
22. Seradyn, www.seradyn.com/. Accessed February 13, 2003.
23. MWG Biotech, www.mwgbiotech.com/html/index.shtml. Accessed February 13, 2003.
24. Beckman Coulter Inc., www.beckman.com/products/pr2.asp. Accessed February 13, 2003.
25. Genetix Ltd., www.genetix.co.uk/. Accessed February 13, 2003.
26. Quail, M. A. (2002) DNA cloning, in *Encyclopedia of the Human Genome*, (Cooper, D. N., ed.) Nature, London.

27. Finnpipette, www.finnpipette.com/. Accessed February 13, 2003.
28. Anachem Ltd., Luton, UK. www.anachemlifescience.com.uk. Accessed February 13, 2003.
29. Matrix Technologies, www.apogentdiscoveries.com/. Accessed February 13, 2003.
30. Tomtec, Connecticut, www.tomtec.com/. Accessed February 13, 2003.
31. Mardis, E. (1994) High-thoughput detergent extraction of M13 subclones for fluorescent DNA sequencing. *Nucleic Acids Res.* **22,** 2173–2175.
32. Washington University Genome Sequencing Center, genome.wustl.edu/tools/protocols/. Accessed February 13, 2003.

12

Isolation of Cosmid and BAC DNA from *E. coli*

Daniel Sinnett and Alexandre Montpetit

1. Introduction

Cosmid and bacterial artificial chromosome (BAC) systems have been developed for the cloning of large DNA inserts averaging 40 kb and 130 kb (range: 90–300 kb), respectively. The resulting clones are more stable than yeast artificial chromosomes (YACs) and rarely chimeric, which makes them excellent tools for the generation of contiguous physical maps. The use of such large-insert clones considerably increases the rate of complex genome mapping and sequencing. Numerous standard protocols have been developed to isolate supercoiled plasmids, but of these, the alkaline lysis protocol remains the most suitable approach to isolate large-insert clones such as cosmids and BACs.

The alkaline lysis procedure relies on a differential precipitation step in which high-molecular-weight chromosomal DNA is precipitated and the relatively small plasmids remain in the supernatant *(1)*. Bacteria are lysed by treatment with a solution containing sodium dodecyl sulfate (SDS) and sodium hydroxide (NaOH), in which SDS denatures bacterial proteins, and NaOH denatures chromosomal and plasmid DNA. The addition of a neutralizing potassium acetate solution causes the covalently-closed plasmid DNA to reanneal rapidly. The bacterial chromosomal DNA and proteins precipitate together with SDS, which form a complex with potassium. The reannealed supercoiled plasmid is then further purified.

The main advantage of using potassium, over sodium, acetate buffer is that it yields plasmids with a lesser amount of RNA and chromosomal DNA contamination. However, the use of SDS in combination with potassium salt can result in a selective loss of large (>100 kb) plasmids *(2)*. It was suggested that the rapid precipitation of SDS with potassium acetate leads to the trapping of large denatured plasmids that cannot renature as fast as small ones. The use of *N*-lauryl sarcosine (SLS) can bypass this problem *(2)*, but this detergent leads to the isolation of increased amounts of RNA.

To circumvent these problems, a modified alkaline lysis protocol was developed *(3)*. In this procedure, detailed in this chapter, SLS was substituted for SDS. Every

From: *Methods in Molecular Biology, Vol. 235:* E. coli *Plasmid Vectors*
Edited by: N. Casali and A. Preston © Humana Press Inc., Totowa, NJ

step was done at 4°C to minimize the nuclease-induced nicking of plasmid DNA during isolation. A proteinase K treatment and a phenol extraction step were added to the procedure in order to reduce the degradation of the DNA. The resulting DNA has little or no contamination with genomic DNA, it is cut by restriction enzymes, it is an efficient template for PCR amplification, and it can be sequenced using cycle-sequencing systems. Furthermore, identical restriction patterns were generated every 2 wk over a period of 6 mo indicating that the DNA preparations were very stable.

This procedure is simple and offers a productive method for the isolation of high-quality and stable large-insert clone DNA suitable for numerous applications.

2. Materials

1. Luria–Bertani (LB) medium: 10 g Bacto tryptone, 5 g Bacto yeast extract, 10 g NaCl; make up to 1 L with double-distilled water (ddH_2O). Sterilize by autoclaving.
2. Terrific Broth (TB): 12 g Bacto tryptone, 24 g Bacto yeast extract, 10 mL of 40% (v/v) sterile glycerol, 17 mL of 1 M KH_2PO_4, 72 mL of 1 M K_2HPO_4; make up to 1 L with ddH_2O. Sterilize by autoclaving.
3. Appropriate antibiotic for selection of cosmid or BAC.
4. Solution I: 50 mM glucose, 10 mM EDTA (pH 8.0), 25 mM Tris-HCl (pH 8.0). Sterilize by ultrafiltration and store at room temperature.
5. 20 mg/mL Lysozyme in 10 mM Tris-HCl (pH 8.0).
6. 10 mg/mL RNase A, DNase-free.
7. Solution II: 0.2 N NaOH, 1% SLS. Prepare fresh.
8. Solution III: 60 mL of 5 M KOAc, 11.5 mL glacial acetic acid; make up to 100 ml with ddH_2O (the final solution is 3 M potassium and 5 M acetate with a pH of approx 4.8). Sterilize by ultrafiltration.
9. Cold (4°C) 100% isopropanol.
10. 95% and 70% Ethanol.
11. TE buffer: 100 mM Tris-HCl (pH 8.0), 1 mM EDTA (pH 8.0).
12. 10% SLS.
13. 20 mg/mL Proteinase K.
14. Phenol : chloroform : isoamyl alcohol (25 : 24 : 1 by volume).
15. Chloroform : isoamyl alcohol (24 : 1 by volume).

3. Method (see Notes 1 and 2)

1. Prepare a 100- to 150-mL culture of the cosmid or BAC clone to be purified in LB or TB media containing the appropriate antibiotic (50 µg/mL kanamycin or 100 µg/mL ampicillin for cosmids and 12.5 µg/mL chloramphenicol for BACs) (see **Note 3**). Grow the bacteria overnight at 37°C in an orbital shaker incubator with vigorous shaking (250 rpm, a g-force of approx 30).
2. Pellet the cells by centrifugation at 10,000g for 5 min. Longer spins than this make it difficult to resuspend the pellet completely. Pour off the media and invert the tube for 5 min on a paper towel. Resuspend the cells by pipetting up and down in 10 mL of chilled solution I containing 2.5 mg/mL lysozyme and 400 µg/mL RNase A (see **Note 4**). Be sure that the cells are completely resuspended. Work on ice to prevent DNA degradation.
3. Add 20 mL of freshly prepared solution II and mix by inversion until lysis is visible. The mixture will become clear and viscous and should contain no particulate matter. Place the tubes on ice.

4. Add 15 mL of solution III, mix rapidly by inversion until a white precipitate appears then incubate on ice for 5 min. Do not vortex at this stage, as this can result in shearing of the DNA.
5. Centrifuge at 10,000g for 15 min. Collect the supernatant in a fresh tube. Transfer the supernatant by pipetting, not by decanting. Do not disturb the white pellet, which contains proteins, RNA, and chromosomal DNA.
6. Precipitate the DNA in the supernatant by adding 0.7 volumes of cold isopropanol. Mix and centrifuge for 15 mins at 10,000g. Carefully discard the supernatant.
7. Wash the pellet with 70% ethanol and centrifuge for 5 min. Carefully remove the ethanol with a pipet and air-dry the pellet for 10 min.
8. To further purify the DNA and to remove contaminants, resuspend the pellet in 500 µL of TE buffer containing 0.1% SLS and 100 µg/mL proteinase K, and incubate for 1 h at 37°C (*see* **Note 5**).
9. Extract the DNA once with an equal volume of a phenol : chloroform solution and then once with an equal volume of chloroform.
10. Precipitate the DNA with 2 volumes of 95% ethanol. Mix well and centrifuge at 15,000g for 5 min in a microcentrifuge.
11. Discard the supernatant and rinse the pellet with 70% ethanol. Centrifuge for 5 min and remove the ethanol with a pipet. Air-dry the pellet for 10 min.
12. Resuspend the DNA pellet in 100 µL of TE buffer. The DNA preparation is stable for several weeks when stored at 4°C (*see* **Note 6**).
13. Estimate the DNA concentration by fluorimetry using a DNA minifluorimeter TKO 100 (Hoefer Scientific Instruments) or by spectrophotometry (*see* **Note 7** and Chapter 8).
14. Check the purity and integrity of the DNA on a 0.8% agarose gel (*see* **Notes 8** and **9** and Chapter 20) and sequence the insert ends if required (*see* **Note 10** and Chapter 22).

4. Notes

1. This protocol can be adapted for minipreparation of DNA *(3)*.
2. Cosmids and BACs can also be prepared using the CONCERT™ High Purity Plasmid Purification System from Invitrogen. However, it is necessary to modify their midiprep protocol as follows. Add RNase A to solution E1 to a final concentration of 400 µg/mL, increase the NaCl concentration in the E5 washing buffer from 0.8 M to 0.9 M by adding 0.58 g of NaCl per 100 ml of E5 and prewarm buffer E6 to 50°C.
3. The BAC and cosmid clones have a very low copy number and are 5–10 times less abundant than conventional plasmids. The use of superrich medium such as TB can be advantageous for culturing plasmids maintained at low copy number.
4. Contaminating RNA may interfere with detection of DNA fragments on agarose gels but can be eliminated by adding DNase-free RNaseA. Note that the use of SLS in **step 3** leads to an increase in the amount of RNA contamination; thus, higher amounts of RNase are needed.
5. DNA prepared without care in removing proteins is not stable, presumably because of nucleases still present in the sample, and will be degraded after a few days stored at 4°C. The use of proteinase K and phenol : chloroform and chloroform extractions will reduce this degradation.
6. Do not store the DNA at –20°C. Freezing and thawing steps will break the large DNA molecules.
7. DNA yields range from 80 ng (in LB broth) to 125 ng (in TB broth) per milliliter of culture. Lower yields may be obtained for BAC DNA, as these large-insert genomic clones are maintained at a single copy per cell. Thus, a larger culture volume may be

needed to achieve a suitable yield. It is critical that the bacterial cell pellet is completely resuspended.
8. A large smear on the agarose gel indicates that host bacterial chromosomal DNA has been isolated along with the recombinant clone. To avoid this problem, use fresh solution II; ensure that the initial bacterial pellet is completely resuspended in solution I before the addition of solution II; do not vortex during **steps 3** and **4**. Additionally, the use of LB growth media, rather than a very rich medium, may help.
9. If a large amount of RNA is still present after the purification, it may inhibit subsequent enzymatic reactions. The use of 2 mM spermidine or 100 µg/mL RNase A in **step 2** can prevent this inhibition.
10. The BAC DNA can be used directly for sequencing with T7 and SP6 BAC end primers. When sequencing, use at least 1 µg of BAC DNA and long (> 20-mer) sequencing primers to ensure specificity.

References

1. Birnboim, H. C. and Doly, J. (1979) A rapid alkaline extraction procedure for screening recombinant plasmid DNA. *Nucleic Acids Res.* **7,** 1513–1523.
2. Thomas, N. R., Koshy, S., Simsek, M. et al. (1988) A precaution when preparing very large plasmids by alkaline lysis procedure. *Biotechnol. Appl. Biochem.* **10,** 402–407.
3. Sinnett, D., Richer, C., and Baccichet, A. (1998) Isolation of stable bacterial artificial chromosome DNA using a modified alkaline lysis method. *Biotechniques* **24,** 752–754.

13

Preparation of Single-Stranded DNA from Phagemid Vectors

W. Edward Swords

1. Introduction

Single-stranded DNA (ssDNA) is the optimal template for most polymerase-based molecular-biology applications, including DNA sequencing and site-directed mutagenesis. Phagemids are chimeric vectors, derived from the ssDNA bacteriophages M13, fd, or f1, that normally replicate as plasmids in bacterial hosts *(1)* (*see* Chapter 2). Phagemids typically lack the genes required for packaging into virions and, thus, "helper" bacteriophage are often required to package phagemids as ssDNA into viral particles, which are subsequently released into the culture medium *(2)*. ssDNA isolated from these virions is suitable for use as a template in a variety of procedures with no further manipulation, as denaturation is not required prior to its use as a template.

Problems associated with the use of phagemids include the relative instability of cloned DNA, compared to plasmid vectors, especially during viral amplification. It is therefore important to minimize the passage of clones. The packaging of DNA into virions is generally inefficient because of competition between the helper phage and phagemid. This is further complicated with larger phagemid clones, as the ratio of packaged phagemid to helper phage is skewed toward the helper as the size of the phagemid increases. Mutant helper phage strains with reduced capacity to interfere with phagemid packaging are often used as helper phages, and they result in a substantial improvement in yield *(3)*. The following procedures allow for the preparation of ssDNA from the pBluescript series of vectors (Stratagene).

2. Materials

2.1. Determination of Helper Bacteriophage Titer

1. Luria–Bertani (LB) broth: 5 g tryptone, 10 g yeast extract, 5 g NaCl. Bring to 1 L with water and autoclave.
2. LB agar: Add 15 g Bacto agar to 1 L of LB broth and autoclave.

3. Semisolid LB top agar: Add 7 g of Bacto agar to 1 L of LB broth and autoclave.
4. Susceptible *Escherichia coli* strain (*see* Chapter 3).
5. K07 helper phage (Invitrogen).

2.2. Preparation of Purified Bacteriophage Particles

1. Susceptible *E. coli* strain harboring phagemid.
2. LB broth (*see* **Subheading 2.1., item 1**).
3. Titered stock of K07 helper phage.
4. Kanamycin.
5. PEG solution: 20% polyethylene glycol, 3.5 M ammonium acetate (pH 7.5).

2.3. Isolation of ssDNA

2.3.1. Triton X-100 Method

1. Triton solution: 0.1% Triton X-100, 100 mM Tris-HCl (pH 7.5), 5 mM EDTA, 1 mg/mL proteinase K.

2.3.2. SDS Method

1. SDS solution: 0.1% sodium dodecyl sulfate (SDS), 100 mM Tris-HCl (pH 7.5), 5 mM EDTA, 1 mg/mL Proteinase K.
2. 5 M NaCl.
3. 3 M Sodium acetate.
4. 100% and 70% Ethanol.
5. TE: 10 mM Tris-HCl (pH 8.0), 0.1 mM EDTA (pH 8.0).

3. Methods

3.1. Determination of Helper Bacteriophage Titer

1. Prepare fifteen 5-mL aliquots of semisolid LB top agar in glass tubes and place at 42°C to keep the agar molten.
2. Add 10^8 colony-forming units of a susceptible bacterial strain to 0.5 mL LB broth, which has been prewarmed to 37°C, in 12 separate tubes. Incubate at 37°C for 2–4 h with vigorous shaking (approx 150 rpm).
3. Prepare dilutions of the phage stock to be tittered, from 10^0 (undiluted) to 10^{-10}, using LB broth.
4. Add 100 µL of the bacteriophage dilutions to the bacterial cultures and incubate at 37°C for 1 h with shaking.
5. Pipet 100 µL of the bacterial cultures into the tubes of top agar, swirl, and layer the molten agar onto an LB agar plate. After the agar has solidified, incubate the plates at 37°C overnight.
6. Count plaques in the bacterial lawn and use the dilution factor to calculate the number of plaque-forming units (pfus) in the original stock.

3.2. Preparation of Purified Bacteriophage Particles

1. Inoculate 5 mL of LB broth with a single colony of the phagemid clone (*see* **Note 1**).
2. Incubate at 37°C with vigorous shaking (approx 150 rpm) and grow the bacteria to late logarithmic phase (optical density [OD_{650}] = 0.5 – 0.7).
3. Add 2×10^9 helper phages, in a small volume (1–2 µL) of LB broth, to 0.25 mL of bacterial culture (*see* **Note 2**). Incubate at 37°C for 1–2 h with vigorous shaking (*see* **Note 3**).

4. Add 0.75 mL of LB broth supplemented with 35 µg/mL of kanamycin, which has been prewarmed to 37°C, and incubate at 37°C for 12–24 h with shaking.
5. Transfer the culture to a 1.5-mL Eppendorf tube and centrifuge at top speed in a microcentrifuge for 5 min.
6. Recover the supernatant and add 300 µL of PEG solution. Incubate at room temperature for 15 min.
7. Centrifuge for 10 min at top speed in a microcentrifuge. Carefully aspirate and discard the supernatant. Be sure that all of the supernatant is removed. A pellet containing the precipitated bacteriophage particles should be visible.

3.3. Isolation of ssDNA

3.3.1. Triton X-100 Method

1. Resuspend the bacteriophage particles in 50 µL of Triton solution and incubate at 37°C for 10 min.
2. Inactivate proteinase K by incubating at 100°C for 10 min.
3. Place on ice for 10 min and then centrifuge at top speed in a microcentrifuge for 5 min.
4. Remove the supernatant containing the ssDNA.
5. The DNA may be used immediately or stored at –20°C for up to 1 yr (*see* **Notes 4–6**). Repeat **steps 3** and **4** before using stored DNA.

3.3.2. SDS Method

1. Resuspend the bacteriophage particles in 300 µL of SDS solution by vortexing vigorously. Incubate at 37°C for 2 h.
2. Add 75 µL of 5 M NaCl and place on ice for 1 h.
3. Centrifuge at top speed in a microcentrifuge for 10 min.
4. Transfer the supernatant to a fresh 1.5-mL Eppendorf tube. Add 50 µL of 3 M sodium acetate and 1 mL of cold ethanol to precipitate the ssDNA and place at –20°C for 1 h or overnight.
5. Centrifuge at top speed in a microcentrifuge for 10 min. Carefully aspirate and discard the supernatant. A faint pellet should be visible.
6. Add 500 µL of cold 70% ethanol to remove salts from the ssDNA and incubate at room temperature for 30 min or at –20°C overnight.
7. Carefully remove the supernatant and air-dry the pellet.
8. Resuspend the ssDNA in 20 µL of water or TE buffer.
9. The DNA may be used immediately or stored at –20°C for up to 1 yr (*see* **Notes 4–6**).

4. Notes

1. It is important to use good bacteriological techniques. Start with single-colony inoculae and carefully monitor the growth of the bacterial cultures.
2. The titer of the helper phage is important and should be determined in plaque assays as described in **Subheading 3.1.**
3. The aeration of bacterial cultures is very important during growth and has dramatic effects on the yield of ssDNA. The use of multiwell plates or other formats with poor aeration is convenient, especially for large sets of clones, but can result in large reductions in yield.
4. Many investigators store phagemid collections in microtiter dishes at –70°C.
5. Quantify the yield and purity of the ssDNA, by measuring the A_{260}/A_{280} ratio (*see* Chapter 8) or by visual analysis on an agarose gel, prior to use of the ssDNA preparations.

6. There is an inverse relationship between packaging efficiency and the size of a phagemid clone. For particularly large clones (>10 kb), packaging may be very inefficient and, therefore, the yield low.

References

1. Zinder, N. D. and Boeke, J. D. (1982) The filamentous phage (Ff) as vectors for recombinant DNA—a review. *Gene* **19,** 1–10.
2. Dotto, G. P., Enea, V., and Zinder, N. D. (1981) Functional analysis of bacteriophage f1 intergenic region. *Virology* **114,** 463–473.
3. Enea, V. and Zinder, N. D. (1982) Interference resistant mutants of phage f1. *Virology* **122,** 222–226.

14

Using Desktop Cloning Software to Plan, Track, and Evaluate Cloning Projects

Robert H. Gross

1. Introduction

Manipulation and analysis of DNA sequences is often a complex task involving many steps, each of which must be carefully planned and executed. To facilitate this process, the number of steps should be minimized and each step analyzed to ensure that it has been completed successfully. Often, this analysis involves restriction-enzyme digestion of DNA constructs followed by gel electrophoresis to examine the results. It is also important to be able to easily track the DNA manipulations. This means that each step must be documented, illustrated, and stored as the project proceeds. It is also desirable to maintain an overview of the cloning strategy and track changes as they are made. Graphical representations of DNA constructs make the process simple to follow and less prone to errors.

Consider the problems that might be encountered in finding appropriate restriction-enzyme sites for use in a cloning project. The sites must be located quickly and without error, both DNA strands must be searched for sites, and sites meeting specific criteria must be identified. It is often required to find a restriction site that cuts only within a certain segment (or only outside of that segment), or an enzyme that cuts a specified number of times, or even sites with blunt ends. During manipulation, the actual staggered cut sites, and not just the recognition sequences, must be tracked. When ligating different fragments of DNA, the compatibility of the "sticky ends" produced by restriction enzyme digestion is important. Other common manipulations include inverting sequences and trimming or filling-in of overhanging ends. In all these manipulations, it is imperative to keep track of the DNA ends. This type of pattern matching is a task well suited to computer analysis.

One of the keys to monitoring cloning projects is to perform restriction enzyme digests and run gel electrophoresis to analyze the results. Predicting fragment sizes and mobilities on the gels is important to be able to interpret the results. Often, there are extra bands on the gel that might be the result of incomplete digestion. A computer analysis will be able to predict gel patterns and also show the partial digest fragments.

The description that follows uses the Gene Construction Kit (GCK) to achieve all of the above objectives. GCK provides a convenient way to keep track of all the various steps in a cloning project while keeping the DNA sequence files readily available for use in other programs. The graphical display allows one to focus on designing the cloning strategy rather than on the details of the individual steps (or on how to run the program). The program handles all of the underlying details of sequence manipulation; thus, the resulting sequences will be accurate. Restriction digests can be conducted *in silico* and compared to actual gel patterns to evaluate the success of each step.

2. Materials

Gene Construction Kit (version 2.5), developed by Textco, is available from SciQuest, Inc. (Research Triangle Park, NC). A demo version is available from www.textco.com/.

3. Methods

3.1. Background

As an example, an actual cloning project will be described. In this project, the coding sequence of a *Drosophila* heat-shock gene (*hsp26*) was cloned into a vector downstream of a regulated promoter. The coding sequence was cloned in both orientations in order to be able to generate both sense and antisense transcripts for downstream applications. The vector plasmid was pRmHa3 (derived from pRmHa1 *[1]*). This plasmid contains an inducible *Drosophila* metallothionein promoter, a polycloning site, a polyadenylation site, and a selectable marker. The architecture of pRmHa3 was unknown and GCK was used to characterize it. The source of the *hsp26* coding sequence was the plasmid pJB1. Plasmid pJB1contains a 2300-nucleotide segment that has the coding region for *hsp26* as well as 176 nucleotides upstream and 1502 nucleotides downstream of the coding sequence.

3.2. Importing a Sequence into GCK

The sequence of pRmHa3 was imported by choosing **File > Import** from the GCK menu. A number of import formats are possible (*see* **Fig. 1**). In this case, the sequence file was a plain text file, so that option was selected from the pop-up menu and the sequence was imported as a linear DNA molecule. A graphical view of pRmHa3 was displayed by choosing **Construct > Display > Display Graphics**. As the plasmid is a circular DNA, it was converted to circular by choosing **Construct > Make Circular**. A label called "junction marker" appeared at the location of the joining of the two ends. GCK will automatically generate such a marker whenever two ends are joined in this way in order to help keep track of each step. In this instance, the junction marker labels were not needed and were removed by selecting them and pressing the "Delete" key.

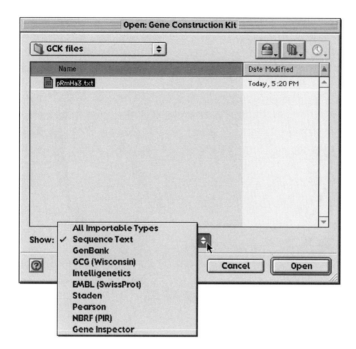

Fig. 1. Importing sequences into GCK. Importing a sequence file is accomplished by selecting **File > Import...**, and then choosing a file format from the pop-up menu. In this case, just a plain text sequence is being imported.

3.3. Annotating a Construct

3.3.1. Identifying Open Reading Frames

1. To characterize pRmHa3, coding regions were identified by highlighting open reading frames (ORFs) using **Construct > Features > Find Open Reading Frames** (*see* **Fig. 2A**). The open reading frames are shown in the plasmid in **Fig. 2B**.
2. ORF1 corresponded to the location of the β-lactamase gene *(1)*. To confirm this, the ORF arrow was selected by double-clicking on it. Double-clicking on a protein sequence (region) will select the corresponding DNA sequence. The corresponding DNA sequence was copied to the clipboard and used to query the NCBI BLAST server (www.ncbi.nlm.nih.gov/BLAST/). This confirmed that ORF1 was indeed the β-lactamase gene that codes for ampicillin resistance.
3. Similar searches using the DNA sequences corresponding to the other ORFs showed them to correspond to known components of the pRmHa3 vector (the metallothionein promoter and alcohol dehydrogenase 3'-untranslated region) and enabled the entire pRmHa3 construct to be annotated. Annotation is accomplished by selecting an object (e.g., the region containing the promoter) and then choosing **Construct > Get Info...** to enter information to store with the construct object.

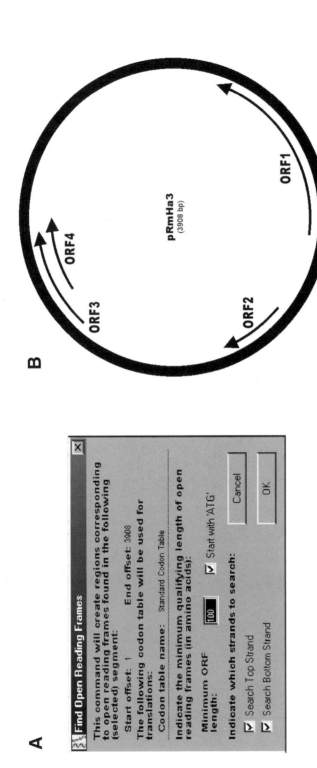

Fig. 2. Identifying ORFs in pRmHa3. (**A**) To find ORFs, the user specifies whether one or both strands should be searched, whether the ORF must start with an ATG, and the minimum size of ORFs to be reported. (**B**) Four ORFs of at least 100 nucleotides were identified in pRmHa3 DNA.

3.3.2. Marking Restriction Sites

There was a polycloning site in the construct that was identified by marking restriction sites using Construct > Features > Mark sites.... Unique sites (those that occur only once in pRmHa3) were marked for all commercially available enzymes (*see* **Fig. 3A**). Note that the dialog box specified that only enzymes that cut less than two times should be marked in order to identify unique sites. The result of this is shown in **Fig. 3B**. Those marked sites not in the polycloning location were removed by selecting them and then pressing the "Delete" key. The final result is shown in **Fig. 3C**.

3.3.3. Color Coding

1. The construct was viewed as a sequence (**Construct > Display > Display Sequence**) and the sequence range to be annotated as the polycloning site was selected.
2. In order to maintain consistency between different projects, a color-coded scheme can be used to identify different functional segments of DNA (e.g., polycloning sites, promoters). The polycloning site was color coded by choosing **Format > Color > Add a Color...** to bring up the dialog box shown in **Fig. 4A**. Pressing "OK" allowed selection of a color from a color palette, which was then defined as the polycloning site color and automatically added to the **Color** menu. Note that all images in this chapter are of necessity shown in grayscale, but in the program, actual colors are chosen to identify segments.
3. This color was then selected by choosing **Format > Color > Polycloning site**.
4. After identifying features, annotating and coloring pRmHa3 looked as shown in **Fig. 4B**.

3.4. Making and Tracking New Constructs

3.4.1. Creating a New Generation

1. The source of the *hsp26* coding sequence was the plasmid pJB1 (*see* **Fig. 5A**). In order to isolate the segment containing the *hsp26* coding sequence, pJ1B was digested with *Eco*RI and *Bam*HI and the smaller of the two fragments generated was isolated using gel electrophoresis and gel extraction. To track this step in GCK, the *Eco*RI–*Bam*HI fragment of pJ1B was selected by dragging the mouse over it (like selecting a word in a word processing program) and copied to the clipboard. Alternatively, the segment could be selected by double-clicking on it.
2. The pRmHa3 file was then opened and a new generation was created through **Format > Chronography > Create New Generation**. By doing this, a new "view" of the construct into which the *Eco*RI–*Bam*HI fragment could be pasted was created. Using "chronography" in this way allows each step in the cloning process to be viewed as a separate "generation."
3. Both *Eco*RI and *Bam*HI produced 5' overhangs. These were end filled with Klenow, and the blunt-ended *Eco*RI–*Bam*HI fragment was inserted into the blunt-ended *Sma*I site of pRmHa3. In GCK, this was done by placing the insertion point at the *Sma*I site in pRmHa3 and then choosing **Edit > Special Paste...** to paste the *hsp26* DNA segment from the clipboard into the *Sma*I site. The special paste option allows one to specify how each of the four ends should be treated during the ligation; in this case, all were made blunt ended. This created one generation of the construct that displayed the original pRmHa3 DNA and the newer generation just created that showed the original plasmid containing the *hsp26* DNA insert. By using chronography, it is possible to track the history of manipulations performed on the construct and (through the Chronography menu that lists each generation) view each of these steps in detail.

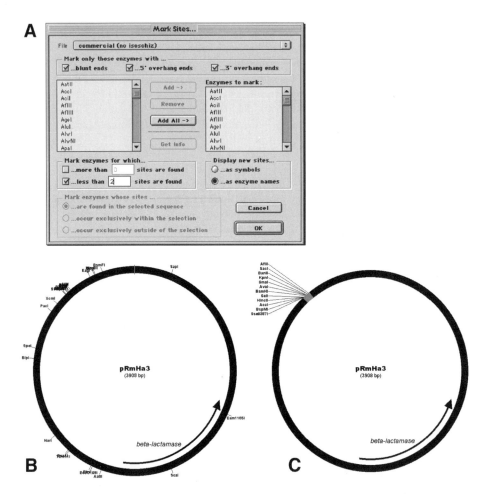

Fig. 3. Marking restriction sites within pRmHa3. (**A**) Marking restriction sites can be customized to show any combination of 5' overhangs, 3' overhangs, or blunt ends, can utilize one or more enzymes, and can be limited to those enzymes that cut only a specified number of times, or inside (or outside) of a selected DNA segment. (**B**) pRmHa3 with all unique restriction sites marked. (**C**) Labels for sites that were not part of the polycloning region were removed and the remaining sites displayed in an orderly fashion by choosing **Format > Site Markers > Automatic Arrangement**. The gray area represents the actual DNA sequence containing the multiple-cloning site.

4. This process was repeated, but this time the *hsp26* segment was pasted into the pRmHa3 *Sma*I site while inverting the segment (this also can be specified in the **Special Paste** dialog box). The end results of these two operations were two constructs that were used to transcribe the *hsp26* DNA in either the sense or the antisense direction (*see* **Fig. 5B,C**).

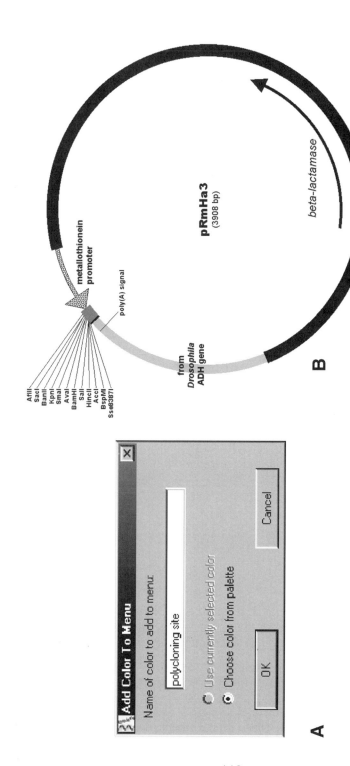

Fig. 4. Color coding pRmHa3. (A) Colors to be added to the Color menu are first named in this dialog box, and then an actual color is chosen from a color palette. This dialog is accessed by choosing **Format > Color > Add a Color....** (B) pRmHa3, completely annotated. As described in the text, various regions of the pRmHa3 DNA were identified and then displayed as shown in this figure.

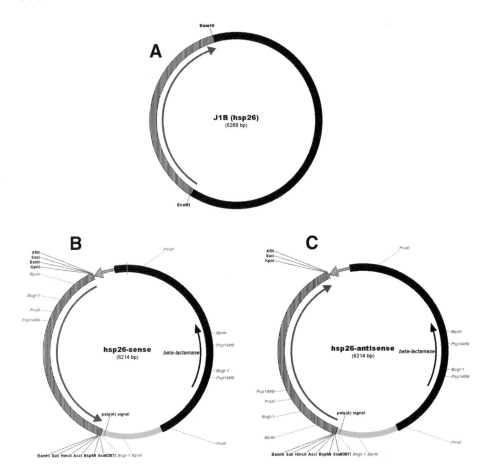

Fig. 5. Cloning the *hsp26* gene into pRmHa3. (**A**) pJ1B, containing the *hsp26* DNA as an *Eco*RI–*Bam*HI fragment (stippled segment). (**B,C**) pRmHa3 after insertion of the *hsp26* gene fragment in either of two orientations. The original features of the vector are still shown and the orientation of the insert is indicated. Enzyme sites marked in gray are those that are used to distinguish orientation of the insert, as indicated in **Fig. 6**.

3.4.2. Restriction Digest Analysis

1. The process of ligating the *hsp26* DNA into the blunt-ended *Sma*I site generated a series of constructs, of which approx 50% would be in the sense orientation and 50% in the antisense orientation. In order to differentiate between the two orientations, restriction digestion followed by gel electrophoresis was used. GCK was used to decide which enzymes were most appropriate for this analysis. All enzymes that cut once in the *hsp26* DNA and twice in the rest of the plasmid were marked (*see* **Fig. 6A**). Those enzymes that cut in an asymmetric manner and in doing so generate different gel patterns for the two orientations were selected (*Bpm*I, *Psp*1406I, *Pvu*II, and *Bcg*I-1).

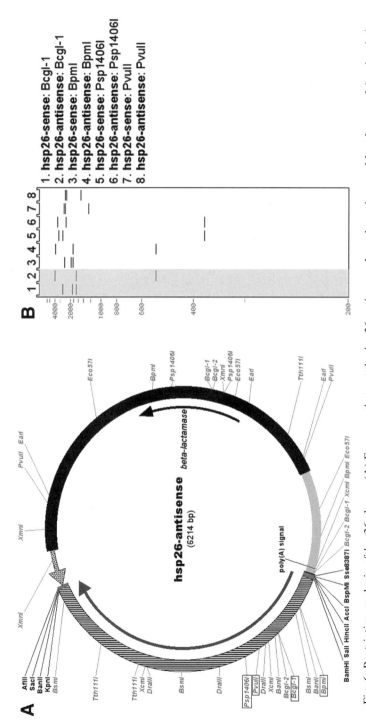

Fig. 6. Restriction analysis of *hsp26* clones. (A) Enzymes that cut the *hsp26*-antisense clone three times, with only one of the sites being in the insert. Enzyme labels that are boxed display an asymmetric distribution. Enzymes labeled in black were part of the original pRmHa3; those in gray were identified in the recombinant plasmid. (B) Virtual gel pattern for the *hsp26*-sense and *hsp26*-antisense constructs digested with the enzymes identified in A. Size markers are indicated along the left edge of the gel. Pairs of adjacent lanes show the digestion pattern for the same enzyme being used to digest different insert orientations. *Bcg*I-1, indicated in lanes 1 and 2, was used in the actual experiments.

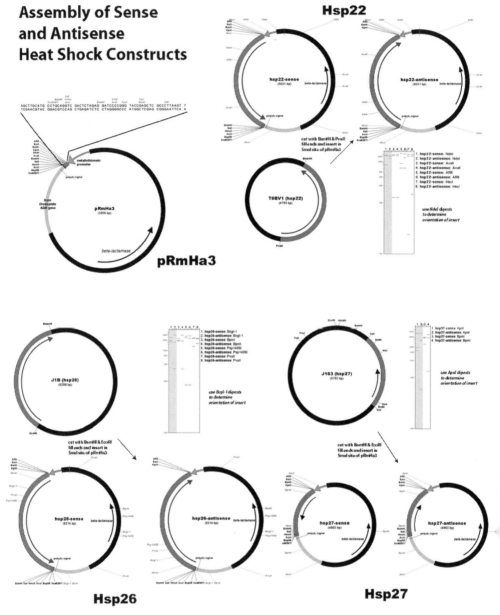

Fig. 7. Creating an illustration of the cloning project. An example Illustration window showing an overview of the entire *hsp*-cloning project. Each construct that *(continued on next page)*

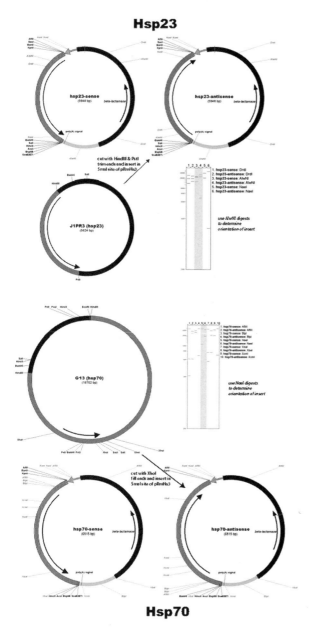

Fig. 7. *(continued)* was utilized in the in this illustration Along with an explanation of each step undertaken are the predicted gel patterns for determining the orientation of the insert DNA.

2. These enzymes were used to perform virtual digests and run on virtual gels in GCK (*see* **Fig. 6B**). Gels can be run with GCK by first selecting the marked enzymes sites to be used in the digest, then copying those sites and pasting them into a Gel window. The digestion pattern is automatically generated. These results identified *Bcg*I-1 as a suitable enzyme to use.
3. The GCK predicted gel patterns were compared to the actual gel patterns generated by *Bcg*I-1 digestion of the *hsp26* clones and enabled the orientation of the *hsp26* DNA in each clone to be deduced (*see* **Note 1**).

3.4.3. Creating an Illustration

An "Illustration" window can hold any object created in GCK (such as constructs and gels) along with any object that can be imported from the clipboard (such as scanned images of gels). Thus, Illustration windows can display both cloning strategies and analytical information. This allows for the creation of a single file and diagram that contains all of the information regarding the construction process. Each step can be annotated as the process continues. An illustration of the cloning project described in this chapter is shown in **Fig. 7**. An Illustration window is created by choosing **File > New > Illustration**. In the Illustration window, any object can be selected by double-clicking on it and edited individually. Thus, double-clicking on a construct within the Illustration will allow pieces of that DNA to be copied or modified, similar to if it were within a Construct window. This not only provides a graphical overview of the process but also stores all the sequences which can be accessed at a later time (*see* **Note 2**).

3.5. Importing Sequences from Genbank

1. GCK's Deluxe Importing feature (**File > Deluxe Import > Search GenBank**) allows for a straightforward importing of GenBank (or EMBL) sequence files directly from the corresponding websites. Sequences can be identified by accession number or keyword searches. Keyword searches will produce a list of sequences matching the search criteria. Once the sequence is identified, it can be directly imported from GenBank into GCK, taking full advantage of the features table of the databank entry (*see* **Fig. 8A,B**).
2. The dialog box in **Fig. 8C** shows how to define how each entry in the GenBank feature table is converted into a GCK feature. Each item in the feature table can be defined individually to have a specific appearance in GCK. This is done by selecting an item in the feature list and then using the items in the **Format** menu to define the appearance of the selected feature in a fashion similar to defining features of a construct (*see* **Subheading 3.3.**). An entire set of conversion specifications can be saved as a default setting, allowing future conversions to follow this default. This provides a consistent way to import any sequence file from GenBank or EMBL into GCK with a uniform feature appearance (*see* **Notes 3** and **4**).

4. Notes

1. Another useful feature of GCK, although not shown here, is the ability to create partial digests. For example, it is possible to specify that during a digest only five of six sites are cut. The resulting digest pattern will show complete digest fragments as solid black lines (bands) and partial digest fragments as dotted blue lines. This feature can be used to distinguish partial restriction digests from contamination in the sample.

Desktop Cloning

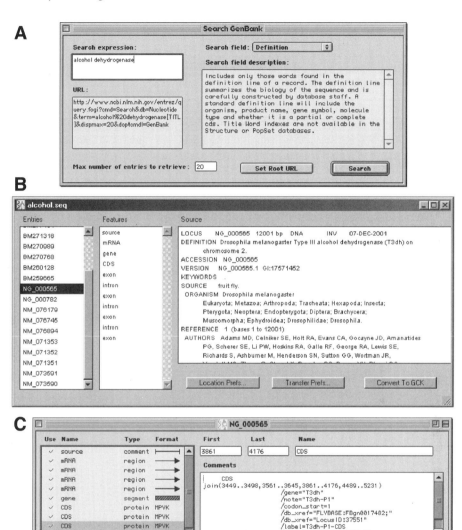

Fig. 8. Importing sequences from Genbank. (**A**) GenBank searches are conducted by typing in a search query in the top left of the dialog, and then specifying the field(s) to be search in the GenBank database using the pop-up menu in the top right (set to "**Definition**" in this dialog box). (**B**) The items listed in the left list are those that are found within GenBank and correspond to the search criteria defined in **A**. Clicking on an item in this "**Entries**" list will show the features of that entry in the middle column (taken from the GenBank features table), and the actual text associated with the selected item in the "**Source**" field on the right. (**C**) The dialog box shows how GenBank features can be mapped to specific GCK features.

2. Illustrations are extremely useful. They can be used to follow the progress of a cloning project, as a way of designing the project, and as a way of illustrating the project for a poster or paper figure. In addition, because each construct in the illustration can be used as a source of actual DNA sequence files, this illustration can be used as a starting point for other cloning projects in the future.
3. The ability to search and retrieve sequences directly from GenBank or EMBL and directly import them into GCK provides a standard way to have a graphical view of any sequence in these databases. Such a standardized graphical view provides a more convenient and intuitive way to view sequences than by looking at textual information.
4. In addition to searching database files, GCK has the ability to search through its own files; thus, if a user (or institution) stores all constructs in a common location, all of those files can be searched from within GCK for sequences or comments of interest. Because comments can be stored as part of any GCK feature (e.g., a marked site, a DNA segment, a protein sequence), a great deal of text can be associated with any construct. Information such as storage location or author is very useful. For example, a search might be used to find all constructs containing an ampicillin resistance gene but not a tetracycline resistance gene.

Reference

1. Bunch, T. A., Grinblat, Y., and Goldstein, L. S. B. (1988) Characterization and use of the *Drosophila* metallothionein promoter in cultured *Drosophila melanogaster* cells. *Nucleic Acids Res.* **16,** 1043–1061.

15

Cloning in Plasmid Vectors

Carey Pashley and Sharon Kendall

1. Introduction

A fundamental step in molecular biology is the cloning of a DNA fragment insert into a plasmid vector. This allows the cloned fragment to be replicated upon transformation of the recombinant molecule into a bacterial cell (*see* Chapters 4 and 5) so that the DNA of interest can be investigated further. Cloning is an essential part of many experiments, including library generation (*see* Chapter 18) and expression studies (*see* Chapter 29).

The vector and insert DNA are usually digested with a type II restriction endonuclease that cleaves at specific sites in the DNA. The two molecules must have compatible ends for cloning to proceed. The generation of compatible ends requires the use of restriction and modifying enzymes, which ultimately results in the generation of either blunt or overhanging ends. The cleaved fragments are mixed in the presence of DNA ligase that produces a mixture of products, some of which should consist of the vector containing the inserted DNA fragment.

Numerous vectors are commercially available to facilitate cloning for various applications (*see* Chapter 2). Chapter 14 presents a detailed description for designing cloning experiments using such vectors. Additionally, specialized vectors are available for cloning polymerase chain reaction (PCR) products generated by *Taq* DNA polymerase (*see* Chapter 17). This chapter focuses on the preparation of DNA fragments for efficient cloning.

1.1. Restriction Digestion

The first step in cloning a DNA insert into a plasmid vector is cutting both vector and insert DNA with the appropriate restriction enzyme(s) to generate compatible ends. This may be a simple single digestion or a double digestion with two enzymes in the case of directional cloning. A partial digestion may be needed in situations where there is a lack of suitable sites to use and the restriction enzyme cuts more than once in the molecule.

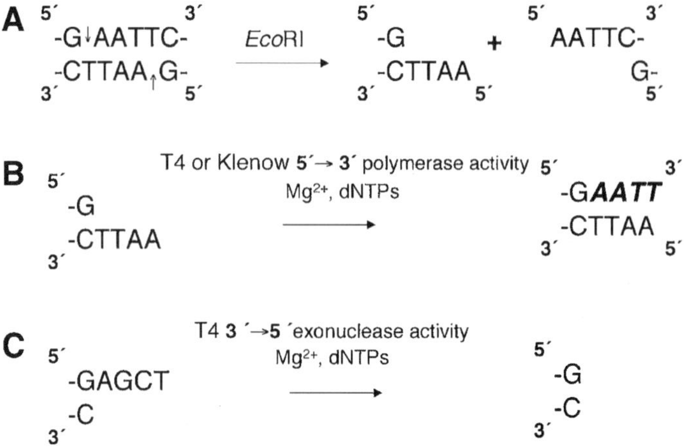

```
A  5'              3'                    5'              5'           3'
   -G↓AATTC-      EcoRI                  -G        +     AATTC-
   -CTTAA↑G-      ———→                   -CTTAA          G-
   3'              5'                    3'       5'    5'


B  5'           T4 or Klenow 5'→ 3' polymerase activity  5'            3'
    -G                    Mg²⁺, dNTPs                    -GAATT
    -CTTAA                ———→                           -CTTAA
   3'                                                   3'            5'


C  5'           T4 3'→5' exonuclease activity            5'
    -GAGCT                Mg²⁺, dNTPs                    -G
    -C                    ———→                           -C
   3'                                                   3'
```

Fig. 1. (**A**) The generation of 5' overhangs by digestion with *Eco*RI. (**B**) The 5'→3' polymerase activity of T4 DNA polymerase or Klenow fills in an overhang 5' end. (**C**) the 3'→5' exonuclease acitivity of T4 polymerase is used to make a 3' overhanging end blunt.

Type II restriction enzymes recognize specific sites within the DNA sequence and cut the DNA at that site. The sequences recognized vary in length and base composition, but the typical type II site is an exact palindrome of 4, 5, 6, 7, or 8 bp with an axis of rotational symmetry (e.g., the *Eco*RI recognition site is GAATTC). Cleavage by a restriction enzyme can generate a number of different ends. Ends with protruding bases, either a 5' or 3' overhang, are known as sticky ends, as they form hydrogen bonds with similar regions on other fragments, whereas ends with no protruding bases are referred to as blunt. For two sticky ends to anneal, they must be compatible. Compatibility occurs when the overhangs generated by the restriction enzyme are complementary to each other. Thus, two pieces of DNA cut with the same enzyme will have compatible ends (*see* **Fig. 1A**). Some enzymes may recognize different sequences but can still be joined together, as they produce identical sticky ends; for example, the overhang produced by *Bam*HI is complmentary to those produced by *Bgl*II, *Bcl*I, or *Sau*3AI.

Cloning is significantly more efficient if complementary overhanging ends, rather than blunt ends, are present on the vector and insert. Directional cloning is considered to be the most efficient approach; in this case, the plasmid DNA is digested by two different restriction enzymes that produce noncomplementary overhangs. If the insert is prepared in the same way, then ligation can only occur only in one orientation. This precludes vector religation and greatly reduces the background level of non-recombinants.

1.2. Introducing a Restriction Site

If the insert DNA does not contain convenient restriction sites, it is possible to generate a site at the desired position by amplifying the insert using the PCR primers designed with the restriction site. After a number of rounds of amplification of the insert, the new restriction site becomes incorporated. An example of a forward primer containing a number of mismatches used to introduce an *Nde* I site (5'-CATATG-3') is as follows:

| Sequence to be amplified | 5'-AGTGGTGCCCTGGTGGTAAAG-3' |
| Sequence of primer with mismatch | 5'-AGTGGTGCCC*CATATG*GTAAAG-3' |

The required sites can also be added by the use of short DNA duplexes, known as linkers or polylinkers. Linkers are blunt-ended DNA fragments that are designed to contain an appropriate restriction-site sequence. They are often phosphorylated at the 5' end, which allows ligation to a dephosphorylated vector.

1.3. Converting an Overhang to a Blunt End

In cloning experiments where compatible ends are not available, it may be necessary to convert a 5' or 3' overhang to a blunt end (*see* **Fig. 1B**). Both bacteriophage T4 DNA polymerase and *Escherichia coli* DNA polymerase I large (Klenow) fragment have 5'ɡ3' polymerase activity and can be used to fill in 5' overhangs. In addition, these enzymes both contain powerful 3'ɡ5' exonuclease activity with single-stranded DNA as the substrate *(1)*. The exonuclease activity of T4 DNA polymerase is 200 times more powerful than that of the Klenow fragment and is thus used to trim 3' overhangs, converting them to blunt ends (*see* **Fig. 1C**). The reaction must contain an excess of dNTPs, as in the absence of dNTPs, the 3'ɡ5' exonuclease activity of T4 DNA polymerase will degrade double-stranded DNA *(2)*.

1.4. Dephosphorylation of Vector DNA

Alkaline phosphatases are commonly used in cloning experiments to dephosphorylate the 5' ends of vector DNA. This prevents self-ligation of the vector, as the enzyme used to ligate the DNA molecules requires a 5'-phosphate group on one of the DNA substrates (*see* **Fig. 2A**). Self-ligation will occur if the ends of the prepared vector are compatible or blunt. Dephosphorylation of the linearized vector DNA, prior to ligation to the insert fragment, decreases the background of the recircularized vector, whereas a DNA insert containing phosphorylated 5' termini can be ligated into the vector. For a linear vector with different 5' ends, alkaline phosphatase treatment is not necessary. Calf intestinal alkaline phosphatase (CIAP) or bacterial alkaline phosphatase (BAP) are most commonly used in these reactions. BAP is more active than CIAP at a higher temperature and is therefore more difficult to heat inactivate than CIAP at the end of the reaction *(2)*. Shrimp alkaline phosphatase can also be used and is easier to inactivate than CIAP.

1.5. Phosphorylation of Insert DNA

Phosphorylation of insert DNA that lacks terminal 5' phosphates, such as PCR products and fragments with synthetic linkers, may be required in preparation for ligation. If the product is to be cloned into a nonphosphorylated vector, it is vital that phosphate groups are added to the insert. In such a situation, T4 polynucleotide kinase can be used; this enzyme catalyzes the transfer of phosphate from ATP to the 5' terminus of dephosphorylated DNA or RNA (*see* **Fig. 2B**) *(3)*.

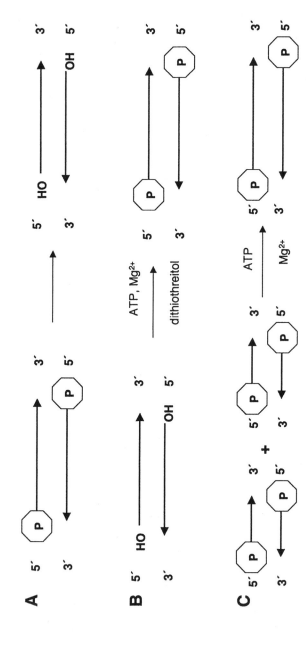

Fig. 2. Schematic representation of the overall reactions catalyzed by enzymes useful in cloning manipulations: (**A**) phosphatase, (**B**) kinase (T4), and (**C**) ligase (T4).

1.6. Ligation of Vector and Insert

The final step in cloning is the joining of the linear DNA fragments together, referred to as ligation. This involves creating a phosphodiester bond between the 3'-hydroxyl group of one DNA fragment and the 5'-phosphate group of another and is equivalent to repairing nicks in a duplex strand. The enzyme most frequently used to ligate fragments is bacteriophage T4 DNA ligase (*see* **Fig. 2C**). T4 DNA ligase (unlike *E. coli* DNA ligase) will join blunt-ended fragments, as well as cohesive-end fragments, efficiently under normal reaction conditions.

2. Materials
2.1. Restriction Digestion

1. Appropriate restriction enzyme supplied with buffer; store at –20°C.
2. 1 mg/mL bovine serum albumin (BSA), acetylated.
3. 0.5 M EDTA (pH 8.0).
4. Phenol : chloroform : isoamyl alcohol (25 : 24 : 1).
5. Chloroform : isoamyl alcohol (24 : 1).
6. 3 M sodium acetate, pH 5.2.
7. 100% Isopropanol.
8. 100% and 70% Ethanol.
9. Sterile double-distilled water.

2.2. Introducing a Restriction Site
2.2.1. Using Polylinkers

1. T4 DNA ligase; store at –20°C.
2. T4 DNA ligase 10X reaction buffer: 300 mM Tris-HCl (pH 7.8), 100 mM MgCl$_2$, 100 mM dithiothreitol (DTT), 10 mM ATP. Store in aliquots at –20°C.
3. Phosphorylated linkers containing the appropriate restriction enzyme site.
4. Sterile double-distilled water.

2.2.2. Using PCR

1. *Taq* polymerase; store at –20°C.
2. *Taq* polymerase 10X reaction buffer: 500 mM KCl, 100 mM Tris-HCl (pH 9.0), 1% Triton® X-100. Store at –20°C.
3. 100 ng/µL Each primer. Store in aliquots at –20°C and avoid freeze–thawing.
4. dNTPs, 10 mM each. Store in aliquots at –20°C and avoid freeze–thawing.
5. 25 mM MgCl$_2$. Store at –20°C.
6. Sterile double-distilled water.

2.3. Converting an Overhang to a Blunt Terminus

1. DNA polymerase Klenow fragment or T4 DNA polymerase. Store at –20°C.
2. Klenow 10X reaction buffer: 500 mM Tris-HCl (pH 7.2), 100 mM MgSO$_4$, 10 mM DTT; or T4 DNA polymerase 10X reaction buffer: 330 mM Tris-acetate (pH 7.9), 660 mM potassium acetate, 100 mM magnesium acetate, 5 mM DTT. Store at –20°C.
3. 1 mg/mL BSA, acetylated.
4. dNTPs, 100 mM each. Dilute to working concentrations of 1 mM and 2.5 mM. Store in aliquots at –20°C and avoid freeze–thawing.
5. Sterile double-distilled water.

2.4. Dephosphorylation of Vector DNA

2.4.1. Using CIAP

1. CIAP, store at –20°C and avoid freeze–thaw cycles. If used infrequently (less than a few times a month), store at –70°C.
2. CIAP 10X reaction buffer: 500 mM Tris-HCl (pH 9.3), 10 mM MgCl$_2$, 1 mM ZnCl$_2$, 10 mM spermidine.
3. 0.5 M EDTA (pH 8.0).
4. Sterile double-distilled water.

2.4.2. Using Shrimp Alkaline Phosphatase

1. Shrimp intestinal alkaline phosphatase (SAP). Store at –20°C and avoid freeze–thawing.
2. SAP 10X reaction buffer: 50 mM Tris-HCl (pH 9.0 at 37°C), 10 mM MgCl$_2$.

2.5. Phosphorylation of Insert DNA

1. T4 polynucleotide kinase, store at –20°C.
2. Kinase 10X buffer: 500 mM Tris-HCl (pH 7.8), 100 mM MgCl$_2$, 50 mM DTT, 1 mM spermidine, 1 mM EDTA. Store at –20°C.
3. 0.1 mM ATP (*see* **Note 1**).
4. 0.5 M EDTA (pH 8.0).
5. Sterile double-distilled water.

2.6. Ligation of Vector and Insert

1. T4 DNA ligase; store at –20°C.
2. Ligase 10X reaction buffer: 300 mM Tris-HCl (pH 7.8), 100 mM MgCl$_2$, 100 mM DTT, 10 mM ATP. Store at –20°C to preserve the ATP (*see* **Note 2**).
3. Sterile double-distilled water.

3. Methods

3.1. Restriction Digestion

3.1.1. Complete Digests

1. Add the following to a 1.5-mL Eppendorf tube on ice (*see* **Note 3**):

DNA 0.1–1 µg (*see* **Note 4**)	x µL
Restriction enzyme (*see* **Notes 5–7**)	1 µL
Restriction enzyme 10X reaction buffer (*see* **Note 8**)	2 µL
BSA 1 mg/mL	2 µL
Sterile double-distilled water to a final volume of	20 µL (*see* **Note 9**)

2. Spin briefly to ensure that the contents are at the bottom of the tube and place in a 37°C water bath for 2–4 h.
3. Stop the reaction by adding 0.5 M EDTA (pH 8.0) to a final concentration of 10 mM (*see* **Note 10**).

3.1.2. Double Digests

3.1.2.1. IF COMPATIBLE BUFFERS ARE AVAILABLE (SEE NOTES 11 AND 12)

1. Add the following to a 1.5-mL Eppendorf tube on ice:

DNA (0.1–1 µg)	x µL
Restriction enzyme 1	1 µL

Cloning in Plasmid Vectors 127

    Restriction enzyme 2   1 μL
    Appropriate 10X reaction buffer   2 μL
    BSA 1 mg/mL   2 μL
    Sterile double-distilled water to a final volume of 20 μL

2. Spin briefly to ensure that the contents are at the bottom of the tube and place in a 37°C water bath for 2–4 h.
3. Stop the reaction by adding 0.5 M EDTA (pH 8.0) to a final concentration of 10 mM (*see* **Note 10**).

3.1.2.1. IF COMPATIBLE BUFFERS ARE NOT AVAILABLE

1. Set up a reaction with one enzyme as described in **Subheading 3.1.1.** and incubate as described.
2. Clean up the reaction using one of the methods described in **Subheading 3.1.4.**
3. Digest with the second enzyme as in **step 1**.

3.1.3. Partial Digests

1. Set up six separate reactions as described in **Subheading 3.1.1.**
2. Place reaction tubes in a 37°C water bath and then stop reactions following 5-, 10-, 15-, 20-, 40-, and 60-min incubations by chilling the tube on ice and adding EDTA (pH 8.0) to a final concentration of 10 mM.
3. Run the reactions on an agarose gel and examine banding pattern (*see* **Note 13**). Purify the desired band from the gel as described in Chapter 16.

3.1.4. Cleaning up Restriction Enzyme Reactions

3.1.4.1. USING PHENOL : CHLOROFORM

1. Make the reaction volume up to 500 μL with sterile double-distilled water (*see* **Note 14**).
2. Add an equal volume of phenol : chloroform : isoamyl alcohol (25 : 24 : 1). Vortex thoroughly and then centrifuge at room temperature for 5 min at top speed in a microfuge (*see* **Note 15**).
3. Transfer the upper aqueous layer to a clean 1.5-mL Eppendorf tube and extract using an equal volume of chloroform : isoamyl alcohol (24 : 1).
4. Precipitate the DNA by adding 1/10 volume of 3 M sodium acetate, pH 5.2, and 1 volume of ice-cold isopropanol. Centrifuge at room temperature for 10 min at top speed in a microfuge to pellet the DNA (*see* **Notes 16** and **17**).
5. Add 1 mL of 70% ethanol to wash the pellet, centrifuge as in **step 4** and carefully pour off the ethanol. Wash again in 100% ethanol and centrifuge as in **step 4**. Pour off the ethanol and leave the pellet to air-dry (*see* **Notes 18** and **19**).
6. Resuspend the DNA in a suitable volume of sterile double-distilled water.

3.1.4.2. USING COMMERCIALLY AVAILABLE KITS

There are a number of kits available to clean up enzyme reactions; these should be used according to the manufacturers' instructions (*see* **Note 20**).

3.2. Introducing a Restriction Site

3.2.1. Using Polylinkers

1. Add the following to a 1.5-mL Eppendorf tube on ice:
    DNA (100–500 ng) x μL

T4 DNA ligase	1 µL
T4 DNA ligase 10X reaction buffer	1 µL
Phosphorylated linkers (see **Notes 21** and **22**)	x µL
Sterile double-distilled water to a final volume of 100 µL	

2. Incubate at 15°C for 6–18 h.
3. Stop the reaction by heating at 70°C for 10 min.
4. Cool the tube on ice and immediately proceed with digestion with the appropriate restriction enzyme as described in **Subheading 3.1.**
5. Once digested, and prior to ligation, the DNA fragment can be run on an agarose gel to remove any nonligated linkers. The fragment can then be purified from the gel as described in Chapter 16.

3.2.2. Using PCR

1. Design suitable forward and reverse primers including the appropriate restriction-enzyme site at the 5' end (see **Notes 23–25**).
2. Set up the following PCR reaction:

DNA template 50–100 ng	x µL
Taq polymerase	1 µL
Taq polymerase 10X reaction buffer	5 µL
100 ng/µL Primer 1	1 µL
100 ng/µL Primer 2	1 µL
dNTPs, 10 mM each	1 µL
25 mM MgCl$_2$	3 µL
Sterile double-distilled water to a final volume of	50 µL

3. Perform PCR in a thermal cycler using optimized cycling parameters.

3.3. Converting an Overhang to a Blunt Terminus

3.3.1. Converting a 5' Overhang to a Blunt Terminus

3.3.1.1. USING KLENOW

1. Following a restriction enzyme digestion that generates a 5' overhang, remove the enzyme as described in **Subheading 3.1.4.**
2. Add the following to a 1.5-mL Eppendorf (see **Note 26**):

Cleaned-up digested DNA	x µL
dNTPs, 1 mM each	4 µL
1 mg/mL BSA	10 µL
Klenow (see **Note 27**)	1–2 µL
Klenow 10X reaction buffer	10 µL
Sterile double-distilled water to a final volume of 100 µL (see **Note 28**).	

3. Incubate at room temperature for 10 min.
4. Stop the reaction by heating at 75°C for 15 min.

3.3.1.2. USING T4 DNA POLYMERASE

1. Following a restriction enzyme digestion that generates a 5' overhang, remove the enzyme as described in **Subheading 3.1.4.**

Cloning in Plasmid Vectors

2. Add the following in a 1.5-mL Eppendorf tube (*see* **Note 26**):

Cleaned-up digested DNA	x µL
dNTPs, 2.5 mM each	4 µL
1 mg/mL BSA	10 µL
T4 DNA polymerase (*see* **Note 29**)	1 µL
T4 DNA polymerase 10X reaction buffer	10 µL
Sterile double-distilled water to a final volume of 100 µL	

3. Incubate at 37°C for 5 min.
4. Stop the reaction by heating at 75°C for 10 min.

3.3.2. Converting a 3' Overhang to a Blunt Terminus

1. Following a restriction enzyme digestion that generates a 3' overhang, remove the enzyme as described in **Subheading 3.1.4.**
2. Add the following to a 1.5-mL Eppendorf tube:

Cleaned-up digested DNA	x µL
dNTPs, 2.5 mM each	4 µL
1 mg/mL BSA	10 µL
T4 DNA polymerase (*see* **Note 29**)	1 µL
T4 DNA polymerase 10X reaction buffer	10 µL
Sterile double-distilled water to a final volume of	20 µL

3. Incubate at 37°C for 5 min.
4. Stop the reaction by heating at 75°C for 10 min.

3.4. Dephosphorylation of vector DNA

3.4.1. Using CIAP

1. Dilute CIAP in CIAP 1X reaction buffer immediately prior to use (do not store diluted enzyme). Add the following to a 1.5-mL Eppendorf tube:

Digested DNA sample	x µL
CIAP (0.01 U/pmol of ends) (*see* **Notes 30–32**)	1–2 µL
CIAP 10X reaction buffer	10 µL
Sterile double-distilled water to a final volume of 100 µL	

2. Depending on the types of ends generated by the restriction digest, incubate the reaction under the following conditions:
   a. For 5'-overhanging ends: Incubate at 37°C for 30 min, add another aliquot of CIAP and repeat the incubation.
   b. For 5'-recessed or blunt ends: Incubate at 37°C for 15 min, add another aliquot of CIAP and incubate for 45 min at 55°C (*see* **Note 33**).
3. To inactive CIAP, add 2 µL of 0.5 M EDTA (final concentration 5 mM) and incubate at 65°C for 20 min.
4. Purify the DNA by either phenol extraction and ethanol precipitation or using commercially available kits (*see* **Subheading 3.1.4.** and **Notes 34–36**).

3.4.2. Using SAP

1. Following a restriction enzyme digestion that generates a 3' overhang, remove the enzyme as described in **Subheading 3.1.4.** (*see* **Note 37**).

2. Add the following to a 1.5-mL Eppendorf tube:

Cleaned-up digested DNA	x µL
SAP (1 unit/µg DNA) (see **Note 38**)	1–2 µL
SAP 10X reaction buffer	5 µL
Sterile double-distilled water to final volume of	50 µL

3. Incubate at 37°C for 10 min.
4. Inactivate SAP by incubation at 65°C for 15 min (see **Note 35**).
5. Spin briefly in a microfuge to collect the contents at the bottom of the tube. Use a 1- to 2-µL aliquot to ligate to the insert (see **Note 36**).

3.5. Phosphorylation of Insert DNA

1. Set up the following reaction in a 1.5-mL Eppendorf tube:

Insert DNA (see **Note 39**)	250 ng
Kinase 10X reaction buffer (see **Notes 40 and 41**)	4 µL
T4 polynucleotide kinase (see **Note 42**)	10–20 U
0.1 mM ATP	2 µL
Sterile double-distilled water to a final volume of	40 µL

2. Incubate at 37°C for 30 min.
3. To stop the reaction, add 2 µL of 0.5 M EDTA.
4. Purify the DNA by phenol extraction and ethanol precipitation or use a commercially available kit (see **Subheading 3.1.4.**).

3.6. Ligation of Vector and Insert

1. Set up the following reaction using the desired vector : insert molar ratio. For example, to obtain a 1 : 3 molar ratio: 50 ng of a 4.0-kb vector requires 37.5 ng of a 1-kb insert (see **Notes 43 and 44**). Example reaction mix:

Vector DNA	50 ng
Insert DNA	37.5 ng
T4 DNA ligase (see **Note 45**)	1 U
Ligase 10X reaction buffer	1 µL
Sterile distilled water to a final volume of	10 µL

It is important to include appropriate controls (see **Note 46**).

2. Incubate the reaction at the following time and temperature as a general guide (see **Notes 47 and 48**).
   a. For cohesive ends, incubate at 22°C (room temperature) for 3 h or at 4°C for 16 h.
   b. For blunt ends, ligate at 22°C for 4–16 h (see **Notes 49 and 50**).
3. Ligation reactions can be frozen and stored at –20°C or transformed into the host strain immediately (see Chapters 4 and 5). The ligation reaction should not exceed 0.5% of the transformation reaction volume (see **Note 50**).

4. Notes

1. ATP should be present at a concentration of at least 1 µM.
2. The buffer is usually provided or prepared by the manufacturer as a 10X concentrate, which, on dilution, yields an ATP concentration of approx 0.25–1 mM. Low concentrations of ATP (0.5 mM) increase the rate of ligation of blunt-ended fragments. The buffer should be completely thawed and thoroughly mixed prior to use.

3. It is important that the reactions be set up on ice and that the restriction enzyme is added last. When using restriction enzymes, always take the enzyme from the freezer and place immediately on ice. Use a separate pipet tip every time the enzyme is dispensed, to guard against contamination of the enzyme stock.
4. DNA quality is one of the most important factors for obtaining complete digestion. Contaminants such as phenol, chloroform, protein, alcohol, and RNA can adversely affect digestion. The purity of DNA can be estimated by ultraviolet (UV) spectrophotometry (*see* Chapter 8), but plasmid DNA isolated from any number of commercially available kits is usually of sufficient quality for complete digestion.
5. Different manufacturers may recommend different digestion conditions and different enzymes may require additional components. Enzymes supplied by Gibco-BRL have sufficient BSA in the storage buffer so that the addition of extra BSA is not required. Certain enzymes supplied by Gibco-BRL, such as *Eco*RII, *Nde*II, and *Rsr*II require the addition of DTT to a final concentration of 1 mM. Always look at the manufacturers' recommended digestion conditions.
6. Most enzymes are supplied at concentrations of 5–10 units/µL, where 1 unit is defined as the amount of enzyme required to fully digest 1 µg of DNA in 1 h. However, volumes smaller than 1 µL are difficult to pipet accurately; thus, 1 µL is routinely used.
7. Adding too much enzyme can cause digestion at sites other than the recognition sequence by some enzymes. This phenomenon is known as "star activity." It can also occur if the incorrect buffer is used or the digestion times are too long.
8. Ensure that all buffers and BSA are thoroughly vortexed after thawing.
9. Reaction volumes smaller than 20 µL are not recommended. The volume can be increased if digesting dilute DNA samples. The recommended maximum volume is 100 µL. It is important that the volume of enzyme used be less than one-tenth of the total reaction volume, as the enzymes are stored in buffer that contains glycerol, which can interfere with the reaction.
10. EDTA addition to stop the reaction is not always necessary and can, in fact, interfere with subsequent reactions. Some enzymes can be heat denatured and information on heat denaturation is usually contained in the manufacturers' instructions.
11. The manufacturers of restriction enzymes supply information on the activity of each enzyme in a particular buffer. This can be used to assess whether or not the two enzymes are both active in a single buffer. For example, if two enzymes each show 100% activity in a particular buffer, then they can both be used with this buffer. If one of the two enzymes shows only 50% activity while the other shows 100% activity in a particular buffer, then you can compensate for this by adding twice as much of the enzyme with only 50% activity. It is not recommended to use a buffer in which activity of either enzyme is below 50%.
12. Some manufacturers supply a buffer that they recommend for use for certain double digestions. The activity of the enzymes in this buffer is usually denoted by the manufacturer.
13. Occasionally, full digestion is reached in as little as 5 min. When this happens, the enzyme can be diluted 1/10 in its appropriate buffer and the timed digestion repeated. Different dilutions can be tried until the partial digest is satisfactory.
14. Volumes smaller than this are difficult to extract.
15. Phenol is highly toxic and should be handled while wearing gloves and goggles.
16. Leaving the tubes at –70°C for 1–2 h after the addition of sodium acetate and ethanol can increase the yield of DNA. However, this also results in the precipitation of salt and should be avoided unless problems with yield are encountered.

17. Do not panic if a pellet is not visible at this stage.
18. Be very careful during the ethanol washes. The ethanol can dislodge the pellet away from the side of the tube and it can be easily lost when decanting the liquid.
19. Ensure that the last traces of ethanol are gone before resuspending the DNA pellet in water. The tubes can incubated at 55°C for 5 min to remove the last traces of ethanol prior to adding water, but be careful not to overdry the pellet because this will make it difficult to dissolve in water.
20. A useful kit for the removal of restriction enzymes is The Wizard® DNA clean-up system (Promega). This should be used according to the manufacturer's instructions. If experiencing problems with ethanol from the washes carrying over, then it is advisable to leave the column on the bench for a few minutes prior to elution. This ensures the total evaporation of all the ethanol used in the washing steps.
21. Linkers are available commercially from Promega and a number of other suppliers of molecular-biology products.
22. The amount of linkers used in the reaction should be 100 times greater in molarity than the substrate DNA.
23. There are a number of websites that are useful in aiding primer design. A particularly good one is Primer 3 (www-genome.wi.mit.edu/cgi-bin/primer/primer3_www.cgi).
24. Primers are available from a number of companies, including Sigma Genosys.
25. Certain restriction enzymes inefficiently cleave recognition sequences located at the end of a DNA fragment so it is advisable to include at least four additional bases in front of the restriction recognition site *(4)*.
26. Keep both Klenow and T4 polymerase on ice during use, but set up the reactions at room temperature.
27. For Klenow, it is recommended to use 1 unit of enzyme per microgram of DNA. Most enzymes are supplied at a concentration of 5 U/μL and we find that using 1 μL works well.
28. The total reaction volume can be between 10 and 100 μL, but 100 μL works well.
29. For T4 polymerase, it is recommended to use 5 units of enzyme per microgram of DNA. Most enzymes are supplied at a concentration of 5 U/μL and we find that using 1 μL works well.
30. The picomoles of ends of linear double-stranded DNA can be calculated using the general formula:

$$(\mu g\ DNA/kb\ size\ of\ DNA) \times 3.04 = pmol\ of\ ends$$

31. Some restriction enzymes, such as *Sal*I, may interfere with subsequent modifications of digested DNA. However, the addition of increased amounts of alkaline phosphatase (1 U CIAP, 1 U TsAP [*see* **Note 35**]), and 150 U BAP) has been shown to reduce this inhibition. Alternatively, the DNA can be purified prior to dephosphorylation *(5)*.
32. In our experience, the addition of 1 μL CIAP works well and is more convenient.
33. The higher temperature of 55°C ensures accessibility of the recessed end to the enzyme.
34. It is necessary to remove CIAP completely from the reaction, as it may interfere with the efficiency of subsequent ligations. As an alternative to the method given, Sambrook and Russell suggest the addition of SDS and EDTA (final concentration 0.5% and 5 mM, respectively) followed by the addition of 100 μg/mL proteinase K and heating to 55°C for 30 min *(2)*.
35. Thermosensitive mutants of bacterial alkaline phosphatase (TsAP) have been developed that have a 40-fold greater activity than wild-type BAP *(6)*. TsAP can be completely inactivated by the addition of EDTA and heating at 65°C, making phenol : chloroform

extractions unnecessary. SAP is also easy to deactivate irreversibly (15 min at 65°C) and does not require extensive dilution. Reactions can be performed in as little as 10 min for dephosphorylation, and 15 min for inactivation. Furthermore, restriction digestion, dephosphorylation, and ligation can be performed in a single tube.

36. In order to assess the success of the dephosphorylation, reaction test ligations can be performed (*see* **Subheading 3.6.**). Ligation reactions containing (1) dephosporylated vector DNA, (2) dephosporylated vector DNA and insert DNA with compatible phosphorylated ends, and (3) linearized vector DNA that has not been treated with alkaline phosphatase should be set up and used for transformation. Dephosphorylation should reduce the transformation efficiency of linear vector DNA by a factor of at least 50. Ligation of insert DNA to dephosphorylated linear vector DNA should increase the transformation efficiency by a factor of at least 5 *(2)*.

37. Alternatively, the restriction digest and dephosphorylation of the vector DNA can be performed in the same tube in 1X restriction digestion buffer. In this case, 15 U of restriction enzyme/µg vector and 10 U of SAP/µg vector should be used in a reaction volume of 30–50 µL.

38. One unit of SAP per microgram DNA should be adequate to dephosphorylate the vector regardless of the type of ends (blunt, 5' or 3' overhang).

39. DNA should not be resuspended in, or precipitated from, buffer containing ammonium salts prior to kinase treatment, as ammonium ions are strong inhibitors of bacteriophage polynucleotide kinase.

40. 10X Ligase buffer can also be used.

41. T4 polynucleotide kinase and buffer are available commercially from many different manufacturers. The conditions recommended by the manufacturer should always be followed.

42. T4 polynucleotide kinase phosphorylates protruding 5' single-stranded termini more efficiently than recessed 5' termini or blunt ends, although the addition of sufficient polynucleotide kinase and ATP allows all termini to be completely phosphorylated.

43. Once the vector and insert have been prepared for ligation, it is necessary to estimate the concentration of the DNA. This may be estimated by agarose gel electrophoresis when run against molecular-weight markers of known concentration or by using a commercial nucleic acid quantification system. The optimal vector: Insert DNA ratio will vary with different vectors and inserts and various ratios should be tried. The equation below can be used to calculate the quantities of DNA necessary for a particular ratio:

$$\frac{\text{ng of vector} \times \text{kb size of insert}}{\text{kb size of vector}} \times \text{Molar ratio of (insert/vector)} = \text{ng of insert}$$

Molar ratios of 1:1, 1:3, and 1:5 (vector:insert) generally work well.

44. Typical ligation reactions use 50–200 ng of vector DNA. Excess DNA may inhibit the transformation. DNA will be less likely to circularize if the concentration of DNA is too high (high concentration of DNA promotes intermolecular ligation resulting in long linear concatemers).

45. Most manufacturers calibrate bacteriophage T4 DNA ligase in Weiss units *(7)*. One Weiss unit is the amount of enzyme that catalyzes the exchange of 1 nmol of ^{32}P from pyrophosphate into [γ, β-^{32}P] in 20 min at 37°C. We usually use 1 µL of T4 DNA ligase for convenience.

46. To assess the efficiency of the transformation, host cells should also be transformed with supercoiled plasmid, as well as linear vector DNA (dephoshorylated and phosphorylated) with which ligase has been included in the reaction. Also, to assess whether the ligation

was a success, samples of linearized vector DNA and samples of ligated products can be analyzed on a 0.8% agarose gel. The banding pattern of the ligation products should be different to that of the unligated sample (shifted to a higher molecular weight).

47. The temperature and time necessary for a successful ligation has considerable range and will vary dependent on the length and base composition of the fragment ends. The optimal temperature is a compromise between that for ensuring annealing of ends (the Tm of ends are generally 12–16°C) and the optimal temperature (25°C) for activity of T4 DNA ligase. Fragments of small length such as linkers may need lower ligation temperatures because of their lower melting temperatures. The temperature may not be so important for blunt-end ligations, which may be incubated at a higher temperature. As a general rule, the lower the temperature, the longer the incubation time required. Reactions performed in the temperature range of 4–15°C generally work well *(8)*.

48. Many manufacturers now produce rapid DNA ligation kits. These kits make it possible to ligate vector and insert in as little time as 5 min and so are useful when time is limiting and the transformation needs to be performed on the same day. In our lab, we have found that the Rapid DNA ligation kit of Boehringer Mannheim very efficiently ligates both cohesive and blunt ends, although many other kits are available.

49. Compared to cohesive ends, the ligation of blunt-ended termini is inefficient, as the ligase is unable to "catch hold" of the molecule to be ligated and has to wait for chance associations to bring the ends together. Blunt-end ligation should be performed at high DNA concentrations to increase the chances of the ends of the molecule coming together in the correct way. Problems in attaining adequate concentrations of blunt-ended DNA can be avoided by adding substances that increase macromolecular crowding of the reaction (e.g., polyethylene glycol) *(2)*. The maximum efficiency of ligation is achieved in reactions containing 15% PEG 8000. The 40% PEG 8000 stock solutions should be prepared in deionized water and stored frozen in small aliquots. A concentration of 15% PEG 8000 stimulates ligation of DNA fragments with cohesive ends and blunt ends (including ligation of synthetic oligonucleotides, which can be very short in length). Alternatively, ligation buffers are commercially available that increase crowding (e.g., the use of 2X ligation buffers rather than 10X).

50. In blunt-end ligations, the restriction sites are not normally reformed if different enzymes are used; therefore, the restriction enzyme may be included in the ligation reaction and only a religated vector should be digested. This should decrease the background of nonrecombinant plasmids on transformation.

References

1. Anderson, S., Gait, M. J., Mayol, L., et al. (1980) A short primer for sequencing DNA cloned in the single-stranded phage vector M13mp2. *Nucleic Acids Res.* **8**, 1731.
2. Sambrook, J. and Russell, D. W. (2001) *Molecular Cloning: A Laboratory Manual*, Cold Spring Harbor Laboratory, Cold Spring Harbor, NY.
3. Richardson, C. C. (1965) Phosphorylation of nucleic acid by an enzyme from T4 bacteriophage-infected *Escherichia coli*. *Proc. Natl. Acad. Sci. USA* **54**, 158.
4. Kaufman, D. I. and Evans G. A. (1990) Restriction endonuclease cleavage at the termini of PCR products. *Biotechniques* **9**, 304–306.
5. Schmidt, B., Natarajan, P., and Fox, D. (1998) Use of alkaline phosphatases in cloning. *Focus* **20(2)**, 52–54.
6. Shandilya, H. and Chatterjee, D. K. (1993) An engineered thermosensitive alkaline phosphatase for dephosporylating DNA. *Focus* **17(3)**, 93–95.

7. Weiss B., Jacquemin-Sablon, A., Live T. R., et al. (1968) Enzymatic breaking and joining of deoxyribonucleic acid VI. Further purification and properties of polynucleotide ligase from *Escherichia coli* injected with bacteriophage T4. *J. Biol. Chem.* **243**, 4543–4555.
8. Ferretti, L. and Sgaramella V. (1981) Temperature dependence of the joining by T4 DNA ligase of termini produced by type II restriction endonucleases. *Nucleic Acids Res.* **9**, 85–93.

16

Extraction of DNA from Agarose Gels

Nicholas Downey

1. Introduction

A common step in cloning experiments is the purification of DNA fragments prior to ligation. Often, both the insert and vector DNA fragments will be derived from restriction endonuclease digests and, thus, will be mixed with enzymes, salts, and possibly other DNA fragments that may inhibit the ensuing ligation reaction. A simple solution to this problem is to purify the DNA fragments of interest. This often involves agarose gel electrophoresis to separate mixtures of DNA fragments, followed by extraction of the DNA fragment of interest from the gel. The importance of this step is obvious from the fact that every vendor of molecular-biology products produces a gel extraction kit. Most of these kits utilize a chaotropic agent, such as sodium iodide, to destabilize the agarose gel. The DNA is subsequently bound to a substrate (e.g., an anionic resin) and washed to remove impurities prior to elution of the DNA from the substrate. Other methods include hot phenol extraction of the DNA from the gel. The use of phenol is best avoided if possible because of safety and waste disposal concerns. Presented here are two alternative techniques that avoid these concerns, do not require the purchase of a specialized kit, and are simple to perform.

2. Materials

2.1. Low-Melt Agarose Protocol

1. Low-melting-point agarose.
2. 5X TBE buffer: 54 g Tris base, 27.5 g boric acid, 20 mL of 0.5 M EDTA (pH 8.0); make up to 1 L with distilled water. Dilute the stock to give a 1X working solution immediately prior to use.

2.1.1. Buffer Exchange

1. G50 Sephadex beads.
2. 23G Needle.
3. TE buffer: 10 mM Tris-HCl (pH 8), 1 mM EDTA.

2.2. Spin Protocol

1. Agarose.
2. 5X TBE buffer (*see* **Subheading 2.1., item 2**).
3. 19G Needle.
4. 3 M Sodium acetate: Dissolve 408.3 g sodium acetate·$3H_2O$ in 800 mL H_2O. Adjust the pH to 5.2 with glacial acetic acid and make up the volume to 1 L with H_2O.
5. 95% Ethanol (ice cold), 70% ethanol.
6. TE buffer (*see* **Subheading 2.1.1., item 3**).

3. Methods
3.1. Low-Melt Agarose Protocol

This protocol makes use of low-melt agarose. In its simplest form, this protocol is more an avoidance of extraction rather than an extraction *per se*. It can be modified to increase the purity of the DNA sample (*see* **Subheading 3.1.1.**).

1. Prepare an agarose gel using low-melt-temperature agarose in 1X TBE (*see* **Note 1**).
2. Perform agarose gel electrophoresis of the DNA sample. Visualize the DNA under ultraviolet (UV) light after ethidium bromide staining (*see* Chapter 15).
3. Use a scalpel to cut a slice from the gel containing the DNA band of interest and transfer to a preweighed 1.5-mL microfuge tube.
4. Weigh the gel slice.
5. Add an equal volume of water (i.e., 1 mL of water per 1 g of gel slice).
6. Incubate at 65°C until the agarose is fully molten.
7. Mix the solution briefly and allow to cool. The agarose will not resolidify.
8. Use this solution in a downstream application.

3.1.1. Buffer Exchange (see **Note 2**)

1. Use a 23-gauge needle to carefully puncture a 0.5-mL microfuge tube at its base.
2. Resuspend some G50 Sephadex beads in TE (or alternative buffer) and load into the prepared 0.5-mL tube.
3. Place the 0.5-mL tube in a 1.5-mL microfuge tube and spin at 300g for 1 min to produce a mini Sephadex column.
4. Transfer the small tube into a fresh 1.5-mL microfuge tube. Load the DNA sample on top of the column and spin for 1 min at 300g.
5. The DNA sample will be eluted in TE and the agarose will remain in the G50 resin.

3.2. Spin Protocol

This protocol is inexpensive, simple, and requires little in the way of specialized materials. The technique is based on the fact that the DNA in an agarose gel slice is, in fact, trapped in solution. Therefore, by applying a centrifugal force, the DNA can be "squeezed" out of the gel.

1. Prepare an agarose gel in 1X TBE and perform agarose gel electrophoresis of the DNA sample as described in Chapter 15 (*see* **Note 1**).
2. After visualization, use a scalpel to cut a slice containing the DNA band of interest from the gel.
3. Use a 19-gauge needle to carefully puncture a 0.5-mL tube at its base. Take care, as this is the thickest part of the tube and the needle might slip.

4. Transfer the gel slice to the tube and close lid. Place the 0.5-mL tube into a 1.5-mL tube.
5. Spin in a microfuge at maximum speed for 5 min.
6. Remove the 0.5-mL tube. Measure the volume of the eluate in the 1.5-mL tube (*see* **Note 3**).
7. Add 0.1 volume of 3 *M* sodium acetate and 3 volumes of 95% ethanol; vortex briefly.
8. Microfuge at maximum speed for at least 15 min.
9. Remove the supernatant taking care not to dislodge the DNA pellet.
10. Wash the DNA pellet with 0.5 mL of 70% ethanol.
11. Resuspend the DNA pellet in water or a suitable buffer for downstream applications (*see* **Notes 4** and **5**).
12. The yield from this protocol varies but is usually between 10% and 25% of the input (*see* **Note 6**).

4. Notes

1. If TBE electrophoresis buffer is inhibitory to the downstream application and the DNA is to be used without further purification, it is possible to use a different buffer system for gel electrophoresis. Use of a modified 1X TAE buffer (10X stock solution: 242 g Tris base, 57.1 mL glacial acetic acid, 100 mL of 0.5 *M* EDTA [pH 8] per liter) will allow the direct use of the eluted DNA directly in most downstream reactions without strongly inhibiting enzyme activities.
2. This buffer-exchange step can be performed if downstream applications require the DNA to be in solution in a particular buffer.
3. If agarose is spun through the hole in the 0.5-mL tube, a small amount of glasswool can be used to "plug" the hole. This will reduce agarose contamination of the eluate.
4. As an alternative to ethanol precipitation, the buffer-exchange step described in **Subheading 3.1.1.** may be employed. This will remove the electrophoresis buffer and leave the DNA in TE or an alternative buffer.
5. If the DNA is in low abundance and isolated in a small gel slice, it may be advantageous to avoid having to precipitate the DNA after elution. In this case, it is best to use a different buffer system for the gel electrophoresis (*see* **Note 1**).
6. An improvement in yield is possible if an extra "freeze" step is included. After the gel slice is cut and transferred to the punctured 0.5-mL tube, store the slice at –70°C for approximately 30 min. Then, proceed immediately to the centrifugation step. This may improve the yield to 50–75% of the input.

17

Cloning PCR Products with T-Vectors

Wade A. Nichols

1. Introduction
1.1. Overview of PCR

Since it was described in 1988 *(1)*, the polymerase chain reaction (PCR) has been a valuable tool for molecular biologists. PCR allows researchers to produce a large quantity of a desired DNA fragment while requiring only a small amount of template. Prior to PCR, isolation of DNA fragments was typically performed by cleavage of the DNA with restriction endonuclease enzymes. The relative abundance or scarcity of appropriate restriction sites within the region of interest greatly affected the ability of researchers to obtain specific DNA fragments. PCR gives researchers the ability to establish the terminal sequences as well as the size of the amplified fragment and, in this way, provides freedom from the problems associated with location and abundance of restriction enzyme recognition sites.

Entire books have been written about the uses and caveats of PCR and it is not the objective of this chapter to cover the same material. Rather, this chapter is directed at the researcher who is already familiar with PCR and is looking for a way to clone PCR products. It is important for two reasons that PCR reactions be optimized before attempting to clone the product. First, the ligation efficiency of T-vectors depends on an abundance of PCR product in the ligation mix and, thus, reactions that produce only small amounts of product are not conducive for cloning of products. Second, T-vectors will ligate to any PCR product present in the reaction, so it is important that reaction conditions do not result in the production of multiple products. The presence of multiple products will increase the effort required to construct a clone and verify that it contains the correct product. PCR optimization kits are commercially available (Invitrogen) and are very effective aids in obtaining high concentrations of a single desired product.

1.2. Strategies for Cloning PCR Products

Although products amplified by PCR can be used directly in many applications, some applications require that the PCR product be cloned into and maintained within a plasmid vector. Cloning of amplified promoter regions for use with reporter genes (*see* Chapters 30 and 31) and cloning of amplified genes for use in an expression system (*see* Chapters 28 and 29) are two examples where insertion of a PCR product into a vector backbone is essential. There are three common strategies for cloning PCR products into plasmid vectors. They are presented in the following subsections.

1.2.1. Blunt-End Ligation

The first approach involves cloning PCR products into a plasmid that has been digested to produce blunt ends. It seems reasonable to assume that the majority of the PCR products will possess blunt ends as determined by the ends of the oligonucleotide primers. However, this is not the case and, therefore, attempts to clone PCR products into plasmids possessing blunt ends are not efficient unless the ends of the PCR product are first modified to produce blunt ends. This additional step involves extra work for the researcher and extra cost.

1.2.2. Addition of Restriction Endonuclease Sites into PCR Product

The second cloning strategy utilizes oligonucleotide primers designed to include restriction enzyme recognition sites near their 5' termini. Following amplification, the resultant product is isolated by precipitation and cleaved by the appropriate restriction endonucleases, resulting in a DNA fragment with cohesive ends that will anneal to complementary sequences on a prepared cloning vector. This technique is widely used with great success. However, it also requires modification of the PCR product prior to cloning. Additionally, oligonucleotide primers designed to include the desired restriction sites are usually larger and, therefore, more expensive than typical primers used for PCR.

1.2.3. TA Cloning

The third, and arguably the easiest of the strategies, takes advantage of a property of *Taq* DNA polymerase to facilitate cloning of PCR products. In 1988, Clark *(2)* described template-independent terminal transferase activity of *Taq* DNA polymerase in which the enzyme adds a single nucleotide to the 3' termini of an amplified fragment. Although this activity may result in the addition of any of the four nucleotide bases, there is a strong bias for the addition of adenylate. PCR products that possess single adenylate 3' overhangs at each end can be cloned into a vector possessing single thymidine 3' overhangs and for this reason, the process is frequently referred to as "TA cloning" and the accepting vector as a "T-vector." A schematic of TA cloning is shown in **Fig. 1**. This chapter will describe the processes of constructing custom T-vectors, the preparation of T-vectors for accepting a PCR product, and, finally, the ligation of PCR products into the T-vector. This method has the advantage that the PCR product is cloned in a single step. In addition, the single 3' nucleotide overhangs prevent self-ligation and concatamerization of both the vector and insert DNA, thus reducing background during the screening stage.

Cloning PCR Products

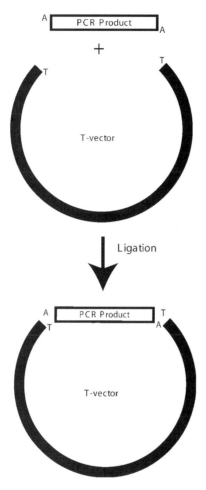

Fig. 1. Schematic of TA cloning. Single 3' T overhangs on the T-vector facilitate annealing to single 3' A overhangs on PCR product.

1.3. T-Vectors

1.3.1. Commercially Available T-Vectors

T-Vectors are available commercially (Invitrogen, Stratagene), and they arrive prepared for immediate use in ligation reactions without additional modification by the researcher. Although these commercially available T-vectors are both convenient and effective, researchers often require features that are not available on commercially available vectors. The construction of custom T-vectors for use in cloning PCR fragments is a simple procedure that can be easily performed by researchers. Even if a commercially available T-vector is suitable for use in a project, constructing and using custom T-vectors is often less expensive when a large number of PCR products are to be cloned.

1.3.2. Custom T-Vectors

The protocols in this chapter describe the preparation of custom T-vectors by (1) the insertion of two *Xcm*I restriction endonuclease sites within an existing vector and (2) by the addition of thymidine 3' overhangs to a blunt-cut vector utilizing the terminal transferase activity of *Taq* DNA polymerase. Methods for the construction of other custom T-vectors are described in **refs. 3–6**.

1.3.2.1. Custom T-Vectors Bearing *Xcm*I Sites

The restriction endonuclease *Xcm*I is isolated from the bacterium *Xanthomonas campestris* and recognizes the sequence 5' CCANNNN^NNNNTGG 3' (the cleavage site is indicated by ^). Cleavage of an *Xcm*I site results in a single 3' overhang on each strand of DNA. Because the bases adjacent to the cleavage site do not determine the recognition of the site by the enzyme, it is possible to design an *Xcm*I site that has an A/T pair at the cleavage site. However, a single site designed in such a way would result in a 3' adenylate overhang on one strand and a 3' thymidine overhang on the opposite strand. For this reason, it is required that two *Xcm*I sites be present to result in 3' thymidine overhangs on each strand (*see* **Fig. 2**). In this fashion, it is possible to construct a custom T-vector by inserting a small oligonucleotide that contains two *Xcm*I sites and a small intervening sequence into an existing vector backbone. Once incorporated into the vector, the sites allow the preparation of the vector by cleavage with *Xcm*I. Although this strategy is useful for a wide array of parent vectors, it should be noted that the strategy would not work well if the parent backbone already possesses *Xcm*I sites. Fortunately, most commonly used vectors lack *Xcm*I sites.

The method described is a slight variation of one described by Borovkov and Rivkin *(7)* and uses pBluescript SK II(+) (Stratagene) as the vector backbone. The procedure is outlined in **Fig. 3**. Two complementary oligonucleotides are designed such that when they are annealed they will possess cohesive ends, allowing them to be inserted into pBluescript SK II(+) which has been digested with *Bam*HI and *Pst*I. The oligonucleotides also bear two *Xcm*I sites and a short intervening sequence that will produce single 3' thymidine overhangs on each DNA strand when cleaved by *Xcm*I. The oligonucleotides are combined at a high concentration, heat denatured, and allowed to cool slowly to facilitate the annealing of the complementary sequences. The resultant double-stranded DNA is then ligated into pBluescript SK II(+) that has been cleaved with *Bam*HI and *Pst*I. The ligation mix is transformed into *Escherichia coli* and transformants are screened for the presence of the new *Xcm*I-containing sequences. Once constructed, *Xcm*I-containing T-vectors must be cleaved with *Xcm*I to produce the desired single 3' thymidine overhangs. Following digestion, the cleaved T-vector is concentrated by ethanol precipitation and is ready for ligation.

1.3.2.2. Custom T-Vectors Prepared with *Taq* DNA Polymerase

As mentioned previously, *Taq* DNA polymerase exhibits terminal transferase activity and attaches nucleotides to the 3' end of blunt fragments. In 1995, Marchuck et al. *(8)* demonstrated that this terminal transferase activity can be utilized to prepare custom T-vectors from any blunt-cut vector. It was also noted that *Taq* DNA poly-

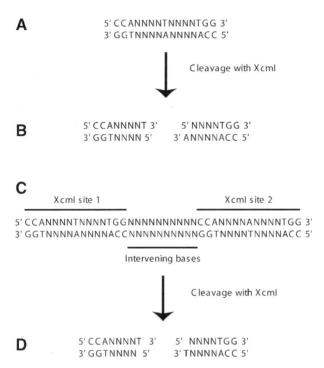

Fig. 2. The use of *Xcm*I sites to construct a site with 3' thymidine overhangs. (**A**) An *Xcm*I recognition site bearing an A/T pair at the cleavage site. (**B**) When this site is cleaved by the enzyme, it results in a 3' adenylate overhang on one strand and a 3' thymidine overhang on the opposite strand. (**C**) A DNA region that contains two *Xcm*I sites and intervening bases designed to produce 3' thymidine overhangs on both strands of DNA. (**D**) When cleaved by *Xcm*I, the region between the two sites is lost and each strand now possesses a 3' thymidine overhang.

merase has a strong bias toward the addition of 3' adenylate when all four dNTPs are present. However, if blunt-ended vectors are combined with *Taq* DNA polymerase and dTTP as the only substrate, the polymerase adds single 3' thymidines to the vector, resulting in a T-vector ready to accept a PCR product with single 3' adenylate overhangs. This chapter will detail the preparation of a T-vector using pBluescript SK II(+) cut with *Eco*RV, although any blunt-cut vector is suitable. An outline of the procedure is shown in **Fig. 4**.

1.4. Cloning PCR Products into T-Vectors

Once prepared, both of the types of T-vectors described are used in a similar fashion. In the ligation reaction, vector DNA is combined with a small volume of a completed PCR reaction, ligase buffer, and T4 DNA ligase. Successful ligation reactions will produce an abundance of plasmids containing the cloned PCR product. The newly formed plasmids can then be introduced into host cells by standard transformation or electroporation techniques (*see* Chapters 4 and 5).

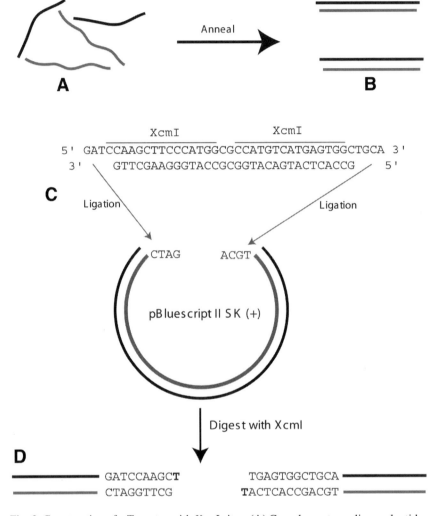

Fig. 3. Construction of a T-vector with *Xcm*I sites. (**A**) Complementary oligonucleotides are combined. (**B**) The oligonucleotide mixture is heated then allowed to cool facilitating the annealing of the complementary sequences. (**C**) The resultant double-stranded structure bears cohesive ends that can be accepted by pBluescript SKII(+) that has been prepared by cleavage with *Bam*HI and *Pst*I. (**D**) Once constructed, the T-vector can be prepared for accepting a PCR product by cleavage with *Xcm*I (3' thymidines in bold).

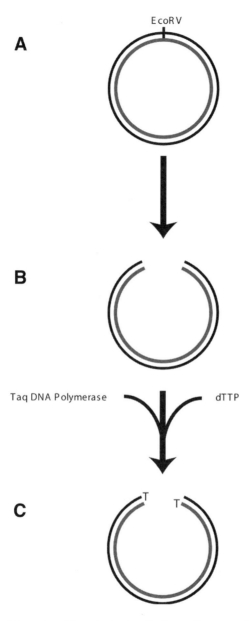

Fig. 4. Preparation of a custom T-vector by terminal transferase modification. An existing vector (**A**) is cleaved with a blunt-cutting restriction endonuclease (*Eco*RV), resulting in a linear fragment possessing blunt ends (**B**). (**C**) The linear vector is incubated with *Taq* DNA polymerase, buffer, and dTTP, resulting in the addition of 3' thymidines to the vector.

2. Materials
2.1. Preparation of an XcmI-Based T-vector
2.1.1. Construction of Custom T-Vector

1. Sterile distilled water.
2. 1 μg/μL Oligonucleotide #1 in water: 5'-GATCCAAGCTTCCCATGGCGCCATGTCAT GAGTGGCTGCA-3'.
3. 1 μg/μL Oligonucleotide #2 in water: 5'-GCCACTCATGACATGGCGCCATGGGAAG CTTG-3'.
4. 1 μg/μL pBluescript SK II(+) in water or TE buffer (available from Stratagene).
5. 10X *Pst*I buffer: 1 M NaCl, 500 mM Tris-HCl (pH 7.9), 10 mM dithiothreitol (DTT), 1 mg/mL bovine serum albumin (BSA).
6. 10X *Bam*HI buffer: 1.5 M NaCl, 100 mM Tris-HCl (pH 7.9), 100 mM MgCl$_2$, 10 mM DTT, 1 mg/mL BSA.
7. *Pst*I restriction enzyme.
8. *Bam*HI restriction enzyme.
9. TE-saturated phenol.
10. TE-saturated phenol : chloroform : isoamyl alcohol (25 : 24 : 1).
11. 3 M sodium acetate (pH 5.2).
12. 95% Ethanol.
13. 10X T4 DNA ligase buffer: 500 mM Tris-HCl (pH 7.5), 100 mM MgCl$_2$, 100 mM DTT, 10 mM ATP, 250 μg/mL BSA.
14. T4 DNA ligase (400 U/μL).

2.1.2. Preparation of XcmI-Based T-Vector for Cloning

1. Sterile distilled water.
2. 1 μg/μL Custom T-vector DNA in water, from **Subheading 3.1.**
3. 10X *Xcm*I buffer: 500 mM NaCl, 100 mM Tris-HCl (pH 7.9), 100 mM MgCl$_2$, 10 mM DTT.
4. *Xcm*I restriction enzyme (*see* **Note 1**).
5. 3 M Sodium acetate (pH 5.2).
6. 95% Ethanol.

2.2. Preparation of T-Vector with Taq DNA Polymerase

1. Sterile distilled water.
2. 1 μg/μL pBluescript SK II(+) in water or TE buffer (Stratagene).
3. 10X *Eco*RV buffer: 1 M NaCl, 500 mM Tris-HCl (pH 7.9), 100 mM MgCl$_2$, 10 mM DTT, 1 mg/mL BSA.
4. *Eco*RV restriction enzyme.
5. TE-saturated phenol.
6. TE-saturated phenol : chloroform : isoamyl alcohol (25 : 24 : 1).
7. 3 M Sodium acetate (pH 5.2).
8. 95% Ethanol.
9. 10 mM dTTP solution.
10. 10X *Taq* reaction buffer: 500 mM KCl, 100 mM Tris-HCl (pH 8.3), 15 mM MgCl$_2$, 2 mg/mL BSA.
11. 1 U/μL *Taq* DNA polymerase.

2.3. Ligation of PCR Product into T-Vector

1. Sterile distilled water.
2. 20 ng/µL Prepared T-vector in water (from **Subheading 3.2.1.** or **Subheading 3.2.2.**).
3. Complete PCR reaction containing amplified target DNA (*see* **Notes 2** and **3**).
4. 10X T4 DNA ligase buffer: 500 mM Tris-HCl (pH 7.5), 100 mM MgCl$_2$, 100 mM DTT, 10 mM ATP, 250 µg/mL BSA (*see* **Note 4**).
5. 400 U/µL T4 DNA ligase.

3. Methods
3.1. Preparation of an XcmI-Based T-Vector
3.1.1. Construction of Custom T-Vector
3.1.1.1. Preparation of Oligonucleotides

1. Combine 25 µL oligonucleotide #1 and 25 µL oligonucleotide #2 in a microcentrifuge tube.
2. Place the tube in boiling water for 5 min, remove, and allow mixture to cool at room temperature for 2 h.
3. The annealed oligonucleotide can be stored at 4°C until needed.

3.1.1.2. Digestion of pBluescript with BamHI

1. Digest pBluescript SK II(+) with *Bam*HI by combining the following: 1 µg of pBluescript SK II(+) DNA, 2.5 µL of 10X *Bam*HI buffer, 5 U of *Bam*HI, water to a final volume of 25 µL. Incubate at 37°C for 1–2 h.
2. Bring the total volume of the reaction to 200 µL with water. Add 200 µL of TE-saturated phenol and mix thoroughly.
3. Centrifuge at full speed for 3 min in a microcentrifuge to separate the phases.
4. Transfer the aqueous phase (top) to a new microcentrifuge tube.
5. Add an equal volume of phenol/chloroform/isoamyl alcohol to the aqueous phase and mix thoroughly.
6. Centrifuge at full speed for 3 min to separate the phases.
7. Transfer the aqueous phase (top) to a new microcentrifuge tube.
8. Add 22 µL of sodium acetate solution and 500 µL of 95% ethanol and place at –80°C for 1 h to precipitate the DNA.
9. Pellet the DNA by centrifugation at full speed for 15 min in a microcentrifuge.
10. Carefully pour off and discard the supernatant, and allow the DNA pellet to air-dry.
11. Resuspend the DNA in 20 µL of water.

3.1.1.3. Digestion of pBluescript with PstI

1. Digest the *Bam*HI-cleaved pBluescript SK II(+) with *Pst*I by adding the following to the DNA solution from **step 11** of **Subheading 3.1.1.2.**: 2.5 µL of 10X *Pst*I buffer, 5 U of *Pst*I, water to a final volume of 25 µL. Incubate at 37°C for 1–2 h.
2. Repeat **steps 2–10** from **Subheading 3.1.1.2.**
3. Resuspend the DNA in 12 µL of water.

3.1.1.4. Ligation of Oligonucleotides and pBluescript

1. Combine 10 µL of oligonucleotide mix (from **Subheading 3.1.1.1., step 2**) and the double-digested pBluescript (from **Subheading 3.1.1.3., step 3**) in a 0.5-mL microcentrifuge tube.

2. Add 2.5 µL of 10X T4 DNA ligase buffer and 0.5 µL of T4 DNA ligase.
3. Incubate at 12°C for 12–18 h.
4. Introduce plasmids into *E. coli* by standard transformation or electroporation techniques (*see* Chapters 4 and 5).
5. Isolate plasmid DNA from transformants by standard techniques (*see* Chapters 8 and 9).
6. Digest candidate plasmids with *Xcm*I by combining the following in a microcentrifuge tube: 0.5 µg of plasmid DNA, 2.5 µL of 10X *Xcm*I buffer, 10 U of *Xcm*I, and water to a final volume of 25 µL. Incubate for 2 h at 37°C.
7. Examine digested candidate plasmids and undigested controls by electrophoresis using a 0.8% agarose gel (*see* Chapter 20). Plasmids containing the desired insert will be linearized by digestion with *Xcm*I, whereas those lacking the insert will remain undigested.
8. Isolate 10 µg of T-vector according to standard techniques (*see* Chapters 8 and 10) and store at 4°C until needed.

3.1.2. Preparation of XcmI-Based T-Vector for Cloning

1. Digest 1 µg of vector DNA by combining the following: 1 µg of vector DNA, 2.5 µL of 10X *Xcm*I buffer, 10 U of *Xcm*I, and water to a final volume of 25 µL. Incubate for 2 h at 37°C.
2. Bring the total volume of the reaction to 100 µL with water. Add 11 µL of sodium acetate solution and 280 µL of 95% ethanol. Place at –80°C for 1 h to precipitate DNA (*see* **Note 5**).
3. Pellet DNA by centrifugation at full speed for 15 min in a microcentrifuge.
4. Carefully pour off and discard the supernatant and allow the DNA pellet to air-dry.
5. Resuspend the prepared vector in 160 µL of sterile distilled water and dispense 10-µL aliquots into freezer-safe 0.5-mL tubes.
6. Store at –20°C for up to 6 mo (*see* **Note 6**).

3.2. Preparation of T-Vector with Taq DNA Polymerase

1. Digest 1 µg of vector DNA with *Eco*RV by combining the following: 1 µg of vector DNA, 2.5 µL of 10X *Eco*RV buffer, 5 U of *Eco*RV, and water to a final volume of 25 µL. Incubate for 2 h at 37°C.
2. Purify the DNA as described in **Subheading 3.1.1.2., steps 2–10**.
3. Resuspend the DNA in 13 µL of water and transfer to a 0.5-µL microcentrifuge tube.
4. Add 2 µL of 10X *Taq* reaction buffer, 4 µL of dTTP solution, and 1 µL of *Taq* DNA polymerase to the resuspended vector.
5. Incubate at 70°C for 2–3 h.
6. Repeat **steps 2–10** of **Subheading 3.1.1.2.** to remove the enzyme and repurify the DNA.
7. Resuspend the prepared vector in 160 µL of sterile distilled water and dispense 10-µL aliquots into freezer-safe 0.5-mL tubes.
8. Store at –20°C for up to 6 mo (*see* **Note 6**).

3.3. Ligation of PCR Product into T-Vector

1. As soon as the PCR reaction has finished, combine the following (*see* **Note 7**): 3 µL of PCR reaction (*see* **Note 8**), 4 µL of prepared T-vector, 1 µL of 10X T4 DNA ligase buffer, 1 U of T4 DNA ligase, and water to a final volume of 10 µL. Incubate for 6–18 h at 8°C (*see* **Note 9**).
2. Introduce plasmids into *E. coli* by standard transformation or electroporation techniques (*see* Chapters 4 and 5).
3. Screen transformants for the desired insert by colony blots (*see* Chapter 21), restriction enzyme analysis (*see* Chapter 20), or PCR (*see* Chapter 19).

4. Notes

1. In the past, researchers have had difficulty with *Xcm*I performance and this hampered preparation of T-vectors. However, higher-quality enzyme is now available and this problem is not as prevalent as it once was. High-concentration *Xcm*I should be purchased from recognized vendors of high-quality enzymes.
2. It is essential that *Taq* DNA polymerase be used for the PCR reaction. TA cloning depends on the addition of 3' adenylate to the PCR product and other DNA polymerases (Vent and *Pfu*) do not possess the necessary terminal transferase activity.
3. The single 3' adenylate overhangs will degrade with time, so it is important to use the PCR product in ligation as soon as possible. If used within 2 h of completion, only a small decrease in ligation efficiency will be seen; however, ligation efficiency decreases sharply after 3 h. To decrease degradation of 3' adenylate overhangs, the thermocycler program should include a final holding temperature of 4°C.
4. T4 DNA buffer contains the chemical DTT, which is essential for efficient ligation. DTT often precipitates when the buffer is stored at –20°C and the precipitate is often visible as small white flakes within the thawed buffer. It is important that ligase buffer be allowed to come to room temperature and mixed vigorously (or vortexed) prior to its use to assure that the DTT is redissolved.
5. It should be noted that phenol extraction is not required for the removal of *Xcm*I and the extra steps involved in extraction may decrease the ligation efficiency of the prepared vector.
6. The single 3' thymidine overhangs on the T-vector will degrade with repeated freeze/thaw cycles and result in decreased ligation efficiency. For this reason, it is recommended that the vector be stored in single-use aliquots at –20°C.
7. Because of the limited efficiency of ligation of DNA fragments with single nucleotide cohesive ends, it is recommended that the PCR product should be present at approximately a 4:1 ratio with respect to the vector in the ligation reaction. Amplification by PCR is sometimes variable, so the researcher should increase or decrease the amount of PCR mix used in the ligation mix according to the approximate concentration of the PCR product. If the ligation reaction is set up before the PCR product is examined (*see* **Note 8**), then a previous PCR reaction performed under identical conditions can be used as an indicator of product concentration.
8. It is best if a PCR product is used in the ligation reaction immediately after the completion of the reaction (*see* **Note 3**); however, it is also important that the remainder of the PCR reaction be examined by gel electrophoresis to determine if the reaction was successful. Researchers have two choices. The ligation reaction can be set up immediately after the completion of PCR. The remainder of the PCR product should then be examined by gel electrophoresis to determine if PCR was successful. If the reaction did not produce a large quantity of a single product, the ligation reaction can be discarded. Alternatively, 5 µL of the PCR reaction mix may be removed and stored at 4°C immediately after completion of the reaction. The remainder of the reaction mix should then be examined by gel electrophoresis to determine if PCR was successful. If the reaction did not produce a large quantity of a single product, the ligation reaction should not be set up. In both cases, a new PCR reaction should be started.
9. The recommended temperature for ligation of single nucleotide cohesive ends is 8°C; however, a slightly higher temperature can sometimes be used with success. If low efficiencies persist, ligations may be performed at 12–14°C.

References

1. Saiki, R. K., Gelfand, D. H., Stoffel, S., et al. (1988) Primer-directed enzymatic amplification of DNA with a thermostable DNA polymerase. *Science* **239**, 487–491.
2. Clark, J. M. (1988) Novel non-template nucleotide addition reactions catalyzed by procaryotic and eucaryotic DNA polymerases. *Nucleic Acids Res.* **16**, 9677–9686.
3. Hawke, N. A., Strong, S. J., Haire, R. N., et al. (1997) Vector for positive selection of in-frame genetic sequences. *BioTechniques* **23**, 619–621.
4. Schutte, B. C., Ranade, K., Pruessner, J., et al. (1997) Optimized conditions for cloning PCR products into an *Xcm*I T-vector. *BioTechniques* **22**, 40–44.
5. Bielefeldt-Ohmann, H. and Fitzpatrick, D. R. (1997) High-efficiency T-vector cloning of PCR products by forced A tagging and post-ligation restriction enzyme digestion. *BioTechniques* **23**, 822–826.
6. Tsang, T. C., Harris, D. T., Akporiaye, E. T., et al. (1996) Simple method for adapting DNA fragments and PCR products to all of the commonly used restriction sites. *BioTechniques* **20**, 51–52.
7. Borovkov, A. Y. and Rivkin, M. I. (1997) *Xcm*I-containing vector for direct cloning of PCR products. *BioTechniques* **22**, 812–814.
8. Marchuk, D., Drumm, M., Saulino, A., et al. (1991) Construction of T-vectors, a rapid and general system for direct cloning of unmodified PCR products. *Nucleic Acids Res.* **19**, 1154.

18

Construction of Genomic Libraries in λ-Vectors

Yilun Wang, Zheng Cao, Darryl Hood, and James G. Townsel

1. Introduction

Lambda (λ) bacteriophages are viruses that specifically infect bacteria. The genome of λ-phage is a double-stranded DNA molecule approx 50 kb in length *(1)*. In bacterial cells, λ-phage employs one of two pathways of replication: lytic or lysogenic. Commonly, λ-phage vectors replicate via the lytic pathway. During lytic growth, the viral DNA is replicated manifold, a large number of phage gene products are synthesized, and progeny phage particles are assembled. The cell is eventually lysed, releasing its many new infectious virus particles; at this time, plaques will form on an infected bacterial lawn. Its high infectivity and clone-style propagation are the inherent basis for the use of λ-phage as a vector. Moreover, the central third of the viral genome is not essential for lytic growth and, thus, can be replaced by a variety of foreign gene segments. λ-Vectors usually contain multiple cloning sites facilitating the cloning of foreign DNA. In addition, λ-phage vectors have the following advantageous features: high cloning efficiency, a relatively large insert-size capacity, and suitability for screening using nucleic acid probes. In terms of gene screening, genomic DNA libraries are often screened by hybridization using a radioactive nucleic acid probe.

The construction of genomic libraries using λ-phage systems has been greatly simplified as a result of the number of commercial ready-to-use λ-vectors that are supplied as predigested and modified arms. In general, the primary considerations in designing a strategy for genomic library construction using the λ-phage vector are as follows:

1. The genomic DNA to be cloned must be digested to give fragments of suitable size and these fragments must be representative of the total genomic DNA *(2,3)*.
2. Ligation should yield the maximum number of recombinants with a minimal background of nonrecombinants;
3. The relative ease of cloning procedures.

In this chapter, the LambdaGEM-12 vector, an EMBL3 derivative developed by Promega *(4,5)* will be used as a representative λ-phage vector. The LambdaGEM-12

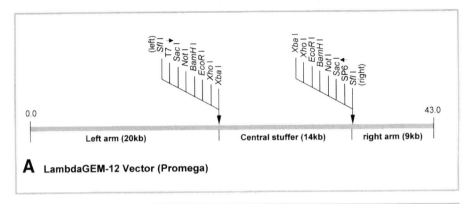

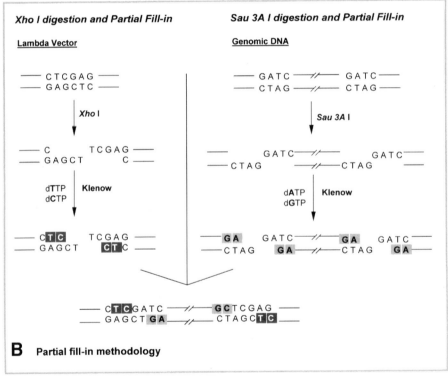

Fig. 1. Schematic diagrams of LambdaGEM-12 vector (**A**) and partial fill-in methodology (**B**).

vector can accommodate large inserts (up to 20 kb). This cloning system is particularly useful for cloning *Sau*3AI digests and permits the use of a partial fill-in strategy because of the availability of a *Xho*I site within the multiple-cloning site (*see* **Fig. 1A**). The partial fill-in of digested-fragment ends prevents vector self-ligation, thus yielding an extra low background *(5)* (*see* **Fig. 1B**). In brief, the main steps involved in genomic library construction are as follows:

1. Purification of genomic DNA.
2. Random dissection of DNA to produce fragments within the size range that can be accepted by the selected vector. This is typically accomplished by partial digestion using a restriction enzyme (RE) with a 4-bp recognition site, commonly *Sau*3AI or its isoschizomer *Mbo*I.
3. Selection of chosen DNA fragments by size fractionation on an agarose gel.
4. Optional modifications, such as partial fill-in of ends or dephosphorylation, to improve cloning efficiency.
5. Ligation of the genomic DNA fragments to λ-phage arms.
6. In vitro packaging, using phage protein extract, to form infectious phage particles.
7. Infection of appropriate *Escherichia coli* host cells and titering of plaques on Luria–Bertani (LB) plates.
8. Extraction of phage particles and purification of λ-DNA for restriction analysis to evaluate the library.
9. Library amplification.

2. Materials
2.1. Preparation of Genomic DNA for Cloning
2.1.1. Purification of Genomic DNA

We suggest the use of the Wizard Genomic DNA Purification Kit (Promega); **items 1–3** are components of that kit. Alternatively, they are available separately from Promega.

1. Nuclei lysis solution. Store at room temperature.
2. RNase A solution, 4 mg/mL. Store at room temperature.
3. Protein precipitation solution. Store at room temperature.
4. Teflon pestle.
5. Isopropanol. Store at room temperature.
6. 70% Ethanol.
7. TE buffer: 10 mM Tris-HCl (pH 8.0), 1 mM ethylenediamine tetraaceticacid (EDTA). Autoclave and store at 4°C.

2.1.2. Partial Digestion of Genomic DNA

1. *Sau*3AI. Store at –20°C.
2. 10X *Sau*3AI buffer: 6 mM Tris-HCl (pH 7.5), 6 mM MgCl$_2$, 50 mM NaCl, 1 mM dithiothreitol (DTT). Store at –20°C.
3. 1 mg/mL Acetylated bovine serum albumin (BSA). Store at 4°C.
4. 0.5 M EDTA (pH 8.0); autoclave.
5. 6X Gel loading buffer: 38% sucrose, 0.25% bromophenol blue, 67 mM EDTA.
6. 0.5% Agarose gel prepared in 0.5X TBE with 0.2 µg/mL ethidium bromide (EtBr) (*see* **Subheading 2.1.3.**, **items 2** and **3**).

2.1.3. Size Fractionation Using Low-Melting-Point Agarose

1. SeaPlaque agarose (FMC).
2. 10 mg/mL EtBr.
3. 0.5X TBE buffer: 45 mM Tris-borate (pH 8.3), 1 mM EDTA.
4. 6X Gel loading buffer (*see* **Subheading 2.1.2.**, **item 5**).

5. Agar*ACE* enzyme (Promega); store at –20°C.
6. 3 M Sodium acetate (NaOAc) (pH 5.8); autoclave.
7. 100% and 70% Ethanol.
8. TE buffer (*see* **Subheading 2.1.1.**, **item 7**).

2.1.4. Partial Fill-In of DNA Ends

1. DNA Polymerase I, Large Fragment (Klenow); store at –20°C.
2. 10X Fill-in buffer: 0.5 M Tris-HCl (pH 7.2), 0.1 M MgSO$_4$, 1 mM DTT, 500 μg/mL acetylated BSA, 10 mM dGTP, 10 mM dATP. Store at –20°C.
3. TE-saturated phenol : chloroform : isoamyl alcohol (25 : 24 : 1). Mix equal parts of TE buffer and phenol and allow the phases to separate. Then, mix 1 part of the lower, phenol phase with 1 part of chloroform : isoamyl alcohol (24 : 1).
4. Chloroform : isoamyl alcohol (24 : 1).
5. 7.5 M Ammonium acetate; sterilize by filtering.
6. 100% and 70 % Ethanol.

2.2. Ligation of Insert to Vector Arms

1. 5 U/μL T4 DNA ligase; store at –20°C.
2. 10X Ligase buffer: 66 mM Tris-HCl, 5 mM MgCl$_2$, 1 mM dithioerythritol, 1 mM ATP, pH 7.5. Store at –20°C.
3. LambdaGEN-12 *Xho*I Half-Site Arms (Promega), store at –20°C (*see* **Note 1**).

2.3. Packaging of Ligated DNA and Titration of Recombinant Phage

2.3.1. In Vitro Packaging

1. Packagene Lambda DNA Packaging Extract (Promega); store at –80°C. Thaw only immediately prior to use.
2. Phage buffer: 100 mM NaCl, 10 mM MgSO$_4$, 35 mM Tris-HCl (pH 7.5). Autoclave and store at 4°C.
3. Chloroform. Store at 4°C and protect from light to reduce photolysis.

2.3.2. Infection

1. *E. coli* KW251 (F$^-$ *supE44 galK2 galT22 metB1 hsdR2 mcrB1 mcrA argA81::Tn10 recD1014*).
2. LB medium: 20 g LB broth base per liter (Sigma). Autoclave and store at 4°C.
3. LB–0.2% maltose–10 mM MgSO$_4$ (LBMM): Add 1 mL of filter-sterilized 20% maltose and 1 mL of autoclaved 1 M MgSO$_4$ to 100 mL cooled sterile LB medium; store at 4°C.
4. LB agar plates: 35 g LB agar per liter (Sigma), autoclave, cool to approx 50°C, add antibiotics if required, and pour 25–30 mL of medium into 90-mm Petri dishes or 65–70 mL into 150-mm dishes. Prepare fresh and allow plates to dry to prevent excess moisture on the surface of the agar.
5. LB top agarose: Add 0.7 g of agarose to 100 mL of LB medium. Autoclave and cool to approx 55°C, add 1 mL of autoclaved 1 M MgSO$_4$, and place in a 50–55°C water bath.
6. Tetracycline: 34 mg/mL in ethanol; store at –20°C.
7. 10 mM MgSO$_4$, ice cold.

2.4. Library Validation

2.4.1. Infection

Items 1–7 of **Subheading 2.3.2.**

λ-Phage Genomic Libraries

2.4.2. Extraction of λ-Phage Particles (see **Note 2**)

1. Phage buffer (*see* **Subheading 2.3.1.**, **item 2**).
2. SM buffer: 50 mM Tris-HCl (pH 7.5), 100 mM NaCl, 8 mM MgSO$_4$, 0.01% gelatin. Autoclave and store at 4°C.
3. Chloroform; store at 4°C protected from light.
4. Nuclease mixture: 0.25mg/mL RNase A, 0.25mg/mL DNase I, 150 mM NaCl, 50% glycerol. Store at –20°C.
5. Phage precipitant solution: 33% Polyethylene glycol (molecular-biology grade, molecular weight [MW] 8000), 3.3 M NaCl.

2.4.3. Purification of Recombinant λ-Phage DNA

1. Proteinase K; store at 4°C.
2. 10% sodium dodecyl sulfate (SDS).
3. 0.5 M EDTA (pH 8.0); autoclave.
4. TE-saturated phenol:chloroform:isoamyl alcohol (25:24:1) (*see* **Subheading 2.1.4.**, **item 3**).
5. Chloroform:isoamyl alcohol (24:1).
6. Isopropanol.
7. 70% EtOH.
8. TE buffer (*see* **Subheading 2.1.1.**, **item 7**).

2.4.4. Restriction Analysis

1. *Sfi*I. Store at –20°C.
2. 10X *Sfi*I buffer: 6 mM Tris-HCl (pH 7.5), 6 mM MgCl$_2$, 50 mM NaCl, 1 mM DTT. Store at –20°C.
3. 0.5% Agarose gel (*see* **Subheading 2.1.2.**, **item 6**).

2.5. Library Amplification

1. Items 1–7 of **Subheading 2.3.2.**
2. Phage buffer containing 0.01% gelatin: Add 500 µL of 2% gelatin to 100 mL of phage buffer (*see* **Subheading 2.3.1.**, **item 2**). Autoclave and store at 4°C.
3. 2% Gelatin; autoclave and store at 4°C.
4. Chloroform; store at 4°C protected from light.
5. Dimethyl sulfoxide (DMSO).
6. Screw-cap glass vials.

3. Methods
3.1. Preparation of Genomic DNA for Cloning
3.1.1. Purification of Genomic DNA

1. Mince 150 mg of tissue in 40 µL/mg of ice-cold nuclei lysis buffer. Homogenize on ice using 10–15 strokes with a Teflon pestle.
2. Transfer the lysate to a high-speed centrifuge tube and incubate at 65°C for 30 min. The solution should become very viscous.
3. Add 5 µL of RNase A solution per milliliter of lysate and mix thoroughly, by inverting the tube, for 2–5 min. Incubate at 37°C for 30 min. Add 330 µL of protein precipitation solution per milliliter of lysate and vortex vigorously for 20 s. Chill the sample on ice for 5 min.

4. Centrifuge at 13,000g for 5 min at room temperature. Transfer the supernatant to a fresh tube and add 1 volume (equal to nucleic lysis buffer added at **step 1**) of room-temperature isopropanol. Gently mix by inverting until the white threadlike strands of DNA form a visible mass.
5. Centrifuge for 1 min at 13,000g at room temperature. Decant the supernatant and add 1 volume 70% ethanol at room temperature. Wash the pellet by gently inverting the tube several times. Centrifuge for 1 min at 13,000g at room temperature. Carefully remove the ethanol using a drawn Pasteur pipet (heat the front end of a glass pipet using a Bunsen burner and pull to form a needlelike tip using forceps; cool to room temperature before use). Air-dry the pellet for about 5 min (avoid overdrying).
6. Add 400–800 µL of TE buffer, resuspend the pellet, and transfer DNA to a 1.5-mL microcentrifuge tube. Redissolve the DNA by incubating at 65°C for 1 h (*see* **Note 3**). Periodically mix the solution by gently tapping the tube. Store at 4°C.
7. Determine the DNA concentration and purity using the A_{260}/A_{280} spectrophotometric ratio method (*see* Chapter 8). The A_{260}/A_{280} absorbance ratio should be greater than 1.8. In addition, visualize the DNA following electrophoresis using a 0.5% agarose gel.

3.1.2. Partial Digestion of Genomic DNA

Intact genomic DNA must be subjected to partial restriction digestion before cloning. Before a large-scale digestion is performed, the optimal conditions for partial digestion are determined in small-scale digests using various ratios of enzyme:DNA.

1. Prepare the following solutions:
   a. 1.5 mL of dilution buffer (DB): 150 µL of 10X *Sau*3AI buffer, 150 µL of acetylated BSA (1 mg/mL), 1.2 mL of distilled H_2O.
   b. 250 µL of assay buffer (AB): 5.5 µg genomic DNA, 25 µL of 10X *Sau*3AI buffer, 25 µL of acetylated BSA (1 mg/mL), 194.5 µL of distilled H_2O.
2. Aliquot 45 µL of AB into five numbered 1.5-mL microcentrifuge tubes. Incubate at 37°C for 60 min.
3. Prepare serial enzyme dilutions in DB, on ice, according to **Table 1**.
4. Start digestion by adding 5 µL of the appropriate enzyme dilutions to the corresponding tubes from **step 2**. Digest at 37°C for 30 min and stop the reaction by adding 1 µL of 0.5 *M* EDTA and 10 µL of 6X gel loading buffer.
5. Load approx 20 µL of each reaction on a 0.5% agarose gel containing approx 0.2 µg/mL EtBr and electrophorese at 2 V/cm for approx 14 h (*see* **Fig. 2A**).
6. Visualize the DNA under ultraviolet (UV) light and determine the optimal ratio of enzyme:DNA that produces the highest intensity of fluorescence in the desired size range. Use half of this amount of enzyme in the following large-scale digestion to obtain the maximum number of fragments with the desired sizes (*2*). For example, in the small-scale digestion shown in **Fig. 2A**, lane 2, 0.1 U of *Sau*3AI/µg DNA produced primary fragments ranging from 15 to 25 kb. Thus, 0.05 U of *Sau*3AI/µg DNA was used in the large-scale digestion.
7. For the large-scale partial digestion, add 20 µg of DNA to 300 µL of DB at 37°C and incubate for 2–4 h. Then, add the appropriate amount of *Sau*3AI and incubate for an additional 30 min. Stop the reaction by the addition of 6 µL of 0.5 *M* EDTA.
8. Visualize 20 µL of the digested products following electrophoresis on a 0.5% agarose gel to confirm digestion (*see* **Fig. 2B**, lane 1).

Table 1
Preparation of Restriction Enzyme Dilutions

Tube No.	Dilution		Final ratio[a] (units of Sau3AI/µg DNA)
1	1/15	3 µL Sau3AI (3 U/µL) + 42 µL DB	1
2	1/150	10 µL 1/15 dilution + 90 µL DB	0.1
3	1/300	10 µL 1/150 dilution + 10 µL DB	0.05
4	1/600	10 µL 1/150 dilution + 30 µL DB	0.025
5	1/900	10 µL 1/150 dilution + 50 µL DB	0.017

[a]After mixing 5 µL of enzyme dilution with 45 µL of AB.

3.1.3. Size Fractionation Using Low-Melting-Point Agarose (see **Note 4**)

1. Prepare a 0.5% low-melting-point (LMP) SeaPlaque agarose gel in 0.5X TBE containing approx 0.2 µg/mL of EtBr. Use a 1.5-mm comb with a widened well (which can be made by taping several teeth together), suitable for loading approx 300 µL of the digested sample, and three regular-size wells for the DNA marker, sample indicator, and blank (see **Fig. 3**).
2. Mix the digested genomic DNA (from **Section 3.1.2., step 7**) with 6X loading buffer. Load 300 µL and 20 µL of the mixture to the widened well and sample indicator well, respectively. Load DNA marker in the outer lane (see **Fig. 3**). Electrophorese at 2 V/cm for 2–3 h.
3. Carefully remove the gel from the buffer chamber. Visualize the DNA under long-wavelength UV light and mark the gel at the position of 14 kb, while shielding the area of gel containing the sample to be recovered by a piece of black paper to protect it from irradiation. Cut the gel from side to side below the 14-kb marker.
4. Return the gel-half containing DNA fragments larger than 14 kb to the running tray in the reverse orientation. Re-electrophorese for about 1–1.5 h or until the "smeared" DNA is squeezed into a relatively compact band. Covering the sample DNA as previously, view the gel under UV light and mark the position of the DNA in the sample indicator well. Excise the sample DNA band from the widened well for extraction.
5. Weigh the gel slice and transfer to a 2-mL microcentrifuge tube. If the gel slice is more than 600 µg, cut it into smaller pieces and load into several tubes. Completely melt the LMP agarose by incubating the gel slice at 65–75°C for 15 min.
6. Transfer the tube to a 42–47°C heating block for 2 min. Add Ager*ACE* enzyme at 1 U per 200 µg of gel and incubate for 15 min.
7. To precipitate the DNA, add 0.1 volume of 3 *M* NaOAc to the digests, mix, and add 2 volumes of room-temperature 100% ethanol. Allow the mixture to stand for 2 h at room temperature, then centrifuge at 13,000g for 15 min at room temperature. Carefully discard the supernatant. Wash the pellet by adding 500 µL of cold 70% ethanol. If the gel slices have been separated into multiple tubes, combine the pellets in one tube. Recentrifuge for 5 min and discard the supernatant.
8. Air-dry the pellet for about 5 min. Add 20–40 µL TE, gently tap the tube and allow the tube to stand at room temperature overnight to dissolve the DNA.

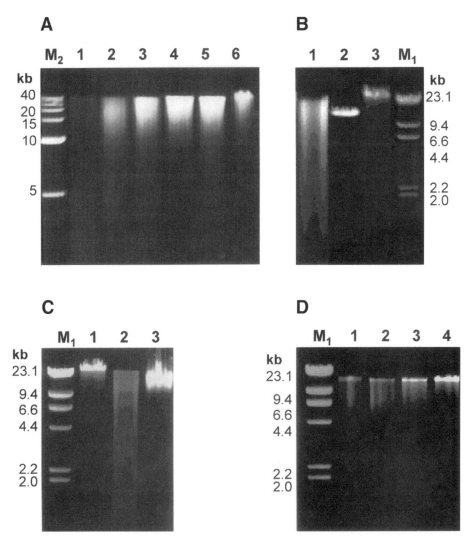

Fig. 2. Genomic DNA preparation for library construction. (**A**) Small-scale partial digestion of genomic DNA using *Sau*3AI. The units of *Sau*3AI/µg DNA are 1, 0.1, 0.05, 0.025, and 0.015 (*lanes 1–5*). *Lane 6*: undigested genomic DNA. (**B**) Large-scale partial digestion of genomic DNA using *Sau*3AI. *Lane 1*: genomic DNA after partial digestion (0.05 U/µg DNA); *lane 2*: insert control DNA standard (16 kb, Promega); *lane 3*: undigested genomic DNA. (**C**) Recovered genomic DNA. *Lane 1*: undigested genomic DNA; *lane 2*: partially digested DNA without size fractionation; *lane 3*: recovered genomic DNA after size fractionation. (**D**) Quantitation of purified genomic DNA. *Lane 1*: purified genomic DNA after partial fill-in; *lanes 2–4*: 50, 100, and 200 ng of diluted DNA standard (16 kb). M1 and M2 represent the sizes of standards of Lambda DNA/*Hin*dIII Marker and 5-kb DNA Ladder, respectively. All DNA samples were resolved by 0.2% EtBr–agarose electrophoresis in 0.5X TBE at 1–2 V/cm.

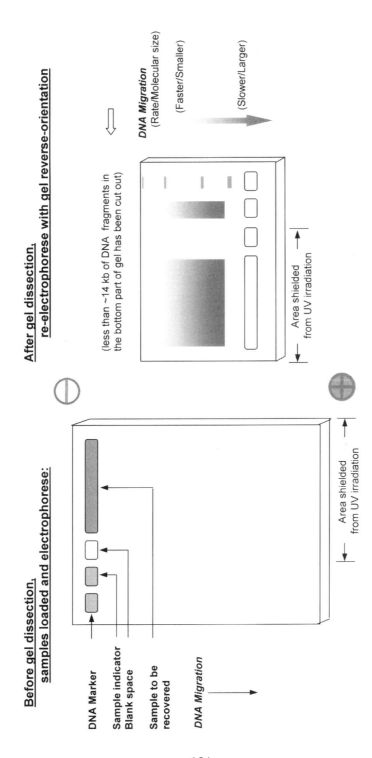

Fig. 3. Schematic illustration of LMP agarose recovery approach for size fractionation of partially digested genomic DNA.

9. To measure the concentration of recovered DNA, analyze a 1- to 2-µL aliquot by 0.5% agarose gel electrophoresis (*see* **Fig. 2C**) and perform an A_{260}/A_{280} ratio determination as described in **Subheading 3.1.1.**, **step 7** (*see* **Note 5**).

3.1.4. Partial Fill-In of DNA Ends

To reduce the self-ligation of genomic fragments, *Sau*3AI-digested ends are partially filled in using dATP and dGTP. The resultant DNA ends are compatible with commercial LambdaGEM-12 *Xho*I half-site arms, which have *Xho*I-digested ends partially filled in with dTTP and dCTP (*see* **Fig. 1B**).

1. Add the following components to a 1.5-mL microcentrifuge tube: 5–10 µg of recovered genomic DNA fragments, 5 µL of 10X fill-in buffer, 2 U of Klenow enzyme, and sterile H_2O to a total volume of 50 µL. Incubate at 37°C for 30 min.
2. Extract the DNA as follows: add 1 volume (50 µL) of TE-saturated phenol:chloroform:isoamyl alcohol (25:24:1), gently mix, and centrifuge at 12,000g for 5 min. Transfer the upper aqueous phase to a fresh tube. Add 1 volume of chloroform:isoamyl alcohol (24:1), gently mix, and centrifuge as above.
3. Transfer the upper aqueous phase to a fresh tube. Add 0.5 volume of 7.5 *M* ammonium acetate and 2 volumes of 100% ethanol, mix, and incubate at –20°C overnight.
4. Centrifuge at 12,000g for 15 min at 4°C. Carefully discard the supernatant, wash the pellet with 500 µL of cold 70% ethanol, and air-dry for about 5 min. Dissolve the DNA in 20 µL of sterile H_2O at room temperature for at least 1 h.
5. Estimate the DNA concentration by comparing serial dilutions of a known concentration standard with the DNA sample by agarose gel electrophoresis and EtBr staining (*see* **Fig. 2D**). Next, read the absorbance of the diluted standards and sample DNA at A_{260}. For the diluted standards, plot the absorbance against DNA concentration and determine the concentration of the genomic DNA directly from the plot (*see* **Note 6**).

3.2. Ligation of Insert to Vector Arms

1. Add the ligation reaction components to five 0.5-mL microcentrifuge tubes, labeled as L, M, H, C_1, and C_2, according to **Table 2**. Incubate at 4°C overnight.
2. Proceed to **Subheading 3.3.** for packaging and titration of phage.
3. Compare the titers from the ligations L, M, and H and determine the optimal ratio of vector arms and insert DNA (V : I). If desired, prepare a scale-up reaction based on the optimal V:I ratio. The total amount of DNA (including insert and vector) should not exceed 5 µg in a 10-µL-volume ligation.

3.3. Packaging of Ligated DNA and Titration of Recombinant Phage

3.3.1. In Vitro Packaging

1. Place the Packagene Lambda DNA Packaging Extract (supplied as 50 µL per tube) on ice. As soon as the extract is thawed, add 5–10 µL of ligation product to a Packagene tube (do not use more than 10 µL of ligation per 50 µL packaging extract) and mix rapidly by tapping the tube.
2. Incubate the packaging mixture at 22°C (room temperature) for 3 h.
3. Add 445 µL of phage buffer and 25 µL of chloroform to each packaging product. Mix gently by inversion and allow the chloroform to settle to the bottom of the tube. The packaged phage particles can be stored at 4°C for 7 d with no significant drop in titer. For long-term storage, *see* **Subheading 3.5.**

Table 2
Preparation of Ligation Mixtures

	L 1:1	M 1:2	H 1:3	C_1 1:1.4	C_2 NA
		(molar ratio of vector : insert)			
Vector DNA[a] (0.5 µg/µL)	2 µL	2 µL	2 µL	2 µL	2 µL
Insert DNA[b]	0.46 µg	0.93 µg	1.39 µg	—	—
Insert control[c] (0.5 µg/µL)	—	—	—	1 µL	—
10X ligase buffer	1 µL	1 µL	1 µL	1 µL	1 µL
T4 DNA ligase (5 U/µL)	1 µL	1 µL	1 µL	1 µL	1 µL
Sterile H_2O to final volume	10 µL	10 µL	10 µL	10 µL	10 µL

[a]The size of vector (LambdaGEM-12 *Xho*I half-site arms) is 43 kb (1 µg = 0.035 pmol).
[b]The average size of insert DNA is about 20 kb (1 µg = 0.075 pmol) (*see* **Fig. 2D**).
[c]The size of insert control is 16 kb (1 µg = 0.095 pmol).

3.3.2. Infection

1. Pick a single colony of KW251 cells from a freshly streaked LB plate containing 15 µg/mL tetracycline. Inoculate 3 mL of LBMM, containing 15 µg/mL tetracycline and shake overnight at 37°C. Subculture 200 µL of the overnight culture into 20 mL of LBMM and incubate at 37°C until the optical density (OD_{600}) reaches 0.6–0.8 (about 2.5 h with shaking at 250 rpm).
2. Centrifuge the cells at 5000*g* for 5 min at 4°C, discard the supernatant, and resuspend the cell pellet in ice-cold 10 m*M* $MgSO_4$ to a final OD_{600} of 1.0. The cells can be used immediately or stored at 4°C for up to 24 h.
3. Make appropriate dilutions of the packaged phage particles in phage buffer. As a general guideline, make a series of threefold dilutions ranging from 1:500 to 1:13,500.
4. Add 100 µL of diluted phage to 100 µL of prepared KW251 cells in a 1.5-mL microcentrifuge tube and incubate for 30 min at 37°C to allow the phage to infect the cells.
5. Transfer the phage–cell mixtures into tubes containing 3 mL of molten LB top agarose (50–55°C). Quickly mix and immediately pour onto LB plates prewarmed to 37°C. Allow the LB top agarose to harden and then incubate the plates inverted at 37°C for 6–8 h or until plaques are present.
6. Count the number of plaques and calculate the phage titer as plaque-forming units (pfu) per milliliter using following formula:

    (Number of plaques × Dilution factor)/Volume of diluted phage = pfu/mL

3.4. Library Validation

A two-stage procedure is designed for library validation. A pilot experiment using a small aliquot of the constructed library is performed (*see* **Note 7**). The purified phage DNA is digested by appropriate RE for analysis and separated on an agarose gel. Typically, the presence of various size inserts produces a smear pattern on the gel. At the second stage, a more detailed analysis is conducted in which a number of individual phage DNAs are purified. The sizes of their inserts are visualized subsequent to agarose gel electrophoresis and the average insert size is calculated (*see* **Note 8**).

3.4.1. Pilot Analysis

3.4.1.1. Infection

1. Prepare four 150-mm LB plates (more if desired). For each plate, mix about $(2–5) \times 10^5$ pfu with 500 µL of KW251 cells, prepared as in **Subheading 3.3.2.**, **steps 1** and **2** (for convenience, all phage and cells required for four plates can be mixed together in one 15-mL conical tube). Incubate at 37°C for 20 min.
2. Transfer the phage–cell mixture into a cell culture tube containing 6 mL of molten top agarose (50–55°C). Mix rapidly and immediately pour onto LB plates that have been prewarmed at 37°C. After the top agarose has hardened, invert the plates and incubate at 37°C for about 6 h or until plaques are nearly confluent.

3.4.1.2. Extraction of λ-Phage Particles

1. Randomly pick 28 well-isolated plaques from the LB agar plates. Transfer the phage "agarose plug" into 1.5-mL microcentrifuge tubes using a blunt 1000-µL pipet tip (simply cut off the end of tip using scissors). These collected samples will be used for phage DNA purification at the second stage (*see* **Subheading 3.4.2.**).
2. Overlay each plate with 10 mL of phage buffer and incubate at 4°C for 1 h. Scrape the top agarose (avoiding the bottom agar) using a spatula and transfer into a 50-mL high-speed centrifuge tube. Break up the agarose with the spatula and incubate at room temperature for 30 min with periodic shaking.
3. Centrifuge at 8000g for 10 min at 4°C. Carefully withdraw the supernatant and transfer to a fresh tube. Add chloroform to 0.3% (v/v); this lysate can be stored at 4°C.
4. To purify the phage particles, add 40 µL of nuclease mixture per 10 mL of phage lysate and incubate at 37°C for 30 min. Then, add 4 mL of phage precipitant, gently mix, and incubate on ice for 1 h (or overnight at 4°C).
5. Centrifuge at 10,000g for 20 min at 4°C. Carefully decant the supernatant, gently resuspend the pellet in 750 µL of phage buffer and transfer to a 1.5-mL microcentrifuge tube.

3.4.1.3. Purification of Recombinant λ-Phage DNA (*see* **Note 9**)

1. Add proteinase K to a final concentration of 0.5 mg/mL and incubate at 37°C for 5 min. Centrifuge at 8000g for 2 min to remove debris.
2. Transfer the supernatant to a fresh 2-mL microcentrifuge tube. Add 75 µL of 10% SDS and 75 µL of 0.5 M EDTA. Incubate at 68°C for 15 min.
3. Extract samples twice with 1 volume of TE-saturated phenol : chloroform : isoamyl alcohol (25 : 24 : 1) and once with 1 volume of chloroform : isoamyl alcohol (24 : 1) as described in **Subheading 3.1.4.**, **step 2**.
4. Carefully transfer the upper aqueous phase to a fresh tube. Precipitate the phage DNA by adding 1 volume of isopropanol. Gently mix and leave at –20°C for 1 h. Centrifuge at 12,000g for 10 min at 4°C.
5. Carefully discard the supernatant. Wash the pellet with 70% cold ethanol. Air-dry the pellet for about 5 min and resuspend in 20 µL TE per 10 mL of starting lysate. Store the purified DNA at 4°C.
6. Estimate the DNA concentration using the A_{260}/A_{280} spectrophotometric ratio method and agarose gel electrophoresis analysis, as described in **Subheading 3.1.1.**, **step 7**.

3.4.1.4. Restriction analysis

1. Analyze the purified phage DNA by restriction digestion with *Sfi*I. To digest the DNA, add the following to a 0.5-mL microcentrifuge tube: 1.5 µg of phage DNA, 15 U of *Sfi*I,

100 µg/mL of BSA, and 1X *Sfi*I buffer. Make up to 40 µL with sterile water and incubate at 50°C overnight (*see* **Note 10**).
2. Resolve the digested products by electrophoresis on a 0.5% agarose gel at 1–2 V/cm (*see* **Fig. 4A**).

3.4.2. Analysis of λ Clones

1. Add 0.5 mL of SM buffer to each agarose plug collected in **step 1** of **Subheading 3.4.1.2.** Place at 4°C for 60 min to allow the phage to elute.
2. Titer each phage eluate as described in **Subheading 3.3.2.** (the pfu from individual plaque plug eluates will vary).
3. Infect prepared KW251 cells with $(2–5) \times 10^5$ pfu of each phage as described in **Subheading 3.4.1.1.**, with the exception that only one 150-mm plate is used for each sample.
4. Extract phage particles and purify DNA as described in **Subheadings 3.4.1.2.** and **3.4.1.3.**, respectively.
5. Analyze each clone by *Sfi*I digestion and electrophoresis, as in **Subheading 3.4.1.4.** Determine each insert size by comparing them to standard DNA markers and calculate the average insert size (*see* **Fig. 4B**).
6. If the library is satisfactory, it may be amplified as described below and used to screen for desired recombinants.

3.5. Library Amplification

1. Mix aliquots of the primary library (from **step 3** of **Section 3.3.1.**), containing approx 5×10^4 phages, with 500 µL of KW251 cells, freshly prepared as described in **Subheading 3.3.2.**, **steps 1** and **2**. Incubate at 37°C for 20 min and transfer to 6 mL of melted (55°C) top agarose. Pour onto 150-mm LB plates. Incubate the plates at 37°C until plaques are pinpoint in size.
2. Cover the plates with 12 mL of phage buffer containing 0.01% gelatin and incubate at 4°C for 8 h. Collect the phage eluate and add chloroform to 10% (v/v). Shake and centrifuge at 5000*g* for 10 min at 4°C. Save the supernatant and add DMSO to 7% (v/v). Determine the titer of the amplified library if desired as described in **Subheading 3.2.2.**
3. Store 100- to 200-µL aliquots of the amplified library in screw-cap glass tubes at –80°C (*see* **Note 11**).

4. Notes

1. A number of similar ready-to-use λ-vectors are commercially available for genomic library construction. For example, LambdaGEM-11 *Bam*HI arm is a similar vector and is currently available (Promega). One advantage of LambdaGEM-11 is that the partial fill-in can be omitted because the vector arms are dephosphorylated and a small fragment is deleted from the central stuffer *(5)*.
2. High-quality reagents should be used for preparing the phage lysate to avoid contamination.
3. Genomic DNA is not readily soluble. DNA sample rehydration can be significantly improved by using a high-purity DNA preparation ($A_{260}/A_{280} \geq 1.80$) and avoiding overdrying the DNA pellet. The sample should also be heated to 65°C for 1 h. Alternatively, the DNA sample can be incubated in water or appropriate buffers (e.g., TE or RE digestion buffer) for 1–4 h at room temperature (or at 4°C overnight) before starting the digestion reaction.
4. It is difficult to produce a particular size of genomic DNA solely by restriction digestion. Non-size-fractionated genomic fragments commonly result in poor cloning effi-

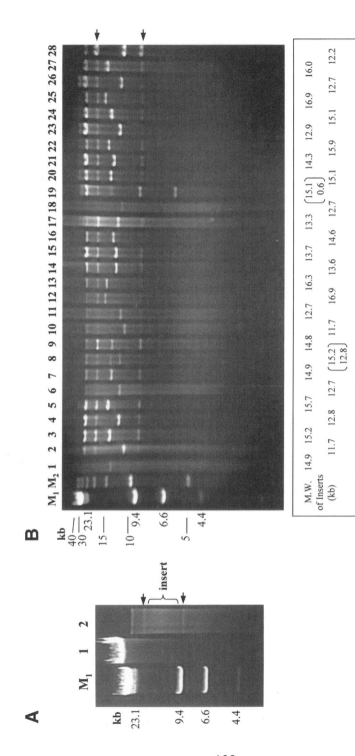

Fig. 4. Library validation by determination of insert size. (**A**) First-stage experiment. *Lane 1*: undigested λ-phage DNA; *lane 2*: *Sfi*I digested λ-phage DNA. (**B**) Second-stage experiment. *Lanes 1–28*: individual λ-phage DNAs digested with *Sfi*I. M1 and M2 refer to Lambda DNA/*Hin*dIII markers and 5-kb DNA ladder, respectively. The 20-kb and 9-kb bands (indicated by arrows) represent the left arm and right arm of the LambdaGEM-12 vector. The molecular weight of each insert was determined by using the Image Acquisition and Analysis program (UVP).

ciency. Therefore, the size fractionation should be regarded as an essential process for genomic library construction. The approach for DNA size fractionation described yields high recoveries.
5. For accurate quantitative determination, we do not recommend that the DNA amount be determined only by the A_{260}/A_{280} UV ratio spectrophotometric method. We suggest that the spectrophotometric determination should be combined with visualization following agarose gel electrophoresis.
6. To obtain the best results in the construction of a genomic library using the presented protocol herein, the quality of the prepared genomic DNA is crucial. The genomic DNA should be of high purity, completely rehydrated, and accurately quantified with respect to insert size.
7. This also serves as a pilot experiment in which all reagents and manipulations are tested. The second-stage experiments (**Subheading 3.4.2.**) should only be started after satisfactory results are obtained.
8. For library validation, a major concern is the fidelity of the genomic DNA sequences retained in the constructed library. In addition, a good library should include all particular sequences of interest. To anticipate the theoretical chance of isolating a desired sequence, use the following formula *(6)*:

$$N = \ln(1-I/G)/\ln(1-P)$$

where N is the number of independent clones that must be screened to isolate a particular sequence with probability P (commonly, $P = 99\%$), I is the average size of the insert (bp), and G is the size of the target genome (bp). For a given species, G is constant. Thus, the average size of inserts and the number of independent clones are the determining factors. In general, the minimal sum of $I \times N$, should be such that the total size of inserts screened represents a 4.6-fold excess over the total size of the genome (G) *(2,6)*.
9. As an alternative, commercial kits for λ DNA purification from phage lysates are available. The Wizard Lambda Preps DNA Purification System (Promega) is recommended.
10. To avoid water evaporating from the reaction, seal the tube with parafilm and immerse the whole tube in a 50°C water bath.
11. The constructed genomic library (either primary or amplified) should be stored in sterile, capped glass vials at −80°C for long-term storage. Do not store in plastic vials because plastic causes the titer of stored libraries to drop dramatically *(7)*. We have also found that smaller aliquots (100–200 µL per vial) store better than the larger aliquots.

Acknowledgments

We thank Dr. York Zhu for the instructive discussion on library construction and Mrs. SeTonia Cook for assistance in the preparation of this manuscript.

References

1. Sambrook, J., Fritsch, E. F., and Maniatis, T. (eds.) (1989) Bacteriophage λ vectors, in *Molecular Cloning: A Laboratory Manual*, Cold Spring Harbor Laboratory, Cold Spring Harbor, NY, pp. 2.3–2.125.
2. Seed, B., Parker, R. C., and Davidson, N. (1982) Representation of DNA sequences in recombinant DNA library. *Gene* **19,** 201–209.
3. Wang, Y., Cao, Z., Reid, E. A., et al. (2000) The use of competitive PCR mimic to evaluate a *Limulus* lambda phage genomic DNA library. *Cell. Mol. Neurobiol.* **20,** 509–520.

4. Frischauf, A. M., Lahrach, H., Poustka, A., et al. (1983) Lambda replacement vectors carrying polylinker sequences. *J. Mol. Biol.* **170,** 827–842.
5. Promega Corporation (1996) *Protocols and Applications Guide*, Promega, Madison, WI.
6. Moore, D. D. (1997) Overview of genomic DNA libraries, in *Short Protocols in Molecular Biology* (Ausubel, F., Brent, R., Kingston, R. E., et al. eds.), Wiley, New York, pp. 5.2–5.4.
7. Klickstein, L. B. (1997) Amplification of a bacteriophage library, in *Short Protocols in Molecular Biology* (Ausubel, F., Brent, R., Kingston, R. E., et al., eds.), Wiley, New York, pp. 5.5–5.6.

19

Rapid Screening of Recombinant Plasmids

Sangwei Lu

1. Introduction

Construction of recombinant plasmid DNA is one of the cornerstones of molecular biology. The ability to clone DNA in a plasmid vector opens doors to downstream applications such as amplification of DNA, expression of desired genes, and construction of DNA libraries. Recombinant plasmids are generally constructed by first isolating the target DNA and linearizing the plasmid vector of choice (*see* Chapter 2). The insert DNA is subsequently ligated to the vector DNA (*see* Chapters 15 and 17) and the ligation mixture transformed into an appropriate *Escherichia coli* host (*see* Chapters 3–5) *(1,2)*. *E. coli* transformed with recombinant plasmid or self-ligated vector will both grow on appropriate selective medium; therefore, screening is almost always necessary to select colonies containing the recombinant plasmid and, in some cases, with the correct orientation of insert.

The conventional method of selecting the desired recombinant plasmid is to randomly select transformants, prepare plasmid DNA, and characterize plasmid clones by restriction enzyme digestion (*see* Chapter 20). This remains the definitive and often necessary method; however, it can be very tedious and labor intensive when a large number of colonies have to be screened in some difficult cloning procedures. Furthermore, it is often necessary to grow the transformants overnight in order to obtain sufficient plasmid DNA for restriction enzyme digestion and sequence analysis. Alternatively, hybridization can be performed, with the insert DNA as probe, to screen for plasmids containing the insert (*see* Chapter 21). This method can be used to screen a large number of colonies; however, a disadvantage is that it is time-consuming. Therefore, it is desirable to use methods to rapidly prescreen transformants for likely candidate recombinant plasmids and, subsequently, confirm the identification by restriction enzyme digestion and/or sequence analysis of purified plasmid DNA.

Multiple goals can be achieved with rapid screening methods. The first and most obvious goal is to determine if any given colony contains plasmid with insert DNA.

From: *Methods in Molecular Biology, Vol. 235:* E. coli *Plasmid Vectors*
Edited by: N. Casali and A. Preston © Humana Press Inc., Totowa, NJ

Second, using primers specific for the insert DNA, it is possible to determine the orientation, as well as the presence, of the insert by using the polymerase chain reaction (PCR) *(3)*. Third, PCR products can be used for DNA sequencing or restriction enzyme digestion to further characterize the insert or the junctions of insert and vector.

Several methods can be used to rapidly screen transformant colonies. This chapter will describe blue–white selection *(1,2,4,5)*, direct gel electrophoresis of colonies *(6)*, and colony PCR. Each method involves a different time commitment and workload, so it is important to balance the effort put into the rapid-screening procedures and the likelihood of obtaining the correct plasmid from the cloning procedure. More difficult cloning procedures with low probabilities of obtaining the correct plasmid call for more extensive prescreening, whereas easy cloning procedures require little or no prescreening.

1.1. Blue–White Selection

Blue–white selection is the easiest screening method of those described in this chapter. This method can be utilized with a large class of cloning vectors, such as pBluecript, pGEM, pUC, and their derivatives (*see* Chapter 2), which contain the coding sequence of the amino-terminal fragment of β-galactosidase (the α-peptide, 146 amino acids). The expression of the α-peptide is often under the control of the *lac* promoter in these plasmids *(2)*. In *E. coli* hosts that support blue–white selection (*see* Chapter 3), the α-peptide associates with the host-encoded carboxyl-terminal fragment of β-galactosidase to produce a functional enzyme. This metabolizes the artificial substrate 5-bromo-4-chloro-3-indoxyl-β-D-galactopyranoside (X-gal) to generate a deep blue. The vector cloning site(s) interrupts the coding sequence, such that insertion of DNA abrogates expression of the α-peptide, the bacteria are unable to produce functional β-galactosidase, and the bacterial colony remains white in the presence of X-gal. If the *lac* promoter is used in an *E. coli* host strain that expresses the *lac* repressor, an inducer is needed to relieve suppression of the *lac* promoter by the repressor. Often, the nonfermentable analog of lactose, isopropyl-β-D-thiogalactoside (IPTG), is used as an inducer (*see* Chapter 29).

1.2. Rapid Screening by Direct Electrophoresis

The method described here is modified from that originally reported *(6)*. Transformant colonies are lysed in the wells of an agarose gel and the released plasmid DNA is separated by direct electrophoresis. The size of the plasmid DNA is compared to that of the parent vector and plasmid DNA that appears to be larger is likely to contain the insert. This method is useful when the insert DNA comprises at least 10% of the recombinant plasmid so that recombinant plasmids and self-ligated vector can be distinguished by gel electrophoresis.

1.3. Direct Colony PCR

Direct colony PCR circumvents the need to culture bacteria overnight in order to obtain sufficient DNA for further analysis. In this method, the PCR reaction achieves multiple goals and can be used to detect the presence of the insert, determine the orientation of the insert (*see* **Fig. 1**), and obtain sufficient DNA for sequencing and restriction digestion analysis.

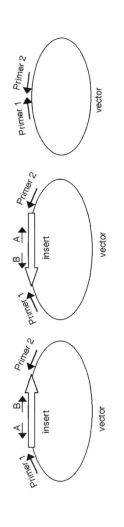

Fig. 1. A scheme for the selection of oligonucleotide primers for use in direct colony PCR. Primers 1 and 2 are vector-based primers that anneal to sites flanking the cloning site; primers A and B are insert DNA-based primers.

2. Materials
2.1. Blue–White Selection
1. Luria–Bertani (LB) agar plates: 10 g tryptone, 5 g yeast extract, 10 g NaCl, 15 g agar. Add water to 1 L and autoclave to sterilize. Cool to approx 50°C, add antibiotics as appropriate, and pour approx 20 mL into each Petri dish.
2. Appropriate antibiotic for the selection of plasmid.
3. 0.1 M IPTG prepared in deionized water. Store at –20°C.
4. 20 mg/mL X-gal, prepared in N,N'-dimethylformamide. Store at –20°C.

2.2. Rapid Screening by Direct Electrophoresis
1. Rapid-screen resuspension buffer: 30 mM Tris-HCl (pH 8.0), 5 mM EDTA, 50 mM NaCl, 20% (w/v) sucrose, 50 µg/mL RNase A and 50 µg/mL lysozyme.
2. Rapid-screen lysis buffer: 89 mM Tris-HCl (pH 8.0), 89 mM boric acid, 2.5 mM EDTA, 2% (w/v) sodium dodecyl sulfate (SDS), 5% (w/v) sucrose, and 0.04% (w/v) bromophenol blue.
3. 1X TAE buffer: 40 mM Tris-acetate, 1 mM EDTA. A 50X stock solution can be prepared by mixing 242 g of Tris base, 57.1 mL of acetic acid, 100 mL of 0.5 M EDTA, and water to a final volume of 1 L.
4. Agarose gel prepared with 1X TAE buffer containing 0.5 µg/mL of ethidium bromide.

2.3. Direct Colony PCR
1. *Taq* DNA polymerase with 10X PCR buffer. Store at –20°C.
2. Deoxynucleotide solution (dNTPs) containing 2 mM each of dATP, dCTP, dGTP, and dTTP. Store at –20°C.
3. 2.5 µM Stock solution of each oligonucleotide primer. Store at –20°C.
4. PCR-grade water.
5. Thermocycler.

3. Methods
3.1. Blue–White Selection
1. Transform *E. coli* with the ligation reaction using methods described in Chapters 4 and 5.
2. Prepare plates by adding 40 µL of 20 mg/mL X-gal and 50 µL of 0.1 M IPTG to each LB agar plate containing appropriate antibiotics and spread with a sterile spreader. Incubate in a 37°C incubator until the plate is dry (*see* **Note 1**).
3. Plate the transformation mixture onto the LB agar plates and allow to dry.
4. Invert the plates and incubate overnight in a 37°C incubator. Usually, at least 15–18 h are necessary for the blue precipitate to be clearly visible.
5. Select white colonies for further characterization (*see* **Notes 2** and **3**).

3.2. Rapid Screening by Direct Electrophoresis
1. Examine transformant colonies. Colonies suitable for rapid screening should be at least 1 mm in diameter. In the case of low-copy-number plasmids, colonies larger than 1 mm in diameter are likely to be necessary. If the colonies are too small, use pipet tips or sterile toothpicks to transfer colonies to a fresh plate and let the bacteria grow at 37°C for 2 h, or until sufficient growth appears on the plate. Label each colony to be screened.
2. Add 10 µL of rapid-screen resuspension buffer to each well of a 96-well "Flexi-plate" (Fisher) or to Eppendorf tubes.

3. Transfer part of a bacterial colony with a pipet tip or toothpick to a well or tube and resuspend it in the rapid-screen resuspension buffer. Make sure that sufficient bacteria to inoculate a broth culture are left on the plate in order to amplify recombinant plasmid-containing clones. Include at least one sample of a bacteria carrying parental vector DNA for each row of samples on the gel. Incubate the bacteria in resuspension buffer at room temperature for 10 min.
4. Prepare an agarose gel. Load DNA marker in one lane of each row. Load 2 µL of lysis buffer into the remaining wells (*see* **Note 4**). Load one lane for each sample.
5. Load the resuspended bacterial colonies into the wells containing lysis buffer (*see* **Notes 5** and **6**).
6. Start electrophoresis at 35 V for 15 min to allow the plasmid DNA to enter the gel and then increase to 100 V and continue electrophoresis for 1 h.
7. Examine the agarose gel on an ultraviolet (UV) transilluminator. Recombinant plasmids should be larger than the vector DNA.

3.3. Direct Colony PCR

The selection of primers for PCR is dictated by the goal of the screening. If the screening is aimed at only determining the presence of insert DNA, vector-based primers that anneal to sequences flanking the cloning site are suitable. For many standard cloning vectors, this allows the use of universal primers (*see* Chapter 2). To determine the orientation of insert DNA, it is necessary to use one primer that anneals to the insert DNA in conjunction with one of the vector-based primers. A sample design scheme for primers is shown in **Fig. 1**.

1. Examine the transformation plates and select suitable colonies as described in **Subheading 3.2., step 1**. Label colonies on the plate so the results of screening can be matched to the colonies.
2. Prepare a master PCR reaction mix allowing 30 µL per sample. The following is the composition of 100 µL of PCR master mix and it can be scaled up as necessary (*see* **Note 7**): 10 µL of 10X PCR buffer, 10 µL of 2 mM dNTPs, 10 µL of 2.5 µM of 5' primer, 10 µL of 2.5 µM of 3' primer, 59 µL of PCR-grade H$_2$O, and 1 µL of *Taq* polymerase. Aliquot 30 µL into each PCR tube.
3. Transfer part of a colony to each tube containing PCR mix using pipet tips or toothpicks and stir to mix. A small amount of bacteria is usually sufficient and the mix should not appear turbid after the addition of bacterial colonies (*see* **Note 8**).
4. Cycle the PCR reactions in a thermocycler. Use conditions suitable for the primer pair selected. Include an initial 2-min denaturing step at 95°C to lyse the bacteria before cycling starts.
5. Examine the PCR products by electrophoresis on an agarose gel. If vector-based primers are used, recombinant plasmid will yield larger PCR products than self-ligated vector. PCR performed with insert specific primers will yield a product only if the insert is in the appropriate orientation.
6. If necessary, the PCR product can be purified using methods such as the Qiagen PCR purification kit and used for restriction enzyme digestion or sequencing analysis (*see* **Note 9**).

4. Notes

1. Instead of spreading IPTG and X-gal on to LB agar plates, they can be added to the agar mixture before plates are poured (final concentration: 6 mM IPTG and 0.3 mg/mL X-gal

in LB agar). However, this approach uses more IPTG and X-gal and these plates have a fairly short shelf life.
2. To ensure that white colonies are not a result of a lack of IPTG or X-gal on the plate, it is preferable to pick white colonies from the vicinity of blue colonies. With some practice spreading IPTG and X-gal, false white colonies rarely occur.
3. If the insert DNA is short, it is possible that colonies containing recombinant plasmid are blue, resulting from read-through of the insert DNA. However, this type of colony usually has a blue center with a white periphery, instead of the uniform blue appearance of self-ligated vector-containing colonies.
4. The lysis buffer must be loaded first into the wells before resuspended colonies are loaded.
5. The blue color from the lysis buffer makes the second loading difficult to track. It is easier to see which lanes have been loaded if the gel is viewed from a 45° angle, instead of from straight above gel. Wells in which the bacterial colonies have been loaded appear white at the top and blue at the bottom.
6. A large number of samples can be analyzed simultaneously using a multichannel pipet for sample processing and gel loading, and a large agarose gel formatted for multichannel pipets. When a large number of colonies are screened at the same time, it is advisable to allow for some evaporation from samples and increase the volume of resuspension buffer to 15 µL.
7. This method can be used to screen a large number of colonies. The size of a batch is only limited by the capacity of the PCR machine. However, cycling should start within 15 min of the first bacterial colony being added to the PCR mix. Otherwise, the efficiency of the PCR reaction decreases, probably the result of interaction between bacteria and components of the PCR mix. With practice, 96 samples can be readily processed within this time frame.
8. The amount of bacteria needed for PCR screening is usually very small. For high-copy-number plasmids such as pGEM and pBluescript, enough bacteria can be obtained by lowering a yellow tip vertically onto a colony and rinsing the bacteria attached to the tip in the PCR mix. Too many bacteria in the mix will inhibit the PCR reaction.
9. The PCR product can also be cloned. However, if restriction digestion needs to be performed before the cloning step, it is important to extract the PCR product with phenol : chloroform at least three times. Otherwise, residual *Taq* polymerase may destroy the ends generated by restriction digestion.

References

1. Ausubel, I. and Frederick, M. (1998) *Current Protocols in Molecular Biology*, Wiley, New York.
2. Sambrook, J. and Russell, D. W. (2001) *Molecular Cloning: A Laboratory Manual*, Cold Spring Harbor Laboratory, Cold Spring Harbor, NY.
3. Mullis, K. B. and Faloona, F. A. (1987) Specific synthesis of DNA in vitro via a polymerase-catalyzed chain reaction. *Methods Enzymol.* **155**, 335–350.
4. Davies, J. and Jacob, F. (1968) Genetic mapping of the regulator and operator genes of the *lac* operon. *J. Mol. Biol.* **36**, 413–417.
5. Horwitz, J. P., Chua, J., Curby, R. J., et al. (1965) Substrates for cytochemical demonstration of enzyme activity I: some substituted 3-indolyl-β-D-glycopyranosides. *J. Med. Chem.* **7**, 574–575.
6. Sekar, V. (1987). A rapid screening procedure for the identification of recombinant bacterial clones. *BioTechniques* **5(1)**, 11–13.

20

Restriction Analysis of Recombinant Plasmids

Joanne Goranson-Siekierke and Jarrod L. Erbe

1. Introduction

A key step in the construction of recombinant plasmids is the verification of the successful cloning of insert DNA into the vector. A number of commonly used plasmids facilitate phenotypic selection and/or screening methods for rapid identification of insert-containing clones. Additional screening methods are required to confirm the presence and/or orientation of the intended insert. A polymerase chain reaction (PCR) strategy may be sufficient to verify the presence and orientation of the intended insert depending on the availability of vector- and insert-specific primers (*see* Chapter 19). In situations where PCR-based analysis is undesirable or additional analysis is required, clear information about the clone can be derived from restriction analysis.

Briefly, this strategy requires four steps:

1. Extraction of DNA from the clone(s) to be tested.
2. Cleavage of the DNA with appropriate restriction enzymes.
3. Electrophoresis of the DNA digests on agarose gels along with molecular-weight standards.
4. Comparison of the sizes of the restriction digest products with those that are expected, based on a restriction map of the desired clone.

This chapter details the steps of the DNA digests *(1–3)* and subsequent analysis *(1,4,5)*. Protocols for the extraction of DNA from *Escherichia coli* clones are given in Chapters 8–11. Alternative protocols are available that may better suit individual needs. For example, the CloneChecker system (Invitrogen Life Technologies) facilitates rapid restriction analysis of plasmid DNA isolated directly from individual colonies, whereas the multiplex technique *(6)* allows for the simultaneous analysis of multiple clones.

Prior to performing restriction analyses, a detailed restriction map of the plasmid in question should be consulted (*see* Chapter 14). Comparison of the observed restriction digestion products with those that are expected, based on a restriction map of the predicted clone, is especially important if cloning was nondirectional because of the use

From: *Methods in Molecular Biology, Vol. 235:* E. coli *Plasmid Vectors*
Edited by: N. Casali and A. Preston © Humana Press Inc., Totowa, NJ

of enzymes resulting in blunt-ended DNA or if a single restriction enzyme was used. The choice of enzymes used for the digestion should be made carefully in order to make the restriction digest pattern clear and simple.

1.1. Determination of Appropriate Restriction Enzymes and Digests

The choice of appropriate enzyme(s) for the restriction analysis of the clone will depend on the plasmid and insert involved. Several criteria may influence this decision. For example, the resulting DNA fragments need to be within a size range detectable on a gel (*see* **Note 1**) and the fragments of interest must be easily distinguishable from each other. Optimally, choose a single enzyme for which there is a unique digestion site in both the multiple-cloning site of the plasmid and close to one end of the insert.

In reality, the ideal choice of enzymes is not always possible. For example, the plasmid may not possess a multiple-cloning site. In this case, a restriction site elsewhere within the plasmid, preferably close to the site of insertion, is used. Digests using two different enzymes that are active in the same reaction buffer are a good choice. Moreover, if a unique restriction site is unavailable, an enzyme site that occurs infrequently may be utilized. It is sometimes necessary to use enzymes that are not functional in the same restriction buffer system, and a single reaction is not possible. **Figure 1** illustrates examples of enzyme choices for different plasmid digestion scenarios. If no usable sites exist either within the plasmid or the insert, a PCR-based analysis may be performed using a primer internal to the DNA insert and a primer specific for the plasmid that is directed toward the site of insertion (*see* Chapter 19).

Before proceeding with the reactions, it is important to refer to the manufacturers' catalogs for vital information regarding the enzymes of choice. Such information will include the specificity of cleavage, optimal reaction conditions, the requirement of the enzyme for other factors such as bovine serum albumin (BSA), blockage of the restriction site by DNA methylation (*see* **Note 2**), and inactivation of the enzyme activity. This information will be necessary to determine which enzymes and reaction conditions will be optimal for analysis of the clones.

2. Materials
2.1. Restriction Enzyme Digestion

1. Predicted restriction map of the plasmid clone. Prepare a map for a clone with the insert in each of the two possible orientations if applicable.
2. Appropriate restriction endonuclease(s) stored at –20°C (*see* **Note 3**).

Fig. 1. *(opposite page)* Example restriction digests for subclone confirmation. A 700-bp blunt-ended fragment was inserted into the *Sma*I site of the theoretical plasmid, pJJ100. Fragment lengths within the insert and vector are shown in base pairs. In (**A**), the insert possesses two restriction enzyme sites, *Eco*RI and *Bam*HI, which also lie within the multiple cloning site of pJJ100. The presence and orientation of the insert can be determined using digests of the subclone with either enzyme. Thus, a subclone having the insert in orientation **1** would yield *Eco*RI fragments of 150 and 3050 bp, whereas a *Bam*HI digest would yield 270- and 2930-bp fragments. A subclone with the insert in orientation **2** would yield *Eco*RI fragments of 650 and 2550 bp and *Bam*HI fragments of 470 and 2730 bp. *(continued on next page)*

Restriction Analysis

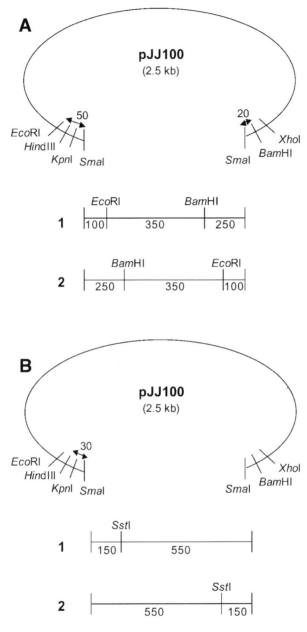

Fig. 1. *(continued)* In (**B**), the insert possesses a single, unique enzyme site, *Sst*I, that is not present in the vector. The presence and orientation of the insert can be determined using this site as well as additional sites within pJJ100. For example, when digested with both *Hin*dIII and *Sst*I, subclones having the insert in orientation **1** would yield fragments of 180 and 3100 bp, whereas subclones with the insert in orientation **2** would give fragments of 580 and 2700 bp.

3. 10X Stocks of the appropriate restriction enzyme buffers (*see* **Note 4**).
4. 10 mg/mL BSA, if applicable.
5. Plasmid DNA, either in water or TE (10 mM Tris-HCl [pH 8.3], 1 mM EDTA).
6. Sterile deionized distilled water.
7. Heat block, water bath, or incubator set at the appropriate temperature (*see* **Note 5**).

2.2. Agarose Gel Electrophoresis

1. Molecular-biology-grade agarose (high melting point).
2. Electrophoresis buffer: Prepare either 5X TBE or 10X TAE stock solutions (*see* **Note 6**); dilute 1:10 with deionized distilled water before use.
    a. 5X TBE: 54 g Tris-HCl, 27.5 g boric acid, 4.65 g EDTA; make up to 1 L with deionized distilled water.
    b. 10X TAE: 48.4 g Tris base, 11.4 mL glacial acetic acid, 20 mL of 0.5 M Na$_2$EDTA (pH 8.0); make up to 1 L with deionized distilled water.
3. Heating plate, microwave oven, or ring stand and Bunsen burner.
4. Gel apparatus and power pack.
5. 5X Sample loading buffer: 50% (v/v) glycerol, 50mM EDTA (pH 8.0), 0.125% (w/v) bromophenol blue, 0.125% (w/v) xylene cyanol. Store stock at 4°C; a <1-mL aliquot can be kept at room temperature for daily use.
6. DNA size marker (*see* **Note 7**).
7. 10 mg/mL Ethidium bromide stock solution, store at 4°C in a dark bottle (*see* **Note 8**).
8. Staining solution: 0.5–1.0 µg ethidium bromide/mL in electrophoresis buffer; keep at room temperature in a covered container.
9. Ultraviolet (UV) light transilluminator (long wave, 365 nm).
10. Photographic documentation system if desired (Polaroid or digital).

3. Methods
3.1. Restriction Enzyme Digestion

1. Thaw all solutions, except the enzyme, and keep on ice.
2. Using a final volume for the digest of 20 µL (or up to 50 µL if the DNA is dilute), add the following into a sterile Eppendorf tube:

    1/10 volume reaction buffer (*see* **Note 9**).
    0.1 mg/mL BSA (if required, see the restriction enzyme manufacturer's catalog).
    0.3–1.0 µg of the DNA to be digested (*see* **Note 10**).

    Add sterile distilled water to the final reaction volume, minus the enzyme volume.
3. Add the restriction enzyme directly from the stock (*see* **Notes 3** and **11**). Mix gently either by pipetting the reaction up and down several times or by lightly flicking the tube (*see* **Note 12**). Centrifuge for 1–5 s to deposit the reaction at the bottom of the tube.
4. Incubate the reaction at the appropriate temperature (*see* **Note 5**) for 1–2 h.

3.2. Agarose Gel Electrophoresis

1. Place an appropriate amount of powdered, molecular-biology-grade agarose into a conical flask, add the desired volume of electrophoresis running buffer, and mix the contents of the flask by swirling (*see* **Note 13**).
2. Melt the agarose in a microwave, on a hot plate or by using a Bunsen burner, being careful to avoid overheating and boiling over of the agarose.

3. Cool the agarose mixture to a temperature of approx 55°C (*see* **Note 14**) and pour into the assembled gel support. Insert the comb and allow the gel to solidify.
4. Carefully remove the dams from the apparatus and place the gel on its support into the electrophoresis apparatus. Fill with sufficient electrophoresis buffer to cover the wells, and remove the comb.
5. Add 5X sample loading buffer, to a final concentration of 1X, to the completed enzyme digestion reactions.
6. Load the samples in separate wells; include a DNA size marker in one of the lanes (*see* **Note 15**). Load all samples quickly to prevent diffusion into the gel.
7. Close the lid to the apparatus and apply the current (*see* **Note 16**).
8. After electrophoresis, soak the gel in staining solution for 20–30 min (*see* **Notes 8** and **17**).
9. View the digestion products using a UV transilluminator and photograph if desired.
10. Using the molecular-weight standards, determine the relative sizes of the fragments resulting from the digest and compare to those expected based on the map(s) of the subclone to determine the presence and/or orientation of the insert.

4. Notes

1. A DNA fragment of a size between 100 and 10,000 base pairs is ideal. DNA fragments that are larger or smaller than this will migrate in the gel, but may blur or fail to resolve *(1,4,5)*. If trying to resolve and analyze fragments larger than 5 kb, a gel that is longer than a mini-gel is required.
2. Certain restriction enzymes cannot recognize their cleavage sites if the DNA is methylated. Be sure to consult the vendor catalog for such information for the enzymes selected. Most *E. coli* strains contain two DNA methylation systems (*dam* and *dcm*); however, various strains of *dam⁻* and *dcm⁻ E. coli* are commercially available (*see* Chapter 3).
3. Restriction enzymes are heat labile and should be kept at −20°C until use. Briefly centrifuge (at 4°C if possible) tubes of enzymes prior to opening for the first time. Store all enzymes at −20°C in a non-frost-free freezer. Keep enzymes on ice while in use. Use only clean, sterile tips when pipetting enzymes.
4. Suppliers of restriction enzymes provide standard restriction enzyme buffers. Each enzyme will perform equally in the designated buffers from different companies.
5. Refer to the manufacturer's catalog to be sure of the correct temperature for the reaction. 37°C is the incubation temperature for most restriction enzymes, but there are a number of exceptions. Using a water bath or heat block is preferable to using an open-air incubator when using short incubation times, as the temperature remains constant throughout the incubation.
6. Both TAE and TBE are low-ionic-strength buffers and are suitable for both preparative and analytical agarose gels *(1,4)*. TAE was the preferred buffer for preparative gels when DNA was purified using glass beads, but methods now accommodate either buffer. Use of TBE is often preferred because of its significantly higher buffering capacity and the sharper resolution of DNA bands. Concentrated solutions of TBE should be stored at room temperature and discarded if a precipitate forms.
7. There are a number of commercially available molecular-weight standards. The 1-kb and 100-bp ladders are very useful although moderately expensive. There are also a number of commercially available DNA samples for which the sizes of the DNA fragments resulting from digestion are known (e.g., λ DNA digested with *Hin*dIII and ϕX174 DNA digested with *Hae*III) *(5)*. These can be purchased as predigested samples or as

DNA stocks that are digested in the laboratory. These DNA stocks are less expensive than the predigested DNAs and ladders.

8. Handle ethidium bromide solutions with extreme care, as ethidium bromide is both a carcinogen and a mutagen. Alternative DNA stains include GelStar nucleic acid gel stain (BioWhittaker Molecular Applications) and methylene blue. Advantages of GelStar include increased sensitivity (4- to 16-fold more sensitive than ethidium bromide), no need for its inclusion in the electrophoresis buffer, and lower background staining than ethidium bromide. GelStar is provided as a 10,000-fold concentrate. GelStar is potentially mutagenic and should be handled with caution. A safe alternative is methylene blue, however, it is 40 times less sensitive than ethidium bromide (the use of 40 ng of DNA/band is recommended). Methylene blue staining is accomplished by soaking agarose gels in a solution of 0.025% methylene blue in distilled water, followed by destaining in distilled water.

9. For digests using multiple enzymes in which the enzymes require different levels of salt in their buffers, perform the reaction first with the enzyme requiring the lowest amount of salt for 1 h. Following the first digest, adjust the salt concentration of the reaction by adding a small volume of concentrated salt solution and the second enzyme and incubate for an additional hour. Universal buffers have been developed using potassium acetate (e.g., One-Phor-All by Amersham Pharmacia Biotech, which we recommend) that make digests with multiple enzymes easier.

10. The smaller the fragment to be detected, the more DNA may need to be used in the reaction to ensure visualization after agarose gel electrophoresis.

11. A unit of enzyme is the amount required to digest 1 µg of a standard DNA sample (typically λ DNA) in 1 h *(1,3,5)*. The number of enzyme sites in the clone to be tested will vary; however, using 5–10 U of enzyme to digest 1 µg of DNA for 1–2 h is a good starting point for most enzymes, assuming that the number of sites expected in the clone is low.

12. Do not vortex reactions containing restriction enzymes, as they may be inactivated by agitation.

13. Gels of an agarose concentration between 0.6% and 1.5% are able to separate double-stranded linear DNA fragments within molecular-weight-ranges of 2–10 kb (0.6% gel) and 0.2–3 kb (1.5% gel) *(1,4,5)*. The concentration of agarose is inversely proportional to the range of molecular weights that they can resolve. A typical percentage of 0.8% of agarose is used for most purposes. Gels with agarose concentrations lower than 0.5% are weak and difficult to handle and should be run in a cold room. Gels with agarose percentages higher than 1.5% become brittle and cannot be cooled to 55°C before pouring. For gels over 1.5%, sieving agaroses (such as New Sieve GTG agarose, FMC BioProducts) are favored, although the resolution for small fragments is not as clear as polyacrylamide electrophoresis (5).

14. Melted agarose can be stored in a 55°C water bath for several days if a conical flask with a screw top is used to prevent evaporation.

15. It is advisable to load a sample of undigested DNA of the clone being analyzed. This is particularly helpful for investigators who are not experienced at analyzing DNA run on agarose gels. It is also advisable to include a control comprising of a digestion of the parental plasmid vector (without insert).

16. For an analytical mini-gel, optimal resolution is obtained at 10 V/cm (referring to the distance between the electrodes). For the resolution of fragments larger than 5 kb, run the gel at 5 V/cm *(1,4,5)*.

17. Ethidium bromide can be included in both the gel (once cooled to 55°C) and electrophoresis buffer (0.5 µg/mL) to avoid the need to stain the gel after electrophoresis. This

allows for faster viewing of the gel on a UV transilluminator. There are several disadvantages to this approach. During electrophoresis, the ethidium bromide migrates in the opposite direction to the DNA. Extended electrophoresis thus removes the stain from the gel and leaves the smaller fragments unstained. Additionally, many investigators report that DNA bands are sharper if ethidium bromide is not included in the gel. A third drawback is that inclusion of ethidium bromide in the gel and buffer results in contamination of the apparatus and a larger amount of ethidium bromide-contaminated waste that must be disposed.

References

1. Sambrook, J., Fritsch, E. F., and Maniatis, T. (1989) *Molecular Cloning: A Laboratory Manual*, 2nd ed., Cold Spring Harbor Laboratory, Cold Spring Harbor, NY.
2. Davis, L. Kuehl, M., and Battey, J. (1994) *Basic Methods in Molecular Biology*, 2nd ed., Appleton & Lange, Norwalk, CT.
3. Smith, D. R. (1996) Restriction endonuclease digestion of DNA, in *Basic DNA and RNA Protocols* (Harwood, A.J., ed.), Humana, Totowa, NJ, pp. 11–15.
4. Sealey, P. G. and Southern, E. M. (1982) Electrophoresis of DNA, in *Gel Electrophoresis of Nucleic Acids: A Practical Approach* (Rickwood, D. and Hames, B. D., eds.), IRL, Oxford, pp. 39–76.
5. Brown, T. A., Ikemura, T., McClelland, M., et al. (1991) *Molecular Biology Labfax* (Brown, T. A., ed.), BIOS Scientific, Oxford.
6. Field, S. J. (1993) Multiplexed minipreps for rapid screening of large numbers of recombinant DNA clones. *BioTechniques* **14,** 530–531.

21

Screening Recombinant DNA Libraries

Wade A. Nichols

1. Introduction

A recombinant DNA library typically represents part or all of an organism's genomic DNA or mRNA (represented as cDNA) cloned into vectors and stored as a collection of thousands of transformants. The construction of a complete library is only half the task; researchers then need to be able to identify the small number of clones bearing the DNA fragment of interest among the numerous transformants within the library. This process is called "screening a library" and it is the molecular equivalent of finding a needle in a haystack. Libraries may be screened by any one of several methods.

1.1. Screening Methods

1.1.1. Phenotypic Screening

In a small number of cases, a cloned fragment of DNA will possess an intact gene that encodes a protein of discernable function. Some examples are genes encoding pigments, secreted enzymes, or assayable metabolic functions. In these cases, it is possible to screen libraries by assaying the expression of these traits. The specific screening method varies according to the properties encoded by the cloned DNA. For example, it may be accomplished by the direct observation of colonies on standard or special indicator media, or by the selection of bacteria on nutrient-limited media. Cases such as these are uncommon and, therefore, phenotypic screening is very limited in its applications.

1.1.2. Immunodetection

Immunodetection relies on the ability of purified antibodies to recognize structures, expressed by the desired fragment of DNA, on the surface of the bacteria. To use this strategy, the protein of interest must be purified, and an antibody that recognizes it must be prepared. If a transformant harboring the target gene, such as an outer-membrane protein, expresses the protein on its surface, it can be identified using the antibodies. This relies on the gene being expressed—the protein being transported to the outer membrane of the host cell and folding to form the epitope that is recognized by the antibody.

This procedure also requires that the host cell does not express a molecule that crossreacts with the antibodies. Although this list of caveats sounds daunting, immunodetection is a powerful tool for screening libraries. However, because the uses for the technique are somewhat limited, it will not be further addressed in this chapter.

1.1.3. Hybridization Screening

Hybridization screening is the most commonly used means of screening recombinant libraries. The technique uses a nucleic acid probe to detect transformants possessing DNA sequences similar to that of the probe. Hybridization screening is an exquisitely sensitive method, requiring only that researchers have a fragment of DNA that will serve as a useful probe. Hybridization screening gets its name because a hybrid DNA molecule is formed when the probe anneals to complementary sequences located on recombinant plasmids within the library. The hybrid structure possesses one strand of DNA from the recombinant plasmid and a partial complementary strand derived from the probe. In many respects, hybridization screening resembles Southern blotting *(1)* because both utilize hybridization of labeled probes to target DNA fixed to a solid matrix. The main advantage of hybridization screening is its sensitivity that allows the detection of very small amounts of DNA that share sequence similarity to the probe. This permits the screening of recombinant clones directly from small colonies of host cells and eliminates the need to first isolate the recombinant plasmids from the bacterial host. This chapter will focus on the conceptual and technical aspects of hybridization screening of recombinant libraries. A flowchart depicting the steps involved is shown in **Fig. 1**.

1.2. Recombinant Libraries

In the past, it was usual to work with pooled libraries in which thousands of library clones were combined into one tube. In recent years, it has become common to separate clones into "discrete" libraries. In such libraries, 96- or 384-well plates are used to store and propagate the library and each well contains a single clone bearing a unique recombinant plasmid. Although construction of a discrete library initially requires more work to separate the individual clones, it allows for the rapid identification and culture of the desired clone. The protocols described in this chapter have been designed for use with a discrete library stored in 96- or 384-well plates.

1.3. Overview of Method

1.3.1. Preparing the Membrane

The first step in hybridization screening is the transfer of bacterial cells bearing the recombinant plasmids to a nitrocellulose membrane. Nitrocellulose membranes possess pores through which nutrients may pass, thus allowing growth of bacterial colonies directly on membranes placed on nutrient agar.

Transferring a library to a nitrocellulose membrane is a simple process and requires only a few steps. First, the membrane is prewetted on the surface of a Luria–Bertani (LB) agar plate and small amounts of bacterial suspensions are removed from a multiwell plate containing the library and transferred onto the membrane. Bacteria serv-

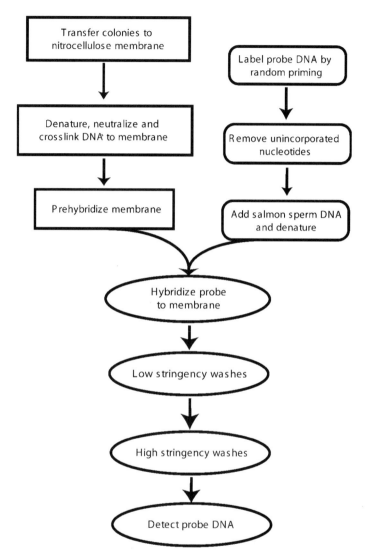

Fig. 1. A flowchart showing the steps for hybridization screening of recombinant libraries.

ing as a negative control are also spotted onto the membrane. The most appropriate negative control is the host bacterial strain bearing the cloning vector used in construction of the library. Once transferred to the membrane/agar support, the plates are incubated overnight at 37°C to allow the growth of bacterial colonies. The following day, the membrane should be covered with bacterial colonies that resemble the pattern of the wells in the multiwell plate. Before proceeding further, a small amount (approx 10 ng) of unlabeled probe DNA is spotted onto the membrane to act as a positive control. **Figure 2** shows what a typical membrane would look like at this point in the procedure.

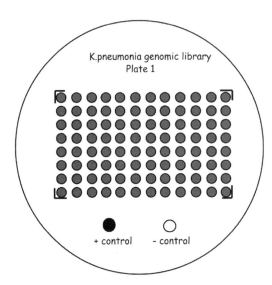

Fig. 2. A graphic representation of a prepared membrane. Note the labeling of library information and of the positions of the corner wells and controls.

Double-stranded plasmid DNA contained within bacterial cells cannot be directly detected by hybridization and must be denatured and directly linked to the nitrocellulose membrane. Exposing the membrane to NaOH results in the lysis of bacterial cells and the denaturation of plasmid DNA into single strands. The membrane is then neutralized with a buffered solution. The DNA is crosslinked to the membrane either by baking or by exposure to ultraviolet (UV) light. At this point, the plasmid DNA is single stranded and tightly bound to the membrane surface in such a way that the bases are able to anneal to complementary probe DNA sequences.

1.3.2. Labeling the Probe

The DNA used as a probe should bear strong sequence similarity to the desired insert DNA so that hybridization between complementary sequences occurs. The most commonly used probes are based on either (1) polymerase chain reaction (PCR) products corresponding to a portion of the desired insert, (2) a homologous fragment of DNA obtained from a closely related organism, or (3) a synthetic degenerate oligonucleotide that has been designed from an elucidated amino acid sequence.

Feinberg and Vogelstein *(2)* first described the production of radiolabeled probes by random oligonucleotide priming in 1983. This protocol uses the large fragment of the *Escherichia coli* DNA polymerase I, also called the Klenow fragment, to synthesize DNA containing radiolabeled nucleotides. The protocol generally utilizes a mixture of random hexanucleotide primers that anneal to complementary sequences on denatured probe DNA that acts as a template. The probe/hexanuclotide duplex is recognized by Klenow fragment that initiates synthesis of a complementary strand of DNA. In the reaction mix, one of the dNTPs (typically dATP or dCTP) is replaced by a radiolabeled analog. The Klenow fragment incorporates the radiolabeled dNTP into the growing strand of DNA, resulting in a radiolabeled DNA fragment. The use of random hexanucleotides to initiate DNA synthesis produces a series of overlapping radiolabeled probes that anneal to overlapping sequences on the target DNA. When using double-stranded template DNA, random priming will produce radiolabeled DNA fragments corresponding to both strands of the template. When using a single-stranded template, DNA random priming will produce a radiolabeled probe that is complementary to the template strand. For this reason, a double-stranded template DNA is best for screening double-stranded target DNA, as the labeled probe can anneal to either strand of target DNA. Following the labeling reaction, unincorporated nucleotides are removed by use of a spin column. Finally, labeled probe is combined with salmon sperm DNA, which decreases the amount of nonspecific binding of the labeled probe to target DNAs, and denatured to produce single-stranded DNA that can anneal to the target sequences.

1.3.2.1. NONRADIOACTIVE ALTERNATIVES

In the last 15 yr, nonradioactive methods of hybridization screening have become popular, as they avoid the safety issues associated with the use of radioisotopes. Initially, the sensitivity of such methods was much lower than those produced by radioactive techniques, however, this is no longer the case and nonradioactive hybridization screening is an excellent alternative to methods requiring radioactive isotopes. Most nonradioactive methods utilize probes that possess nucleotides tagged with the molecule digoxigenin (DIG). Detection of a DIG-labeled probe is ultimately accomplished by immunodetection using anti-DIG antibody Fab fragments. Nonradioactive hybridization screening with DIG-labeled probes is complex and requires that all of the reagents be optimized for use together. For this reason, it is strongly recommended that anyone interested in performing nonradioactive screening buy a prepared kit from a commercial vendor such as those sold by Roche or Fermentas.

1.3.3. Prehybridization

The next step in the protocol is called prehybridization. This step involves the wetting of the membrane with the hybridization solution in the absence of a labeled probe. Prehybridization usually takes only 1 h and it allows the membrane to reach the optimal hybridization condition before the introduction of the probe.

1.3.4. Hybridization

Hybridization is the step in which the labeled probe anneals to complementary sequences in target DNA. For this to occur, the hybridization solution must contain components that facilitate precise hybridization of complementary sequences at relatively low temperatures (3). Dextran sulfate is added to the hybridization buffer to increase the frequency of specific annealing between probe and target, and the inclusion of formamide decreases the effective melting temperature of the probe allowing the hybridization to occur at reduced temperatures. To assure maximum annealing, hybridization reactions are usually performed for 16–24 h at 42°C.

1.3.5. Stringency Washes

During hybridization, the labeled probe anneals to complementary sequences on target DNA. However, some probe DNA will also anneal weakly to target DNA with which it shares low levels of homology. Additionally, some labeled probe will weakly adhere to the nitrocellulose membrane. It is critical to selectively remove the probe adhering to the membrane and to low-homology target DNAs while retaining the probe that is annealed to the homologous target sequences. This is achieved with a series of stringency washes.

Low-stringency washes are performed at room temperature to remove labeled probe that is bound directly to the nitrocellulose membrane. Unlike the target DNA, the probe was not crosslinked to the membrane and, thus, it is possible to remove probe bound directly to the membrane without disrupting the crosslinked target DNA. High-stringency washes are performed at an elevated temperature to facilitate the removal of probe that is weakly annealed to low-homology sequences in target DNA. Careful selection of an appropriate temperature at which to perform high-stringency washes is very important (*see* **Note 1**). An additional consideration for stringency washes is the salt concentration of the wash solution. High salt concentrations in the wash solution stabilize the probe/target hybrid and, therefore, there exists an inverse relationship between salt concentration and stringency. Following the completion of high-stringency washes, there should be little probe annealed to low-homology sequences.

1.3.6. Detection

The final step in hybridization screening is the detection of candidate clones by autoradiography of the membrane. The decay of the radioactive isotope causes dark spots on the X-ray film when it is developed, indicating the presence and location of library clones that contain recombinant plasmids that are homologous to the probe. It is then possible to use a template of a multiwell plate to determine the wells in which the candidate clones are stored.

2. Materials
2.1. Preparing the Membrane

1. Recombinant library stored in multiwell plates.
2. LB agar: 10 g/L tryptone, 10 g/L yeast extract, 5 g/L NaCl, and 15 g/L bacto-agar. Sterilize by autoclaving.
3. 150-mm-diameter LB agar plates containing appropriate antibiotics (*see* **Note 2**).
4. Nitrocellulose membrane (*see* **Note 3**).
5. Pin replicator or multichannel pipettor.
6. India ink pencil (available at art stores).
7. Four pieces of filter paper cut slightly larger than nitrocellulose membrane.
8. Denaturing solution: 0.5 M NaOH.
9. Neutralizing solution: 1 M Tris-HCl (pH 7.5).
10. Rinse solution: 0.5 M Tris-HCl (pH 7.5), 1.5 M NaCl.
11. Vacuum oven or UV crosslinker.

2.2. Labeling the Probe

1. Sterile distilled water.
2. 100 ng Gel-purified probe DNA (see **Note 4**).
3. Random hexanucleotide mix (approx 1.0 mg/mL).
4. 5 U/µL of Klenow fragment.
5. 10X Klenow reaction buffer: 0.5 M Tris-HCl (pH 7.5), 100 mM MgCl$_2$, 10 mM dithiothreitol (DTT), 0.5 mg/mL bovine serum albumin (BSA).
6. 10 mCi/mL [α-^{32}P] dCTP (specific activity 3000 Ci/mmol) (see **Note 5**).
7. dNTP mix: 5 mM with respect to dATP, dGTP, and dTTP.
8. Bio-spin nucleotide removal column (Bio-Rad) or similar.

2.3. Prehybridization

1. Formamide.
2. 20X SSC: 3 M NaCl, 0.3 M sodium citrate. Adjust the pH to 7.0 with HCl.
3. 100X Denhardt's solution: Combine 2.0 g of Ficoll 400, 2.0 g of polyvinyl pyrrolidone, and 2.0 g of BSA in 10 mL of water.
4. 10% (w/v) sodium dodecyl sulfate (SDS).
5. 50% (w/v) Dextran sulfate.
6. 2 M Tris-HCl (pH 7.5).
7. Hybridization solution: 5 mL water, 48 mL formamide, 24 mL 20X SSC, 1 mL 2 M Tris-HCl (pH 7.5), 1 mL 100X Denhardt's solution, 20 mL 50% dextran sulfate solution, 1 mL 10% SDS. Note that the SDS must be added as the final reagent.
8. Hybridization bottle or sealable plastic bags.

2.4. Hybridization

1. Labeled probe (prepared as in **Subheading 3.3.**).
2. 2.5 mg/mL Sonicated salmon sperm DNA: Add 75 mg of salmon sperm DNA to 30 mL of water. Sonicate on ice for approximately 40 min (pulse for 1 min, rest for 1 min). Continue until the solution can be easily pipetted.

2.5. Stringency Washes

1. 1 L Wash solution 1: 2X SSC (see **Subheading 2.3., item 2**), 0.1% SDS.
2. 1 L Wash solution 2: 0.2X SSC, 0.1 % SDS (see **Note 6**).

2.6. Detection

1. X-ray film (film specifically made for autoradiography is available from vendors).
2. Cardboard cut to the same size and shape as the X-ray film.
3. Film cassette.
4. X-ray developer.
5. Clean multiwell plate.

3. Methods

3.1. Preparing the Membrane

1. Remove a multiwell plate containing the library from the freezer and allow the bacterial suspensions to thaw on ice.
2. Label the edge of a nitrocellulose membrane with the specific information about the library (plate number, date, etc.) using India ink. Be sure to not touch the membrane with bare hands.

3. Using forceps, carefully place the nitrocellulose membrane on top of an LB agar plate. Moisture on the agar should wet the membrane as it is laid on the surface of the agar.
4. Using a pin replicator or multichannel pipettor, carefully transfer 1 µL of each of the bacterial suspensions to the nitrocellulose membrane. If using a multichannel pipet, try to retain the spacing between rows when transferring samples to the membrane (*see* **Note 7**).
5. Pipet a small amount of negative control suspension near one end of the membrane and mark its position with India ink.
6. Replace the lid of the Petri dish and incubate, upside down, at 37°C overnight.
7. The next day, examine the membrane for bacterial growth. Bacterial colonies should be present in a pattern that mimics the positions of wells in the multiwell plate.
8. Carefully mark, with India ink, the four corners of the grid that has been formed by the bacterial colonies. This will show the position of the grid on the membrane when the cells are no longer visible and will be important for the identification of candidate clones during the detection stage.
9. Transfer 10 ng of unlabeled probe DNA to the membrane to serve as a positive control. Mark the location of the control with India ink (*see* **Fig. 2**).
10. Using forceps, carefully transfer the membrane (bacteria side up) onto filter paper that has been saturated with denaturing solution (*see* **Note 8**).
11. Using forceps, carefully transfer the membrane (bacteria side up) on to filter paper that has been saturated with neutralizing solution. Allow the membrane to sit on the filter paper for 10 min.
12. Using forceps, carefully transfer the membrane (bacteria side up) on to filter paper that has been saturated with rinse solution. Allow the membrane to sit on the filter paper for 5 min.
13. Crosslink DNA to the membrane by drying in a vacuum oven at 80°C for 2 h, or in a UV crosslinking chamber, following the manufacturer's instructions. If necessary, the membrane may be transferred to a plastic bag and stored for several days following this step.

3.2. Labeling the Probe

1. Combine the following in a microcentrifuge tube:

    2.5 µL dNTP mix
    2.5 µL 10X Klenow reaction buffer
    5.0 µL [α-^{32}P] dCTP
    1.0 µL Klenow fragment (5 U)

    Store on ice until needed
2. Combine the following in a screw-cap microcentrifuge tube:

    100 ng Probe DNA
    1 µg Hexanucleotide mix
    Water to a final volume of 14 µL.

    Seal the lid tightly and place in a boiling water bath for 5 min. Chill briefly on ice.
3. Add the reaction mix from **step 1** to the mix from **step 2**. Incubate at room temperature for 2–18 h.
4. Remove unincorporated nucleotides by use of a spin column, according to the manufacturer's instructions (*see* **Note 9**).
5. Store the labeled probe at –20°C until needed.

3.3. Prehybridization

1. Wearing gloves, carefully transfer the membrane to a hybridization bag or bottle (*see* **Note 10**).
2. Slowly add sufficient hybridization solution to cover the membrane.
3. Prehybridize with agitation at 42°C for 1 h (*see* **Note 11**).

3.4. Hybridization

1. While the membrane is prehybridizing, remove the labeled probe from the freezer and allow it to thaw on ice.
2. Combine 1 mL of sonicated salmon sperm DNA and the entire labeled probe in a 15-mL screw-cap tube.
3. Denature in a boiling water bath for 10 min. Chill briefly on ice.
4. Add the entire mixture to the membrane/hybridization solution. Agitate gently to ensure that the probe is evenly distributed in the solution.
5. Reseal the hybridization bag or bottle and hybridize for 12–16 h at 42°C.

3.5. Stringency Washes

1. Remove the hybridization solution from the bag or bottle (*see* **Note 12**).
2. Add 100 mL of wash solution 1 to the bag or bottle containing the membrane. Incubate at room temperature for 20 min. Discard the solution and perform three more washes.
3. Add 100 mL of wash solution 2 to the bag or bottle containing the membrane. Incubate at the appropriate wash temperature for 20 min (*see* **Notes 1**, **6**, and **13**). Discard the wash solution and perform three more washes.

3.6. Detection

1. Carefully wrap the membrane in plastic wrap. Tape the membrane to a cardboard support, with the bacteria side facing up. Place the cardboard with membrane in a film cassette of the appropriate size (*see* **Note 14**). The use of intensifying screens in the cassette will decrease the exposure time.
2. In a dark room, remove a piece of X-ray film from the box and place on top of the membrane. Mark the top right corner of the film either by cutting this corner or by marking it with a permanent marker. Close the film cassette and expose the film to the membrane for 1 h.
3. In a dark room, remove the X-ray film from the cassette and develop in an X-ray film developer (alternatively, the film may be developed by hand).
4. Place the developed film over the membrane inside the film cassette. Use the cassette to properly align the film to the cardboard support and the membrane. Use a permanent marker to mark the outline of the membrane, the position of the four corners of the grid, and the position of the positive and negative control on the X-ray film (marked with India ink on the membrane).
5. Examine the X-ray film to make sure that there is a dark spot at the position of the positive control but no dark spot at the position of the negative control. If either of these controls did not perform as expected, it is recommended that the experiment be repeated with a new probe.
6. If any dark spots are present within the limits of the four corners of the grid, the position of the candidate clone within the multiwell plate must be determined. Place the film over a multiwell plate (or a template of a multiwell plate) and align the marks on the film that denote the four corner wells with the corner wells of the plate. Examine the position of any dark spots and record the address on the plate (e.g., A7 or D4). The bacteria in those wells of the library plate contain candidate clones.

7. Potential problems include the following: no signal from any of the library clones (*see* **Note 15**); clusters of positive signals (*see* **Note 16**); a positive signal for all of the clones (*see* **Note 17**); and spotting, indicated by a developed film that is covered with numerous dark spots that are smaller than bacterial colonies on the membrane (*see* **Note 18**).
8. Candidate clones may be screened by a secondary hybridization. To perform the second screen, pipet 1 µL of each candidate clone onto a nitrocellulose membrane on the surface of a new LB agar plate. Repeat the entire procedure. A clone that gives a signal in this second screen is a strong candidate for containing the DNA of interest. Such a clone may be analyzed using standard techniques described in this volume, such as restriction analysis (*see* Chapter 20) and sequencing (*see* Chapter 22) of purified plasmid DNA.

4. Notes

1. As a rule of thumb, the optimal wash temperature can be calculated by the following equation:

$$T_s = 70 - (100 - h)$$

where T_s is the calculated wash temperature (°C) and h is the percent identity of the probe and target DNA. For example, if the probe has 80% identity to the target DNA, the wash temperature would be $T_s = 70 - (100 - 80)$ or $T_s = 70 - 20 = 50°C$. Determining the extent of sequence identity shared by the probe and target DNA can be difficult. For example, a PCR product amplified from genomic DNA will be identical to target DNA within the library if both PCR template DNA and the library DNA were derived from the same organism. Thus, a probe generated from the PCR product template will have 100% identity to the target DNA in the library. However, if the template DNA used to generate the probe and the library DNA are derived from related organisms, then the approximate degree of identity between known gene homologs that are shared between the two organisms should be used. Thus, if gene homologs from two species of bacteria typically demonstrate 82–85% sequence homology, then assume that the probe will bear a similar degree of homology to the target. Finally, synthetic oligonucleotide probes designed from amino acid sequences can vary from 65% to 80% identity, as the degree of identity is affected by the accurate prediction of codon usage within a gene.
2. The larger 150-mm Petri dishes plates are needed to accommodate the size of 96- or 384-well plates.
3. Nitrocellulose membranes are available that have been precut to the size and shape of 150-mm Petri dishes. Although nylon membranes are often used in Southern blotting techniques, the use of nylon is not recommended in hybridization screening of libraries. Cell debris from the bacterial cells adheres to nylon resulting in an increase of background signal on the blot.
4. Ideally, the probe DNA should be at least 200 bp in length. It is essential that no contaminating DNA be present in the probe template because the contaminant DNA will also be labeled. The use of a contaminated probe results in additional work for the researcher, as there is no way to determine if a library clone hybridized to the desired probe or contaminant DNA. It is particularly important that probe DNA does not bear sequence similarity to the cloning vector used in the construction of the library. Even a small amount of a vector-based sequence in a probe will result in the probe hybridizing to every library clone. To avoid possible probe contamination, it is recommended that probe DNA, obtained by restriction endonuclease cleavage or PCR, be purified by agarose gel electrophoresis followed by extraction from the gel (*see* Chapter 16).

5. All researchers working with isotopes should be aware of proper institutional guidelines for purchasing, storing, using, and discarding radioactive material. All researchers working with isotopes should always wear proper safety equipment, including gloves, lab coats, face shields, film badges, and so forth.
6. An inverse relationship exists between salt concentration in wash solutions and the effective stringency of the wash performed with the wash solution. Therefore, stringency may be adjusted by varying the salt concentration of wash solutions. High-stringency washes are usually performed in the presence of 0.2X SSC; moderate washes in 1X SSC; and low-stringency washes are performed in the presence of 3–5X SSC (carried out at 55°C instead of 65°C).
7. The transfer of small volumes of bacterial suspension from the multiwell plate to the membrane is best accomplished by the use of a pin replicator specifically designed for this purpose. Pin replicators have metal pins projecting from a platform and the pins are arranged in such a way that each pin corresponds to the position of a single well on a multiwell plate. The end of each pin has been etched, allowing a small volume of bacterial suspension to adhere to the pin. If a pin replicator is not available, a multichannel pipettor may be used for libraries stored in 96-well plates. It is extremely difficult to properly transfer and align bacterial suspensions from a 384-well plate with a multichannel pipettor and this practice is discouraged.
8. It is essential that the denatured DNA remain in the same position as the bacterial colony; therefore, the membrane must not be submerged in solutions during these steps.
9. Failure to remove unincorporated nucleotides will result in increased background signal on the blot, which makes accurate detection of hybridizing library clones more difficult.
10. Hybridization reactions are best performed in glass bottles within a hybridization oven; however, sealable plastic bags or watertight plastic containers in a shaking water bath or incubator can also be used.
11. Formamide is combustible at low temperatures and should never be used at temperatures exceeding 50°C.
12. It is important that the nitrocellulose membrane not be allowed to dry between hybridization and stringency washes, as drying of the membrane causes irreversible crosslinking of the probe to the membrane and results in extremely high background levels.
13. If using a probe <200 bp in length, it is necessary to reduce the temperature of the stringency washes. It is recommended that initial washes are performed at 27°C and then detection performed to determine if a dark spot, corresponding to the negative control, is observed. If there is a dark spot corresponding to the negative control, the membrane is rewashed at 32°C and redeveloped. Continue increasing the high-stringency wash temperature until no signal is detected from the negative control.
14. It is important that the membrane not be allowed to move while in the cassette. For this reason, it is strongly recommended that the membrane be taped to a piece of cardboard that has been cut to the same size and shape as a piece of X-ray film.
15. Unfortunately, candidate clones are not always detected by the first screening of a library. If the controls are working properly, this could indicate that the amount of library DNA from the colonies is insufficient for detection. In this situation, the X-ray film is left exposed to the membrane for a longer period of time than needed to expose the positive control in order to overexpose the film and see if any signals appear. When performing long exposures, it may be necessary to cover the area of the membrane containing the positive control with foil. This will reduce the intensity of the signal from the positive control.

16. False positives are typically seen when high-stringency washes have failed to properly remove the probe that is bound weakly to target DNA. If several positives are present in a cluster, it usually means that the membrane in that area was not washed thoroughly during the stringency washes. It is highly unlikely that a random library contains several adjacent clones with the same insert.
17. False positives can be caused by contaminated probe DNA, particularly if the contaminant DNA shares sequence identity with the cloning vector used in the construction of the library. If every library clone hybridizes to the probe, then it is likely that the probe is hybridizing to the vector sequences. If this is the case, a new probe should be designed.
18. Spotting is the result of radiolabeled DNA directly adhering to the membrane. Extreme spotting indicates failure to remove unincorporated labeled dATP following the labeling of the probe. Spin columns remove more of the unincorporated radiolabeled dATP than cleanup of the probe by ethanol precipitation. Less severe spotting suggests failure of the stringency washes to remove excess probe.

References

1. Southern, E. M. (1975) Detection of specific sequences among DNA fragments separated by gel electrohoresis. *J. Mol. Biol.* **98,** 503–517.
2. Feinberg, A. P. and Vogelstein, B. (1983) A technique for radiolabeling DNA restriction endonuclease fragments to high specific activity. *Anal. Biochem.* **132,** 6–13.
3. Denhardt, D. (1966) A membrane filter technique for the detection of complementary DNA. *Biochem. Biophys. Res. Commun.* **23,** 641–646.

22

Sequencing Using Fluorescent-Labeled Nucleotides

Allison F. Gillaspy

1. Introduction

The most widespread method used for DNA sequencing today is the Sanger dideoxy method that was first described in 1977 (1). This method takes advantage of the requirement for a free 3' hydroxyl group to form the necessary phosphodiester bridge between two nucleotides during DNA polymerization. In short, DNA sequencing reactions using this method are performed on a DNA template using a specific primer, DNA polymerase, and a mixture of deoxynucleotidetriphosphates (dNTPs) and dideoxynucleotides (ddNTPs) as a substrate. When a dNTP is incorporated into a growing DNA strand, chain elongation proceeds normally. However, when a ddNTP is incorporated by the polymerase, the process of chain elongation is aborted because of the absence of the 3' hydroxyl group on the nucleotide analog. The resulting DNA polymerization products are of varying lengths depending on when a ddNTP was incorporated into the growing strand. Theoretically, there should be a ddNTP incorporated at least once at each of the base positions for any particular DNA template. This pool of DNA fragments is then separated by size on high-resolution polyacrylamide gels. The resolution of these gels is so high that the difference in size resulting from a single nucleotide can be determined.

Although the overall strategy for DNA sequencing has not changed much in the past several years, the methods for separation, detection, and analysis of data have changed dramatically. Originally, the ddNTPs used for DNA sequencing were tagged with a radiolabel so that the sequencing products could be detected by autoradiography (1). Because each of the four ddNTPs was labeled with the same radionucleotide, four separate reactions had to be performed for each template to be sequenced. In addition to the obvious hazards to the individual performing these experiments, this method had other drawbacks. These include the fact that all labware involved was contaminated with radioactivity and that the reaction set up and data analysis were often extremely laborious.

Advancements in the field of DNA sequencing now make it possible to set up a sequencing reaction in a single tube. This is the result of the discovery that the four different ddNTPs could be labeled using chemically distinct fluorescent dyes *(2)*. The different dyes are designed so that they fluoresce at different wavelengths. In addition to ease of reaction setup, this new method also eliminated the danger related to the use of radioactive chemicals.

The advent of the fluorescent-based sequencing method required the development of new detection systems as well. When a fluorescent sequencing reaction is complete, the products are still separated by size on an acrylamide gel matrix. However, the products are allowed to run off the end of the gel and a detector records the fluorescent emission of each of the DNA fragments as they leave the gel *(3)*. The data are automatically recorded and transferred to a computer so that it can be analyzed by the use of sequencing analysis software.

Advancements in DNA sequencing technology continue to develop at an astonishing rate. Among the most recent is the development of capillary electrophoresis machines that can be used to run large numbers of samples simultaneously *(4)*. Although the principle of the procedure is the same, these new machines make it possible to increase the number of samples analyzed in 1 d into the thousands. This increased throughput has facilitated large-scale genome sequencing projects that have revolutionized many areas of molecular research.

2. Materials

2.1. Sequencing Reaction Setup

1. Purified plasmid (100–500 ng/μL), polymerase chain reaction (PCR) product (10 ng per 100 bp), or bacterial artificial chromosome (BAC) clone (200–600 ng/μL) (*see* **Note 1**).
2. Sequencing primer: 3.2 pmol/μL for plasmids and PCR products, or 12 μM for BAC clones (*see* **Note 2**).
3. Applied Biosystems Big Dye terminator ready reaction mix (www.appliedbiosystems.com). For alternatives, *see* **Note 3**.
4. 5X Sequencing reaction buffer: 400 mM Tris-HCl (pH 9.0), 10 mM MgCl$_2$. Aliquot and store at –20°C (*see* **Note 4**).
5. Deionized ultrafiltered (DIUF) water.
6. Thin-wall PCR tubes or 96-well PCR plate.

2.2. Ethanol-Precipitation Cleanup of Single Reactions

1. 3 M Sodium acetate (pH 5.6): Dissolve 204.15 g of sodium acetate in 250 mL of double-distilled water (ddH$_2$O). Adjust pH to 5.6 using glacial acetic acid. Add quantity sufficient (q.s.) ddH$_2$O to 500 mL. Sterilize by autoclaving and store at room temperature.
2. 95% Ethanol, chilled to –20°C.
3. 70% Ethanol, chilled to –20°C.
4. DIUF water.

2.3. Cleanup of Multiple Reactions

1. Sephadex G50-150 solution: 7 g Sephadex G50-150 in 100 mL DIUF water (*see* **Note 5**). Make up solution in a sterile container at least 24 h prior to use to ensure complete hydration of sephadex beads. Store for up to 3 wk at 4°C.

2. Millipore 96-well filter plate (0.65 µm hydrophilic, low protein binding) (www.millipore.com) and 96-well V-bottom polystyrene plate.

2.4. Pouring and Loading Sequencing Gels (see Notes 5 and 6)

1. 10X TBE solution: Dissolve 54 g Tris base, 28 g boric acid, 4 g ethylene diaminetetraacetic acid (EDTA) disodium salt in 350 mL DIUF water. Once chemicals are in solution, add enough DIUF water to q.s. solution to 500 mL. Verify that pH is between 8.2 and 8.3. Filter sterilize using a 0.4-µm bottle-top filter unit and store at room temperature for up to 1 mo.
2. Polyacrylamide gel mix such as Long Ranger (Applied Biosystems) or PAGE-PLUS (AMRESCO, www.amresco-inc.com). Store acrylamide solutions at 2–6°C for up to 1 mo.
3. Ultrapure urea.
4. 10% Ammonium persulfate (APS) solution: Dissolve 0.1 g of APS in 1 mL of DIUF water. Make the solution fresh each day of use. Store the APS compound in a desiccant at room temperature for up to 2 mo.
5. Tetramethylethylenediamine (TEMED). Store at room temperature for up to 2 mo.
6. Sequencing gel mix: Dissolve 18 g of ultrapure urea in 26 mL DIUF water by stirring gently at room temperature. Add 5 mL of 10X TBE buffer and 6 mL PAGE PLUS or 5 mL FMC Long Ranger acrylamide mix. Add enough DIUF water to q.s. solution to 50 mL. Filter sterilize through a 0.2-µm nylon bottle-top filter and degas for 15 min at room temperature. For optimum results, this solution should be made fresh on each day of use.
7. Blue dextran solution: Dissolve 500 mg blue dextran in 10 mL of DIUF water. Aliquot and store at 4°C.
8. Deionized formamide: Add 1 g of Amberlite 150L resin to 20 mL of ultrapure formamide. Cover and stir the mixture gently for 15–20 min. Filter through a 2-µm nylon filter unit. If the pH is not >7.0, then add 1 g of fresh resin and gently stir for an additional 15–20 min. Filter and check the pH. Store deionized formamide in small (100 µL) aliquots at –20°C for up to 6 mo. Avoid repeated freeze–thaw cycles by using a new tube of deionized formamide each day (*see* **Note 7**).
9. Loading dye for slab gel sequencing: Add 20 µL of blue dextran solution to 100 µL of deionized formamide and mix. Make this solution fresh daily.
10. Materials needed to pour a sequencing slab gel: 48- or 36-cm glass sequencing gel plates, six clamps, sequencing gel comb, Saran wrap, and low-lint Kimwipes®.

3. Methods
3.1. Sequencing Reaction Setup (see Note 8)
3.1.1. Reaction Setup for Slab Gel Sequencing of Plasmids and PCR Products

1. Add 1–5 µL of plasmid DNA or 10 ng per 100 bp PCR template DNA to a sterile, thin-wall PCR tube or a 96-well PCR plate. The volume should not exceed 6 µL.
2. Add 1 µL of sequencing primer and mix gently.
3. Add 2 µL of 5X sequencing reaction buffer and mix gently.
4. Add 1 µL of ABI Big Dye terminator ready reaction and mix gently.
5. Add sufficient ultrapure water to bring reaction to a final volume of 10 µL.
6. Mix sample gently, cover tube, and keep on ice until ready for cycling in the thermal cycler.

3.1.2. Reaction Setup for Slab Gel Sequencing of BAC Clones

1. Add 5 µL of BAC clone template DNA to a sterile PCR tube.
2. Add 1 µL of sequencing primer.

3. Add 2 μL of 5X sequencing reaction buffer and mix gently.
4. Add 2 μL of ABI Big Dye terminator ready reaction mix.
5. Gently mix the sample, cover the tube, and place on ice until all samples are ready to be placed in the thermal cycler.

3.1.3. Reaction Setup for Capillary Sequencing of Plasmids and PCR Products

1. Add 500–1000 ng of DNA template (in a maximum volume of 3 μL) to a 96-well reaction plate.
2. Add 1 μL of sequencing primer to each sample and mix.
3. Add 1 μL of ABI Big Dye terminator ready reaction to the sample and mix.
4. Add sufficient ultrapure water to each sample to q.s. the reaction to 5 μL.
5. Mix gently, cover, and place on ice until all samples are ready to be placed in the thermal cycler.

3.2. Reaction Conditions

3.2.1. Cycling Conditions for Slab Gel Reactions

1. If the thermal cycler does not have a heated lid, place a single drop of mineral oil over the top of each reaction to avoid evaporation.
2. Denature samples for 5 min at 95°C followed by 60 cycles of 95°C for 30 s, 50°C for 20 s, and 60°C for 4 min. For sequencing of BACs, increase the number of cycles to 100. Program the thermal cycler to hold at 4°C indefinitely after cycling if samples are to be processed at a later time. This cold-temperature hold is sufficient for storage of samples for several hours with little or no adverse effects on signal intensity.

3.2.2. Cycling Conditions for Capillary Sequencing Reactions

1. If the thermal cycler does not have a heated lid, place a single drop of mineral oil over the top of the reaction to avoid evaporation.
2. Denature samples for 2 min at 95°C followed by 60 cycles at 95°C for 30 s, 50°C for 30 s, and 60°C for 4 min. Complete the reaction with a final extension step at 60°C for 10 min. Program the thermal cycler to hold at 4°C indefinitely after cycling if samples are to be processed at a later time. This cold-temperature hold is sufficient for storage of samples for several hours with little or no adverse effects on signal intensity.

3.3. Cleaning Up Sequencing Reactions (see Note 9)

3.3.1. Ethanol-Precipitation Cleanup of Single Reactions

1. Add DIUF water to the completed sequencing reaction to adjust the volume to 20 μL.
2. Add 2 μL of 3 *M* sodium acetate (pH 5.6) and mix well.
3. Add 50 μL of ice-cold 95% ethanol. Invert the tube several times to mix.
4. Incubate the sample on ice for 10 min to precipitate the DNA.
5. Centrifuge the sample at 4°C, 10,000*g* for 30 min to pellet the DNA.
6. Decant the supernatant and wash the pellet by adding 250 μL of ice-cold 70% ethanol, being careful not to dislodge the pellet from the side of the tube.
7. Centrifuge the sample at 4°C, 10,000*g* for 10 min and decant the supernatant.
8. Air-dry the pellet for 10–20 min or dry in a Speed vac, at low setting, for 10 min.
9. Samples that have been dried in this manner can be stored at –20°C for up to 2 wk without any adverse effects (*see* **Note 10**).

DNA Sequencing

3.3.2. Cleanup of Multiple Reactions

1. Tape a Millipore 96-well filter plate firmly on top of a 96-well V-bottom plate.
2. Dispense 250 µL of hydrated Sephadex G50-150 to each well of the 96-well filter plate.
3. Centrifuge in a swinging bucket rotor for 3 min at 500g.
4. Remove the bottom plate, discard the flow through, and then reattach the V-bottom collection plate to the filter plate that contains the Sephadex.
5. Add an additional 250 µL of Sephadex G50-150 to each well.
6. Centrifuge in a swinging bucket rotor for 3 min at 500g.
7. Discard the flow through and attach the filter plate to a new V-bottom collection plate. The filter plate can be used immediately or made 3–4 h ahead of time and stored covered at 4°C.
8. Remove the plate of sequencing reactions from the thermal cycler and spin briefly to ensure the condensation is off of the sides of the wells.
9. Add sufficient DIUF water to each well to q.s. solution to 15 µL total volume. This addition of water is to ensure optimum sample recovery from the Sephadex. Sample recovery is increased when volumes from 10 to 15 µL are placed on the column.
10. Transfer the entire sequencing reaction to the appropriate well of the filter plate.
11. Centrifuge the filter plate in a swinging bucket rotor for 5 min at 500g.
12. Remove the 96-well V-bottom collection plate that now contains the cleaned-up samples and place in a vacuum oven that has been preheated to 60°C to dry the samples completely (approx 30–45 min) (*see* **Notes 10** and **11**).

3.4. Pouring and Loading Sequencing Gels (see Note 6)

3.4.1. Assembly of Slab Gel Apparatus

1. Place clean, dry sequencing plates onto a low-lint Kimwipe® on a flat work surface. Orient the plates with the inside surface facing upward.
2. Gently wipe the inside of the plates with ddH$_2$O and a clean low-lint Kimwipe.
3. Gently wash the plates with 95% ethanol, making sure to use a fresh Kimwipe.
4. Lightly wet the 0.2-mm-thick spacers with ddH$_2$O and place them at the edges of the back sequencing plate (the longer of the two plates). Be sure to align the spacers flush with the sides and bottom of the plate to prevent any leakage of the gel.
5. Before placing the front sequencing plate on top of the back plate, quickly brush over the surface with a Kimwipe that has been slightly dampened with 95% ethanol. This step will remove any additional dust or lint that may have settled onto the plate surface during assembly.
6. After the gel plates have been assembled and the plates are aligned flush at their side and bottom edges, place three extra-large binder clamps on each side of the plate assembly (one on each side about 1 in. from the bottom, one on each side in the middle and one on each side directly below the notch at the top of the spacers).
7. Slightly elevate the loading end of the plate assembly.

3.4.2. Pouring Slab Gels

1. Add 50 mL of gel mix, 35 µL of TEMED, and 250 µL of 10% APS solution to a clean, dry glass beaker (*see* **Note 12**).
2. Mix the solution by gently swirling the beaker, being careful not to generate bubbles. Introduction of air into the gel mix at this point will directly affect polymerization of the gel.

3. Using a 50-mL syringe, slowly dispense the gel mix into the space between the glass plates. While dispensing the gel mix, it may be necessary to gently tap the surface of the top plate to remove bubbles and to help the gel travel toward the bottom of the plate assembly.
4. Once the gel mix has reached the bottom of the plates, remove the slight elevation at the loading end and continue pouring the gel until the area between the plates is completely filled. Removing the elevation will prevent the gel from simply passing through the plates and spilling onto the work surface.
5. Insert the comb (straight edge down) between the plates at the top of the loading area until the notches on the comb come into contact with the top of the front sequencing plate.
6. Place three bulldog clamps across the top of the gel sandwich to ensure a tight fit between the plates and the comb.
7. After the gel has polymerized (15–20 min), place a damp cloth or Saran wrap over the top and bottom of the gel sandwich to prevent the gel from drying out in these areas.
8. Allow the gel to polymerize an additional 2 h before use. Gels should be loaded no more than 6 h after pouring to avoid low-quality data.

3.4.3. Loading Samples

1. Resuspend slab gel samples in 2 µL of sample loading buffer by gently pipetting several times. Load 1–2 µL on the gel as per the manufacturer's protocol. For capillary reactions, resuspend each sample in 20 µL of formamide and load the entire plate directly onto the machine.
2. Sequencing gels should be run according to the protocol provided with the sequencer.

4. Notes

1. Template DNA must be of extremely high quality. When using the ABI dye chemistries, purify plasmid templates using Qiagen mini-spin kits (www.qiagen.com) or Promega Wizard preps (www.promega.com) for the most reproducible results. For PCR purification, gel extraction using the QIAquick method (Qiagen) is the best option. For BAC clone purification, the ProPrep BAC system (Ligo Chem, www.ligochem.com) or Qiagen tip-100 or tip-500 are recommended.
2. In the author's experience, the length of the primer used for sequencing is not as important as is the uniqueness of the primer sequence. Most sources suggest the use of primers between 18 bases and 22 bases for the best results. However, the author routinely uses sequencing primers between 21 bases and 35 bases without problem.
3. For optimal results an alternative method must be used to sequence GC-rich or AT-rich templates. Although the ABI Big Dye chemistry will occasionally give good results on these types of template, ABI suggests using the dRhodamine dye terminator mix. In addition to using the dRhodamine dye set, the author routinely increases the reaction volume to 20 µL. In this case, use the same template concentration as in **Subheading 2.1.**, but use 6.4 µL of primer (3.2 pmol/µL), 4 µL of a 1 M betaine solution, and 8 µL of the dye terminator mix. Cycling conditions and other downstream procedures are the same as described.
4. 5X Sequencing buffer is used only when setting up reactions to be run on a sequencing slab gel. Also, this buffer is not used when sequencing GC- or AT-rich templates or when using the dye terminator mix at full strength, as suggested by the manufacturer.
5. Use ultrapure reagents when possible. Lower-grade chemicals can adversely affect multiple factors, including gel polymerization and overall sequence quality. Pay special attention to using DIUF water when setting up sequencing reactions and making solu-

DNA Sequencing

tions for acrylamide gels. The author has experienced problems with background fluorescence when using other water sources.

6. Reagents for a sequencing gel mix are not needed when running sequencing reactions on a capillary electrophoresis unit. When using this type of equipment use only the polymer that is recommended by the manufacturer of the sequencer.
7. Several companies, including Amersham Pharmacia (www.amershambiosciences.com) and Applied Biosystems, sell formamide solutions specifically for use in fluorescent-based sequencing applications. Purchase of these solutions is highly recommended in order to reduce the likelihood of the sequencing quality being affected by poorly made or expired formamide stock. In the author's experience, the majority of failures that are not the result of poor template quality or improper template concentration are because of a problem with the formamide used during sample loading.
8. As with any solution that contains enzyme, the fluorescent-labeled terminator mix should be kept on ice during use and all reactions should be kept chilled before placing in the thermal cycler.
9. It is imperative that some method of cleanup be used before running sequencing reactions on a gel. This is because there will be some unincorporated dye terminator components in the reaction even after cycling to completion. If cleanup is insufficient, then the data quality will be poor because of excess dye "blobs" that will run off the gel during the collection of data at the beginning of the run. Signal resulting from excess dyes will overshadow any "real" signal in the first 100–200 bp of sequence. Although in this chapter it is suggested that ethanol precipitation is used for the cleanup of small numbers of reactions, the author routinely uses the Sephadex cleanup method regardless of the number of samples. If sequencing reactions are performed on a regular basis, the Sephadex cleanup method is the more cost-effective with regard to both reagents and time.
10. Following cleanup and drying, reactions can be stored covered at –20°C for up to 2 wk with minimal loss of signal intensity. Samples that have been resuspended in solutions containing formamide should not be stored for more than 18 h at –20°C before running on a gel, as this routinely results in extremely low signal intensity or a complete loss of data.
11. Although some samples will require longer drying times, it is highly recommended that drying times do not exceed 1 h at 60°C. Drying times in excess of this not only contribute to difficulty with resuspension of samples but also affect signal strength. Excessive drying of the ABI Big Dye terminators can result in a complete loss of signal from the TAMRA and R110 dyes that are coupled to the ddTTP and ddGTP nucleotides, respectively.
12. The amounts given in **Subheading 3.4.2.** for sequencing gel mix are for running samples on a 48-cm slab gel. Most systems allow for shorter gels to be run, but in the author's experience, the shorter gel length and run times are not worth the reduction in the length of sequence and quality of data obtained.

Acknowledgments

The protocols discussed in this chapter are based in part on those developed in the laboratory of Dr. Bruce Roe at the University of Oklahoma Advanced Center for Genome Technology (ACGT) (www.genome.ou.edu) and at Dr. David Dyer's Laboratory for Microbial Genomics at the Oklahoma Health Sciences Center (www.microgen.ouhsc.edu). Special thanks to Jenny Gipson, Mandy Gipson, Cindy Maddera, and Allen Gies for their technical advice and suggestions.

References

1. Sanger, F., Nicklen, S., and Couldon, A. R. (1977) DNA sequencing with chain-terminating inhibitors. *Proc. Nat. Acad. Sci. USA* **92,** 6339–6343.
2. Prober, J., Trainor, G., Dam, R., et al. (1987) A system for rapid DNA sequencing with fluorescent chain-terminating dideoxynucleotides. *Science* **238,** 336–341.
3. Ansorge, W., Sproat, B., Stegemann, J., et al. (1987) Automated DNA sequencing: ultrasensitive detection of fluorescent bands during electrophoresis. *Nucleic Acids Res.* **15,** 4593–4602.
4. Zagursky, R. and McCormick, R. (1990) DNA sequencing separations in capillary gels on a modified commercial DNA sequencing instrument. *BioTechniques* **9,** 74–79.

23

Site-Directed Mutagenesis Using the Megaprimer Method

Zhidong Xu, Alessia Colosimo, and Dieter C. Gruenert

1. Introduction

Site-directed mutagenesis (SDM) is used to introduce a defined mutation into target DNA of known sequence to study, for example, gene expression or protein structure–function relationship. A number of polymerase chain reaction (PCR)-based mutagenesis methods have been developed *(1)*. Among them, the "megaprimer" method is probably the simplest and most flexible *(2–5)*. The megaprimer method utilizes two external oligonucleotide primers and one internal mutagenic primer in two rounds of PCR with a DNA template that contains the sequence to be mutated. The first round of PCR is carried out using one of the external primers and the mutagenic primer containing the desired mutation. This amplifies an intermediate PCR product that is purified and used as a "megaprimer" for the second round of PCR, along with the other external primer. The final PCR product is cloned into appropriate vectors and used in downstream applications.

Different modifications of the megaprimer method for SDM have been reported *(2,3,6,7)*, each of which aims to increase the efficiency of mutagenesis. For example, the use of an excess of the purified "megaprimer" over the template DNA in the second round of PCR increases the frequency of mutant clones. Typically, these modifications result in an overall efficiency of mutagenesis greater than 50%.

This chapter describes a simplified version of the megaprimer method based on a dilution of the products from the first-round PCR and their use as a template for the second-round amplification (*see* **Fig. 1**) *(8)*. In the first PCR amplification, a wild-type sequence is mutated with a forward primer (A) and a mutagenic internal primer (M). The desired mutation(s) give rise to a new restriction enzyme cleavage site and/or a functional mutation in the amplified product. A diluted aliquot of the first-round PCR reaction is used in a second round of PCR with external forward and reverse primers (A and B, respectively). During the first cycle, the mutated fragment is

From: *Methods in Molecular Biology, Vol. 235:* E. coli *Plasmid Vectors*
Edited by: N. Casali and A. Preston © Humana Press Inc., Totowa, NJ

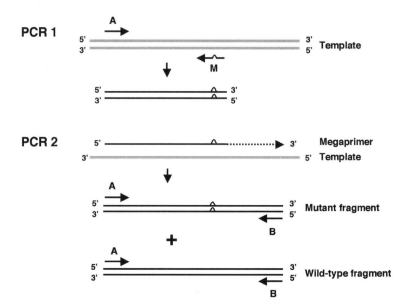

Fig. 1. Schematic illustration of the megaprimer method for site-directed mutagenesis. The first-round PCR (PCR1) is performed using external forward primer (A) and mutagenic reverse primer (M) to amplify a mutagenic "megaprimer" fragment from a DNA template. The second-round PCR (PCR2) is performed using external forward (A) and external reverse (B) primers using a dilution of the first-round PCR product as the template. The megaprimer product formed in the first PCR is extended with DNA polymerase. In the subsequent cycles of PCR2, because of the large excess of mutant amplification products, the mutant fragment is preferentially amplified with respect to the original wild-type sequence.

extended with DNA polymerase using the wild-type sequence DNA as the template. In subsequent cycles, because of a large excess of mutant fragment over the original wild-type sequence, the mutagenic sequence is preferentially amplified with primers A and B. This modified version of the megaprimer method was successfully implemented in our previous study investigating allele-specific oligonucleotides for cystic fibrosis gene therapy (9). This modification further simplifies the overall procedure for SDM and results in a relatively high frequency (>60%) of mutated fragment compared to the original wild-type sequence.

2. Materials

1. DNA template and plasmid carrying the gene sequence to be mutated.
2. Oligonucleotide primers: Two external primers (forward and reverse) and one internal mutagenic primer.
3. 5 U/µL *Pfu* DNA polymerase and 10X reaction buffer (Stratagene, La Jolla, CA).
4. Deoxyribonucleotide triphosphate (dNTP) mix, containing 10 mM each of dATP, dCTP, dGTP, and dTTP.
5. Light mineral oil.

6. QIAquick gel purification kit (Qiagen, Chatsworth, CA).
7. Phenol : chloroform : isoamyl alcohol (25 : 24 : 1).
8. 100% ethanol; 70% ethanol.
9. Cloning vector, such as pCR-Script Amp SK(+) vector (Stratagene).
10. *Eschericia coli* XL1-Blue MRF' Kan Supercompetent cells (Stratagene).

3. Methods
3.1. Primer Design
3.1.1. Mutagenic Primer

We design primers that are 22–24 bp in length. This gives sufficient length for incorporating the required base-pair change and to give the desired annealing temperature (T_m) > 60°C. Mutagenic primers are designed as described by Kuipers et al. *(10)* such that the 5'-end immediately follows a T residue in the same strand. This consideration is particularly important when *Taq* DNA polymerase is used to generate megaprimer fragments. *Taq* DNA polymerase adds a non-template-coded A residue onto the 3' ends of PCR fragments. The addition of this extra base effectively introduces an extra mutation into the second-round PCR product and undermines extension of the megaprimer in the initial cycle(s) of the secondary amplification, because of the mismatch between the 3' end of the megaprimer and the template.

3.1.2. External Primers

Mutagenic fragments generated by SDM can be directly cloned into vectors without modification, such as the pCR-Script cloning vector as described in this chapter. However, when mutagenic fragments are to be cloned into vectors at specific restriction sites, both external primers must be designed to contain the appropriate restriction sites. Restriction sites should be located at the 5' end of primers with a few additional bases to the 5' of the site to facilitate restriction enzyme digestion of the PCR products.

3.2. PCR Mutagenesis
3.2.1. First-Round PCR

1. Assemble the PCR reaction in a 0.2-mL tube as follows:

    1 ng template plasmid DNA (*see* **Note 1**)
    10 pmol forward external primer
    10 pmol mutagenic primer
    0.5 µL dNTP stock solution (0.25 m*M* final concentration)
    1 U *Pfu* DNA polymerase (*see* **Note 2**)
    2 µL of 10X *Pfu* reaction buffer

    Make up to a total volume of 20 µL with deionized water.
2. Overlay the reaction mixture with 20 µL of light mineral oil.
3. Perform PCR amplification using the following parameters: 94°C for 3 min for initial denaturation; 25 cycles of 94°C for 20 s, 58°C for 20 s, and 72°C for 1–2 min, followed by a final extension at 72°C for 5 min.
4. Verify the primary PCR amplification by agarose gel electrophoresis.

3.2.2. Second-Round PCR

1. Make a 1 : 25 and a 1 : 50 dilution of the primary PCR reaction by adding 2 µL of the PCR product to 48 or 98 µL of deionized water, respectively. Use 2 µL of each diluted sample for the second-round PCR (*see* **Note 3**). Do not add additional template DNA.
2. Perform the second-round amplification as described in **Subheading 3.2.1.**, except substitute the internal mutagenic primer with the reverse external primer (i.e., use forward and reverse external primers).
3. Verify the second-round PCR product by agarose gel electrophoresis.
4. If only the expected size DNA fragment is amplified in the second-round PCR, then this fragment can simply be precipitated from the PCR reaction (*see* **Subheading 3.2.3.**).
5. If the second-round PCR amplifies nonspecific bands, then excise the desired band from the agarose gel using a scalpel blade. Purify the DNA fragment from the gel using the QIAquick gel extraction kit according to the manufacturers' protocol.

3.2.3. Ethanol Precipitation

1. Transfer the PCR product into a 1.5-mL Eppendorf tube; add water to bring the volume to 50 µL.
2. Extract the PCR product with 50 µL of phenol : chloroform : isoamyl alcohol. Centrifuge for 5 min at top speed in a microfuge to separate phases.
3. Transfer the aqueous fraction to a new tube, add 2.5 volumes of 100% ethanol, and mix well by vortexing for a few seconds. Place the sample at –20°C for 30 min and then centrifuge for 15 min at top speed in a microfuge at 4°C.
4. Rinse the pellet with 200 µL of 70% ethanol, recentrifuge, and remove ethanol completely; air-dry.
5. Redissolve the pellet in 20 µL of deionized water.

3.2.4. Subcloning of Mutagenic Fragments

1. Clone the purified DNA fragment into pCR-Script Amp SK(+) vector according to manufacturer's instructions (*see* **Note 4**).
2. Transform the ligation products into *E. coli* XL1-Blue MRF' Kan Super competent cells according to the manufacturers' instructions.
3. Analyze five clones from each transformation (*see* **Note 5**). If the mutation changes or creates a restriction site, restriction analysis will distinguish mutated clones from the wild-type sequence clones (*see* Chapter 20). However, it is advisable to sequence the entire cloned fragment to confirm the mutation and to ensure that no other mutations have been introduced (*see* Chapter 22).

4. Notes

1. Template DNA should be kept at a low concentration (e.g., around 1 ng). Excess template leads to high levels of wild-type sequence being carried over into the second-round PCR, which results in a high level of wild-type sequence in the second-round PCR products.
2. *Pfu* DNA polymerase has a 3'g5' exonuclease activity that increases the fidelity of DNA replication. *Pfu* DNA polymerase also lacks terminal transferase activity that catalyzes the addition of a non-template-coded A residue to the 3' end of the PCR product. These two features render it a good choice for SDM experiments. However, regular *Taq* enzyme (such as Ampli*Taq* DNA polymerase; Perkin-Elmer, Norwalk, CT) can also be used as long as the mutagenic primer is designed such that its 5' end immediately follows a T

residue in the same strand (*see* **Subheading 3.1.1.**). Our previous study showed that this gives similar results when either the *Pfu* or *Taq* enzyme is used *(8)*.
3. Dilution of the first-round PCR product, which is used as a template in the second-round PCR, improves the yield of the desired mutant clones relative to those carrying parental wild-type sequences. When 1–2 µL of undiluted first-round PCR product is used as the template, nonspecific amplification products are observed after the second-round PCR. Using our experimental conditions, 3.2 or 1.6 ng of mutated product and 4 or 2 pg, respectively, of wild-type plasmid DNA, are present in the second-round PCR reaction. Under these conditions, the mutation-containing product is in approx 800-fold excess over the wild-type plasmid DNA and is, thus, preferentially amplified in the second-round PCR.
4. Second-round PCR products can be cloned into any appropriate DNA cloning vector. The pCR-Script Amp SK(+) vector allows cloning of PCR-amplified DNA fragments without restriction enzyme digestion and allows the use of blue–white screening (*see* Chapter 19) to identify clones containing insert DNA. In the case that SDM fragments are to be cloned at specific restriction sites, both the vector and PCR products must be digested appropriately before ligation (*see* Chapter 15).
5. The number of clones to analyze depends on the efficiency of mutagenesis. In our experience using the current protocol, at least two-thirds (>60%) of subclones contain the desired mutation.

References

1. Ling, M. M. and Robinson, B. H. (1997) Approaches to DNA mutagenesis: an overview. *Anal. Biochem.* **25,** 157–178.
2. Sarkar, G. and Sommer, S. S. (1990) The "megaprimer" method of site-directed mutagenesis. *BioTechniques* **8,** 404–407.
3. Kammann, M., Laufs, J., Schell, J., et al. (1989) Rapid insertional mutagenesis of DNA by polymerase chain reaction (PCR). *Nucleic Acids Res.* **17,** 5404.
4. Barik, S. and Galinski, M. (1991) "Megaprimer" method of PCR: increased template concentration improves yield. *BioTechniques* **10,** 489–490.
5. Picard, V., Ersdal-Badju, E., Lu, A., et al. (1994) A rapid and efficient one-tube PCR-based mutagenesis technique using *Pfu* DNA polymerase. *Nucleic Acids Res.* **22,** 2587–2591.
6. Giebel, L. B. and Spritz, R. A. (1990) Site-directed mutagenesis using a double-stranded DNA fragment as a PCR primer. *Nucleic Acids Res.* **18,** 4947.
7. Herlitz, S. and Koenen, M. (1990) A general and rapid mutagenesis method using polymerase chain reaction. *Gene* **9,** 143–147.
8. Colosimo, A., Xu, Z., Novelli, G., et al. (1999) Simple version of "megaprimer" PCR for site-directed mutagenesis. *BioTechniques* **26,** 870–873.
9. Goncz, K. K., Kunzelmann, K., Xu, Z., et al. (1998) Targeted replacement of normal and mutant CFTR sequences in human airway epithelial cells using DNA fragments. *Hum. Mol. Genet.* **7,** 1913–1919.
10. Kuipers, O. P., Boot, H. J., and de Vos, W. M. (1991) Improved site-directed mutagenesis method using PCR. *Nucleic Acids Res.* **19,** 4558.

24

Site-Directed Mutagenesis by Inverse PCR

Clifford N. Dominy and David W. Andrews

1. Introduction

Site-directed mutagenesis has revolutionized the study of protein structure and function by enabling the controlled and systematic production of mutant proteins. Early methods of site-directed mutagenesis involved the use of a mutated oligonucleotide primer to prime synthesis of a target single-stranded DNA template. These approaches were very inefficient, yielding success rates of 1–5% *(1)*. A dramatic improvement in the efficiency of generating mutations resulted from the use of single-stranded, uracil-containing DNA molecules isolated from *ung⁻ dut⁻ Escherichia coli* strains (*see* Chapter 3). Again, the mutation is introduced in a mutated oligonucleotide primer. Selection against the wild-type sequence parent DNA occurs on transformation into wild-type *E. coli*. Mutagenesis by this method was relatively efficient, with rates of 15–35%, but required a number of subcloning steps involving single-stranded M13 phage clones *(2)*. It was only following the development of the polymerase chain reaction (PCR) that the two concepts were combined, dramatically improving the efficiency of the whole procedure.

Early PCR-based mutagenesis strategies usually involved multiple subcloning steps. With the advent of techniques that facilitate the amplification of large PCR products, site-directed mutagenesis using complementary, mutagenic primers to amplify entire constructs was developed as a method for mutagenesis involving minimal cloning steps. This approach has been termed "inverse PCR" and represents an efficient, reliable, and, consequently, cost-effective way of producing site-specific mutations. The technique relies only on the gene of interest being carried in a vector small enough that the entire plasmid can be amplified by PCR. As long as this criterion is satisfied, the target DNA is mutated by one or more mutagenic oligonucleotides without the need for further subcloning of the altered product. Because of the inherent processivity and fidelity of the various thermostable DNA polymerases, it is prudent to restrict the size of the construct to less than 10 kilobase pairs (kb).

Classical inverse PCR *(3)* involves a pair of complementary primers that anneal to the same region of the opposing DNA strands of the construct; one of the primers has a mutation relative to the wild-type or parental sequence. After amplification of the entire plasmid using these oligonucleotides, the linear product is phosphorylated and circularized by ligation. The great advantage of this technique over the original M13-based method *(1)* is the ability to rapidly generate mutated amplified products. Successive rounds of amplification result in an exponential increase in the linear mutated construct relative to the parental template sequence. The principle drawbacks to this approach are that it depends on faithful amplification of the termini of the DNA fragments (often PCR leaves a single undesirable A residue at the end, which disrupts the reading frame of the final product) and the relatively low efficiency of ligation of blunt-ended products.

This chapter covers the three main types of inverse PCR currently in use today. All have been used in the author's laboratory to generate site-directed mutants. The types of mutation made include point mutations, insertions, and large deletions within genes. The three techniques differ from one another in subtle ways and each has strengths and weaknesses.

1.1. Enzymatic Inverse PCR
Using Type IIS Restriction Endonucleases

Enzymatic inverse PCR using Type IIS restriction endonucleases (EIPCR-IIS) is a significant improvement over the classical method *(4)*. In this technique, the 5' termini of both primers contain a unique Type IIS restriction site, such as *Sap*I. Type IIS restriction enzymes cleave DNA at defined sites outside of their recognition/binding sequences. This enables removal of the ends of a PCR product by the appropriate enzyme and avoids problems that may arise from PCR-mediated addition of extra residues to the PCR product's termini. The overhanging ends that result from digestion with the Type IIS enzyme are efficiently ligated without the need for prior phosphorylation while retaining the original reading frame of the coding sequence (*see* **Fig. 1**).

A potential drawback to the use of this strategy for introducing point mutations is that unless a restriction site has been deleted or created, distinguishing differences in the mutated and parental DNA requires DNA sequencing. EIPCR-IIS, as outlined in **Fig. 2**, is therefore particularly useful for rapidly generating site-specific deletions of genes where a change in restriction fragment size is detectable by gel electrophoresis.

1.2. Enzymatic Inverse PCR
Using Type II Restriction Endonucleases

The author's laboratory has adapted the original EIPCR protocol for the use of class II restriction enzymes *(5)*, thereby extending the versatility of the technique. The principle difference from EIPCR-IIS is that in stage 1 of this process, a unique Type II enzyme recognition site is artificially introduced into the construct. In stage 2, the intermediate construct is digested at this novel site, and the overhang is removed to bring the original coding sequence back into frame. The cloning intermediate, containing the novel restriction site, is very useful for selecting positive clones and also

Mutagenesis by Inverse PCR

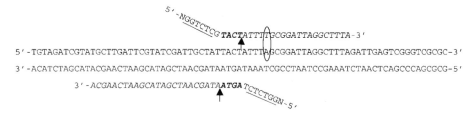

Fig. 1. Primer design for EIPCR-IIS. Primers are italicized and the point mutation is circled. Arrows highlight the cleavage sites of *Eco*31I, whose recognition sequence is underlined. The four-nucleotide overhang generated by *Eco*31I is shown in bold.

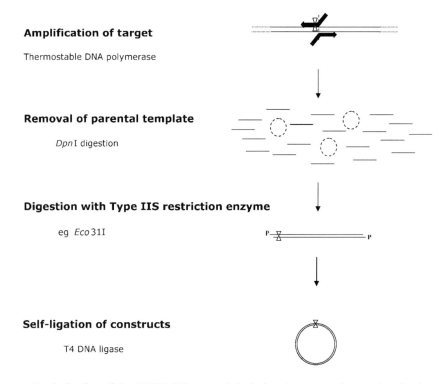

Fig. 2. Outline of the EIPCR-IIS protocol depicting the enzymatic steps involved.

provides a useful means to map the mutants without introducing a silent restriction site. **Figures 3** and **4** outline the principle of EIPCR-II.

Advantages of EIPCR-II are an increased efficiency and fidelity of ligation compared to classical inverse PCR. Unlike EIPCR-IIS, the intermediate can be used for optimizing conditions for the creation of the final mutant construct. The restriction

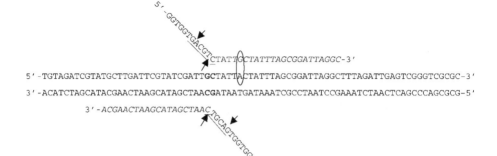

Fig. 3. Primer design for EIPCR-II. Primers are italicized. Each includes a unique *Aat*II restriction site (underlined) with the additional bases required for digestion of the PCR product at the 5' end. Arrows highlight the cleavage sites of *Aat*II. The introduced point mutation is circled.

site used in EIPCR-II can also be used to conveniently mix and match mutants. A potential disadvantage of this procedure is the additional enzymatic steps required to remove the restriction site overhang to bring the coding sequence back into the correct coding frame.

1.3. Quikchange Site-Directed Mutagenesis

QuikchangeT™ site-directed mutagenesis, by Stratagene (La Jolla, CA), represents a subtle departure from the above-outlined protocols (6). The method takes advantage of the lack of strand-displacement activities of some thermostable DNA polymerases such as *Pfu*Turbo DNA polymerase. In this procedure, two complementary mutagenic oligonucleotides anneal to opposing stands of the target template. Amplification of the DNA template using these mutant primers extends the product on opposing strands in an inverse manner around the circular construct (*see* **Figs. 5** and **6**). On completion of the amplification step, the *Pfu* polymerase is displaced from the template rather than displacing the mutagenic primers. This results in the formation of a circular template with nick sites at the 5' terminus of each primer. Using this system, the amplification reaction is very inefficient and the product consists of a mixture of mutated and parental sequences. Subsequent endonuclease digestion with *Dpn*I removes methylated and hemimethylated parental sequences, which greatly improves the selection of the mutated template over parental constructs. The reaction mixture is transformed into *E. coli*, and host enzymes repair the nicked junction sites to generate an intact circular supercoiled plasmid.

The primary advantage of this technique is its relative speed in generating new mutants. The major disadvantages are the inefficiency of the polymerization reaction and the relatively limited choice of primers and enzyme that can be used. In addition, a highly efficient transformation system is essential, as the nicked double-stranded product formed by the polymerase does not transform as efficiently as the parental wild-type supercoiled plasmid.

Mutagenesis by Inverse PCR

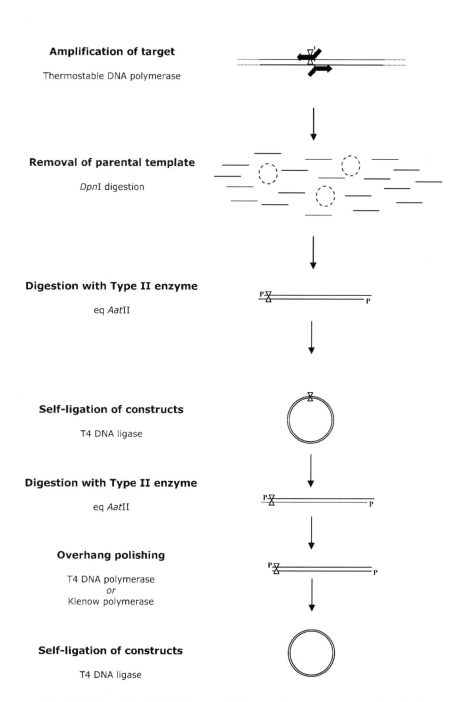

Fig. 4. Outline of the EIPCR-II protocol showing the enzymatic steps involved.

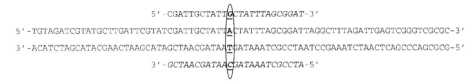

Fig. 5. Primer design for Quikchange site-directed mutagenesis. Complementary mutant primers are italicized. The desired (G to A) point mutation that is incorporated into the primers is circled.

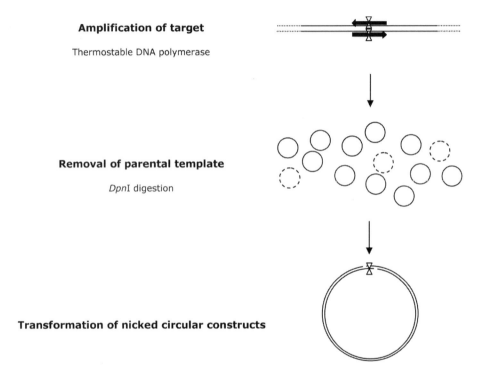

Fig. 6. Outline of the Quikchange site-directed mutagenesis protocol showing the enzymatic steps involved.

1.4. Choosing a Method

All three of these techniques are useful to quickly and efficiently generate either site-directed mutations or specific additions or deletions within genes of interest. EIPCR-IIS should be considered when the introduced mutation alters the restriction pattern of the plasmid as it facilitates screening of mutants.

EIPCR-II is useful for generating large numbers of mutants in a cost effective way, as it uses commonly available enzymes, permits the combining of mutations via the intermediates, and *Dpn*I digestion is usually not necessary. EIPCR-II is the method of choice when particular mutations prove difficult to generate as the efficiency of each step can be quantified. Furthermore, when generating large numbers of mutants, the intermediate screening step is useful for identifying which reactions are likely to give positive results quickly.

Quikchange site-specific mutagenesis is the quickest of the methods presented here, with the mutagenesis step being completed in a few hours. As there is no ligation step, transformation can proceed by the end of the first day and mutant constructs are selected and analyzed on d 2 and 3. Cost, specifically of the *Pfu* polymerase, can be a hindrance to any large-scale mutagenic experiments. It is the most difficult of the three procedures to optimize, as there are no analyzable intermediates.

2. Materials
2.1. Amplification Reaction
1. 10X Amplification reaction buffer: 100 mM Tris-HCl (pH 8.8), 500 mM KCl, 0.8% Nonidet P40, MgCl$_2$ (*see* **Note 1**).
2. 10 mM dNTP stock solution (2.5 mM each of dGTP, dATP, dCTP, and dTTP).
3. Mutagenic primers (*see* **Subheading 3.** for design guidelines).
4. Thermostable DNA polymerase (*see* **Note 2**).

2.2. Restriction Digestion
1. 10X *Dpn*I reaction buffer: 500 mM potassium acetate, 200 mM Tris-acetate (pH 6.9), 100 mM magnesium acetate, 10 mM dithiothreitol (DTT).
2. *Dpn*I restriction endonuclease.
3. Appropriate Type IIS (**Subheading 3.1.**) or Type II (**Subheading 3.2.**) restriction endonuclease and corresponding 10X reaction buffer.

2.3. Overhang Polishing Reaction
2.3.1. Using E. coli Klenow Fragment
1. 10 U/μL *E. coli* DNA polymerase Klenow fragment.
2. 10X Klenow fragment reaction buffer: 500 mM Tris-HCl (pH 8.0), 50 mM MgCl$_2$, 10 mM DTT.
3. 10 mM dNTP stock solution.

2.3.2. Using T4 DNA Polymerase
1. 3 U/μL T4 DNA polymerase.
2. 10X T4 DNA polymerase reaction buffer: 500 mM NaCl, 100 mM Tris-HCl (pH 7.9), 100 mM MgCl$_2$, 10 mM DTT.
3. 1 mg/mL Bovine serum albumin (BSA).
4. 10 mM dNTP stock solution.

2.4. Phenol Extraction and Ethanol Precipitation
1. Phenol:chloroform (1:1).
2. 0.3 M Sodium acetate (pH 5.2).
3. 100% Ethanol.
4. TE: 10 mM Tris-HCl (pH 7.6), 1 mM EDTA.

2.5. GENECLEAN® (see Note 3)

1. Glassmilk™ (Q-BIOgene).
2. NaI solution: Dissolve 90.8 g of sodium iodide and 1.5 g of sodium sulfite in 100 mL of water. Filter through a 0.45-μm filter, then add another 0.5 g of sodium sulfite. Store at 4°C in dark bottles.
3. Wash solution: 50% Ethanol, 0.1 M NaCl, 10 mM Tris-HCl (pH 7.5), 1 mM EDTA. Store at –20°C.
4. TE: 10 mM Tris-HCl (pH 7.6), 1 mM EDTA.

2.6. Ligation and Transformation

1. T4 DNA ligase.
2. 10X Reaction buffer: Usually provided with enzyme. Typical composition: 400 mM Tris-HCl (pH 7.8), 100 mM MgCl$_2$, 100 mM DTT, 5 mM ATP.
3. *E. coli* DH5α competent cells (*see* Chapter 4).

3. Methods
3.1. EIPCR-IIS

The EIPCR-IIS protocol is outlined in **Fig. 2**.

3.1.1. Primer Design

Careful primer design is crucial for the success of any DNA amplification experiment and is particularly critical when designing primers for site-specific mutagenesis. Considerations for general primer design are outlined in **Notes 4** and **5**. It is useful to incorporate a novel restriction site within the primer sequence to facilitate identification of mutants. This is a good practice because of the occasional incomplete degradation of the parent DNA strands by *Dpn*I digestion (the authors have found that for some templates, the rate of recovery of the parent plasmid can be unexpectedly high). There is an abundance of software available to assist with the successful design of mutagenic primers and to introduce silent mutations (*see* **Note 6**). The degeneracy of the genetic code often needs to be employed to obtain an optimum primer sequence for PCR that also introduces a silent site for screening purposes. Finally, the primers must be designed to include a Type IIS restriction site at the 5' end (*see* **Note 7**).

3.1.2. Amplification Reaction

1. To amplify the target DNA set up the following reaction (*see* **Note 8**):

10X Reaction buffer	5 μL
Target DNA	10–50 ng
Mutant primer 1	125 ng
Mutant primer 2	125 ng
10 mM dNTP stock	1 μL
2.5 U Thermostable polymerase	1 μL
Double distilled deionized water to	50 μL

2. Perform the PCR reaction in a thermocycler at conditions determined empirically according to the thermodynamic properties of the oligonucleotide primers used (*see* **Note 9**).

The following cycling conditions are typically used in the author's laboratory (see **Note 10**):

Step	Cycles	Temperature	Time
1	1	94°C	45 s
2	25–35	94°C	45 s
		53°C	45 s
		72°C	60 s/kb
3	1	72°C	5 min
4	1	4°C	Indefinitely

3. Analyze PCR reactions by electrophoresis on a 0.8% agarose gel to ensure that a single product of the correct size is obtained.

3.1.3. Restriction Digestion with DpnI

1. Add 10 U of *Dpn*I to the amplification reaction in order to digest methylated and hemimethylated parental DNA (*see* **Note 11**).
2. Incubate at 37°C for 1 h.

3.1.4. Phenol Extraction and Ethanol Precipitation

1. To remove the DNA polymerase from the PCR reaction, add an equal volume of phenol : chloroform (1 : 1) and mix by vortexing for 5 s. Centrifuge at 12,000g for 1 min and recover the aqueous top phase (*see* **Note 12**).
2. Precipitate the DNA by adding 0.1 volume of 0.3 M sodium acetate (pH 5.2) and 2 volumes of 100% ethanol.
3. Pellet the DNA in a microcentrifuge at 12,000g for 10 min, aspirate the supernatant, and allow the pellet to air-dry. Resuspend in 10 µL of TE.

3.1.5. Restriction Digestion with Type IIS Enzyme

1. To digest the mutated DNA at the restriction site incorporated into the primer, set up the following reaction:

*Dpn*I-digested PCR product	1 µg
10X Reaction buffer	2 µL
Restriction enzyme	2 U
Distilled deionized water to	20 µL

2. Incubate for 1 h at 37°C. Terminate the reaction by heating at 65°C for 20 min to inactivate the enzyme.

3.1.6. GENECLEAN®

1. To purify the digested PCR product, run the reaction on a 0.8% agarose gel and excise the desired band.
2. Weigh the gel slice in a microcentrifuge tube and add 200 µL of NaI solution per 0.1 g of gel.
3. Heat at 50–55°C for 2–5 min to dissolve the agarose, vortex to mix.
4. Add 2 µL of Glassmilk matrix suspension per 1–10 µg of DNA and vortex to mix. Incubate for 5 min at room temperature with periodic vortexing.
5. Spin for 1 min at full speed in a microcentrifuge and aspirate the supernatant.
6. Resuspend the pellet in 200 µL of NaI solution.
7. Spin for 1 min at full speed in a microcentrifuge and aspirate the supernatant.
8. Resuspend the pellet in 200 µL of wash solution.

9. Spin for 1 min at full speed in a microcentrifuge and aspirate the supernatant.
10. Repeat the wash and carefully remove all excess liquid.
11. Resuspend the pellet in 10 μL of TE buffer and incubate at 50°C for 5 min.
12. Centrifuge briefly and transfer the buffer, containing the extracted DNA, to a clean tube.

3.1.7. Ligation and Transformation

1. Set up the ligation reaction as follows (*see* **Note 13**):

Amplified, restriction-digested template	300 pmol
10X Ligation buffer (*see* **Note 14**)	3 μL
T4 DNA ligase	3 Weiss units
Distilled deionized water to	30 μL

    Incubate at 16°C overnight.
2. Using standard methods, transform 10% of the ligation product into *E. coli* DH5α competent cells. Plate onto Luria–Bertani (LB) agar supplemented with the appropriate antibiotic to select for the mutant constructs (*see* Chapter 4 and **Note 15**).

3.1.8. Analysis of Mutants

Isolate plasmid DNA from candidate clones (*see* Chapters 8–10). Mutants may be identified by screening for a change in size of the construct in the case of deletion mutants or by screening for an introduced silent restriction site if appropriate (*see* Chapter 20 and **Note 16**).

3.2. EIPCR-II

The EIPCR-II protocol is outlined in **Fig. 4**. In stage 1, an intermediate is constructed containing the desired mutation and an introduced restriction site. In stage 2, the introduced restriction enzyme site is removed, to restore the correct reading frame of the gene of interest and generate the final construct.

3.2.1. Primer Design (see **Notes 4 and 5**)

An important difference with designing primers for EIPCR-II is that, by definition, it uses restriction sites in which the enzyme cuts at the recognition site. These enzymes usually do not have a degenerate cleavage site and, thus, the overhangs must be removed so that only two nucleotides from the recognition site remain in the final PCR product (*see* **Table 1**). To facilitate the accurate removal of the overhang, use restriction sites with a 3' overhang and primers that incorporate the novel restriction site at their 5' end (*see* **Notes 6** and **17**).

3.2.2. Stage 1

1. Amplify the target DNA by PCR and analyze the products by agarose gel electrophoresis as described in **Subheading 3.1.2**.
2. Degrade template DNA by endonuclease digestion with *Dpn*I, as described in **Subheading 3.1.3**.
3. Purify the PCR product by phenol extraction and precipitate with ethanol as described in **Subheading 3.1.4**.
4. Digest the mutated DNA with the Type IIS restriction enzyme, whose recognition site was incorporated into the primer, as described in **Subheading 3.1.5**.

Table 1
Restriction Enzymes That Generate
Four-Base 3' Overhangs Suitable for EIPCR-II

Enzyme[a]	Recognition sequence[b]	Dinucleotide[c]
AatII	GA*CGT*C	GC
ApaI	G*GGCC*C	GC
BbeI	G*GCGC*C	GC
Bme1580I	G*KGCM*C	GC
BmtI	G*CTAG*C	GC
BsiHKAI	G*WGCW*C	GC
HaeII	R*CGCG*Y	AT/GC
KpnI	G*GTAC*C	GC
NsiI	A*TGCA*T	AT
NspI	A*CATG*T	AT
PstI	C*TGCA*G	CG
SacI	GA*GCT*C	GC
SphI	G*CATG*C	GC

[a]Most commonly used name.

[b]The recognition sequences of suitable enzymes include six bases and a four-nucleotide 3' overhang (italics). End repair with the Klenow fragment of DNA polymerase removes the overhang and subsequent ligation leaves the dinucleotide indicated in bold at the junction site.

[c]Enzymes that permit junction sites of GC, CG, and AT are commercially available.

5. Purify the digested PCR product using GENECLEAN®, as described in **Subheading 3.1.6.**.
6. Ligate the restricted PCR product, as described in **Subheading 3.1.7.**, to yield the intermediate construct.
7. Transform 10% of the ligation reaction into *E. coli* DH5α competent cells using standard methods (*see* Chapter 4).
8. Screen colonies for the introduced restriction site by restriction analysis (*see* Chapter 20). Typically, approx 80% of the resultant clones will be mutants.

3.2.3. Stage 2

1. Digest the intermediate construct with the restriction enzyme whose recognition site was incorporated into the primer (*see* **Subheading 3.1.5.**).
2. Remove the 5' overhangs to restore the reading frame of the mutated coding sequence, using either T4 DNA polymerase or the Klenow fragment of *E. coli* DNA polymerase.

    a. Using *E. coli* Klenow fragment
    Combine the following components:

Digested DNA (*see* **Note 18**)	100–250 ng
10X Klenow polymerase reaction buffer	2 µL
10 mM dNTP mix	1 µL
10 U of Klenow polymerase	1 µL
Distilled deionized water to	20 µL

    Incubate at 37°C for 10 min and then heat inactivate the enzyme at 75°C for 10 min (or according to the manufacturer's recommendation).

b. Using T4 DNA polymerase
   Combine the following components:

Digested DNA (see **Note 18**)	100–250 ng
10X T4 DNA polymerase reaction buffer	2 µL
10 mM dNTP mix	1 µL
1 mg/mL BSA	1 µL
3 U of T4 DNA polymerase	1 µL
Distilled deionized water to	20 µL

   Incubate at 24°C for 20 min and heat to 75°C for 10 min to inactivate the enzyme.

3. Precipitate the template with ethanol (**Subheading 3.1.4.**, **steps 2** and **3**) and resuspend in 15 µL of distilled water.
4. Ligate the blunt-ended product to yield the final construct as described in **Subheading 3.1.7.** (see **Note 19**).
5. Transform 10% of the ligation product into *E. coli* DH5α competent cells using standard methods (see Chapter 4).
6. Isolate plasmid DNA (see Chapters 8–10) and screen mutants for the final in-frame site-directed mutant by looking for clones that now lack the unique restriction site used in the cloning process (see Chapter 20).

3.3. QuikChange Site-Directed Mutagenesis

The QuikChange site-directed mutagenesis protocol is shown in **Fig. 6**.

3.3.1. Primer Design (see **Notes 4–6**)

When designing primers for QuikChange site-directed mutagenesis, care should be taken to synthesize complementary mutagenic oligonucleotides that are between 25 and 30 nucleotides in length, have a GC content that is as close to 50% as possible, and have a melting temperature of at least 78°C. Ideally, the mutation site should be in the center of the oligonucleotide, as indicated in **Fig. 5**.

3.3.2. Mutant Construction

1. Set up the following PCR reaction:

10X Reaction buffer	5 µL
Template DNA	10–50 ng
Mutant primer 1	125 ng
Mutant primer 2	125 ng
10 mM dNTP stock	1 µL
2.5 U *Pfu*Turbo DNA polymerase	1 µL
Double-distilled deionized water to	50 µL

2. Perform the PCR reaction in a thermocycler using the following conditions (see **Note 20**):

Step	Cycles	Temperature	Time
1	1	94°C	30 s
2	18	94°C	30 s
		55°C	45 s
		68°C	120 s/kb
3	1	68°C	5 min
4	1	4°C	Indefinitely

3. Digest methylated parental DNA by adding 2 units *Dpn*I per 50-μL reaction. Incubate at 37°C for 1 h.
4. Heat inactivate *Dpn*I by incubation at 65°C for 20 min.
5. Transform 10% of the amplification reaction into competent *E. coli* DH5α (*see* Chapter 4).
6. Isolate plasmid DNA from candidate *E. coli* clones (*see* Chapters 8–10) and sequence the region that contains the mutation (*see* Chapter 22). This is a practical approach because this technique gives a very high frequency of mutant constructs (80–90% of constructs) although the authors typically obtain small numbers of colonies (*see* **Note 21**) *(4.5)*.

4. Notes

1. Magnesium chloride is required for the activity of the DNA polymerase and is typically used at 1.5 m*M* final concentration, although variation of Mg^{2+} levels between 1.0 and 2.5 m*M* $MgCl_2$ can increase the specificity of the amplification reaction *(7)*.
2. The choice of enzyme depends on the length of the construct and the GC content of the target sequence. It is important to choose a thermostable polymerase that has both the processivity to amplify the entire construct as well as the fidelity to amplify the target sequence without introducing mutations in addition to those directed by the mutagenic primers. The authors use an enzyme cocktail comprising *Taq* DNA polymerase that has high processivity and Vent DNA polymerase that has high fidelity because of its proofreading activity. These enzymes are typically combined at a *Taq*/Vent ratio of 20:1 enzyme units. A list of DNA polymerase fidelities (error rates) for the enzymes used in this chapter illustrates the advantages of each enzyme. The data are technical information supplied by the manufacturer of the relevant enzyme.

	Error frequency (errors per nucleotides incorporated)
Taq Polymerase (Fermentas MBI)	8.0×10^{-6}
Vent Polymerase (New England Biolabs)	2.8×10^{-6}
Pfu DNA polymerase (Stratagene)	1.3×10^{-6}

3. Any similar commercial kit may be used, following the manufacturer's instructions. An alternative protocol is given in Chapter 16.
4. Conventional wisdom dictates that both primers of a primer pair should have a similar melting temperature and that they should ideally have a GC content of 50–60%. Runs of Gs and Cs should be avoided to minimize false priming in regions of high GC content. A primer should not readily form dimers, either with itself or with its amplification partner. Entropically favorable ($\Delta G < 0$) hairpin loops should be avoided, especially when stable at temperatures approaching the annealing conditions of the amplification reaction.
5. There are a number of methods available for calculating the melting temperature of an oligonucleotide with the most common method being the (4+2) rule, whereby guanidine and cytosines contribute approx 4°C to the melting temperature, whereas adenosines and thymidines account for 2°C. A more accurate method to calculate the thermodynamic T_m is to use the formula

$$T_m = 81.5 + 0.41(\%GC) - (675/N) - \%(\text{mismatch})$$

in which *N* is the length of the primer in nucleotides.
6. Software in the public domain such as Primer Generator (www.med.jhu.edu/medcenter/primer/primer.cgi) *(8)* and the authors' laboratory's own silent restriction-site program (available for free use at http://www.dwalab.com) may be used to engineer silent restric-

tion sites into one of the primers. This is most often accomplished by replacing one particular codon with another to generate a nucleotide change that is required to create a restriction site while maintaining the amino acid sequence of the protein. The new site need not be introduced at the mutation point and can be in either of the primers. In rare cases, silent mutagenesis can introduce codon bias that may adversely affect expression levels of the mutant gene.

7. Until recently, commercially available Type IIS restriction enzymes were inefficient, expensive, and rather limited in number compared to Type II enzymes. However, many high-efficiency, inexpensive, Type IIS enzymes are now available. The best characterized for use in EIPCR-IIS are those with six-base recognition sites that generate four nucleotide 5' overhangs such as *Bpi*I, *Eco*31I, and *Esp*3I that are used in the Dovetail™ PCR cloning kit (MBI Fermentas). These enzymes require only three nucleotides 5' of the recognition site to efficiently cleave the PCR product. Only the four-nucleotide overhang remains on the PCR product, and as it does not contribute to the recognition site, any sequence can be accommodated. As the recognition sites are six nucleotides, they occur infrequently in most plasmids and coding sequences. *Sap*I is even less likely to occur elsewhere in the plasmid, as it has a seven-nucleotide recognition site. However, ligation of *Sap*I overhangs is less efficient because *Sap*I generates only a three-nucleotide overhang.

8. The selection of the thermostable polymerase is important only in terms of the quantity and fidelity of the product. Enzymes, such as *Taq* polymerase, that insert additional A residues on the 3' termini of constructs can be used, because the ends of the PCR product are removed prior to ligation.

9. Annealing conditions depend on the thermal properties of the primer pairs used. Typically, annealing temperatures between 45°C and 60°C are used (*see* **Note 5**). As with all PCR techniques, the chosen melting temperature should be high enough to induce DNA strand separation but maintain the stability of the polymerase as judged by its half-life at that temperature. Generally, the annealing temperature should not be more than 5°C below the melting temperatures of the primers to ensure the specificity of primer binding to the template. Primer extension is carried out between 68°C and 72°C. When a polymerase with proof-reading activity is included in the reaction, 25 or more cycles of amplification are performed. On completion of the amplification reaction, a final extension step of 5 min at 72°C is carried out to deplete any unincorporated nucleotides and complete any unfinished product.

10. Reaction conditions are dependent on both the type of thermocycler and the brand of tubes used. Tube thickness plays an important role in heat transduction during amplification and affects the ramp times between steps in the program.

11. *Dpn*I generally exhibits 100% activity in 1X amplification buffer. To ensure that the enzyme is active in the amplification buffer, which may contain dimethyl sulfoxide (DMSO) as an additive, perform a digestion of a control plasmid in amplification buffer.

12. *Taq* DNA polymerase must be completely removed from the reaction mixture, as even traces of the enzyme can fill in the 5' overhangs that are generated in the digestion reaction.

13. For maximal efficiency of the ligation step, precipitate the purified PCR product as described in **Subheading 3.1.4., steps 2** and **3**.

14. The maximum efficiency of ligation occurs in the ligase buffer supplied with the ligase enzyme, but the use of this requires an extra precipitation step to isolate the DNA. Ligations are often reasonably efficient in 1X restriction buffer.

15. As a shortcut, after 10 min of ligation the reaction can be precipitated as described in **Subheading 3.1.4., steps 2** and **3** and transformed in its entirety.

16. In the authors' experience, mutagenesis by classical inverse PCR is generally 40–55% efficient, whereas the group that pioneered the technique report an efficiency of 82% when mutating the α-complement portion of the *lacZ* gene in *E. coli* *(3)*. By contrast, using EIPCR-IIS, the reported mutagenesis rate (95%) using a similar *lacZ* system is similar to that experienced by the authors *(4)*. In general, this method works well for deletion mutants, but screening for point mutations without the aid of introduced (or deleted) restriction sites in the primers requires the sequencing of several clones and, thus, can be tedious. For reasons that are not clear, this approach can occasionally generate a number of inexplicable mutations. Some of these can be unwanted mutations at the ligation point apparently the result of damaged ends on the PCR product.
17. Add additional nucleotides 5' of the introduced restriction site to aid in digestion of the PCR products. For example, *Aat*II requires at least six nucleotides 5' to its recognition site in order to cut. By comparison, *Kpn*I requires only three additional nucleotides. Enzyme manufacturers usually provide this information.
18. Care should be taken to use nanogram amounts of DNA in order to not saturate the enzyme.
19. Selection for positive clones can be improved by digesting the ligation reaction prior to transformation to linearize any plasmids that contain the introduced site. Selection is improved because linear DNA is a much less efficient substrate for transformation than is circular DNA.
20. In practice, the extension time and number of cycles can be shortened for smaller templates (under 5 kb). Occasionally, primer displacement is observed, resulting in a duplication of the sequence covered by the primers. In our hands, this usually occurred in cases where the melting temperature of the primers was within 10–12°C of the extension temperature. Lowering the extension temperature in subsequent amplification reactions is the most cost-effective way of solving the problem.
21. In many cases, the mutational frequency is very high, but in other cases, the mutation may be very difficult to generate. This appears to be a property of the specific sequence of the DNA region in question and may be difficult to avoid because of restrictions in primer placement.

References

1. Smith M. (1985) In vitro mutagenesis. *Annu Rev Genet.* **19**, 423–462.
2. Kunkel, T. A., Benebek, K., and McClary, J. (1991) Efficient site-directed mutagenesis using uracil-containing DNA. *Methods Enzymol.* **204**, 125–139.
3. Hemsley, A., Arnhem, N. Toney, M. D., et al. (1989) A simple method for site-directed mutagenesis using the polymerase chain reaction. *Nucleic Acids Res.* **17**, 6545–6551.
4. Stemmer, W. P. C. and Morris, S. K. (1992) Enzymatic inverse PCR: a restriction site independent, single fragment method for high efficiency site directed mutagenesis. *BioTechniques* **13**, 214–220.
5. Hughes, M. J. G. and Andrews, D. W. (1996) Creation of deletion, insertion and substitution mutations using a single pair of primers and PCR. *BioTechniques* **20**, 188–196.
6. *Quikchange Site-Directed Mutagenesis Kit. Instruction Manual.* (1998) Stratagene, La Jolla, CA.
7. Kunkel, T. A. and Loeb, L. A. (1979) On the fidelity of DNA replication. Effect of divalent metal ion activators and deoxyribonucleoside triphosphate pools on in vitro mutagenesis. *J. Biol. Chem.* **254**, 5718–5725.
8. Turchin, A. and Lawlor, J.F. (1999) The Primer Generator: a program that facilitates the selection of oligonucleotides for site-directed mutagenesis. *BioTechniques* **26**, 660–668.

25

Creating Nested DNA Deletions Using Exonuclease III

Rosamund Powles and Lafras M. Steyn

1. Introduction

DNA fragments cloned into plasmids are frequently greater than 500 base pairs in length and thus may be too long to sequence from a single primer-binding site in the vector. An efficient way to sequence such large DNA inserts is to generate a nested set of deletions in the target DNA, effectively moving the priming site closer to the sequence of interest. Similarly, nested deletions can be used to delineate a feature of interest (e.g., a replicon *[1]* or promoter) or to subclone a region of DNA devoid of restriction enzyme sites.

Exonuclease III catalyzes the stepwise removal of mononucleotides from the recessed or blunt-ended 3' hydroxyl termini of double-stranded DNA. A 3' overhang of greater than or equal to four nucleotides is protected from exonuclease III activity. Exonuclease III will, however, act on nicks in duplex DNA or 3' overhangs of less than four nucleotides *(2)*. The enzyme also carries three other activities: an endonuclease specific for apurinic DNA, an RNase H activity *(3)*, and a 3' phosphatase activity, which removes 3' phosphate termini but does not cleave internal phosphodiester bonds.

*Bal*31 nuclease or pancreatic DNase I can also be used to generate nested deletions, but, of the three enzymes, exonuclease III is the best. Exonuclease III digestion proceeds at a uniform and predictable rate. Thus, deletions of predetermined lengths can be obtained by simply removing aliquots from the reaction at different time intervals. In addition, the entire series of enzymatic reactions required can be carried out in a single set of tubes without purification of intermediate products *(4,5)*.

In contrast, *Bal*31 degrades both ends of double-stranded DNA simultaneously; thus, both the target fragment and the vector sequences are degraded. In addition, *Bal*31 nuclease is processive and tends to digest the double-stranded DNA in an asynchronous manner. To obtain viable deletion mutants, it is therefore necessary to purify the truncated target fragments and reclone them into an appropriate vector *(5)*.

DNase I behaves like an endonuclease in the presence of Mg^{2+}, introducing nicks into double-stranded DNA. However, in the presence of Mn^{2+} or Co^{2+}, the enzyme cuts both strands of DNA at approximately the same place *(6)*. This property can be used to create deletions. However, the reaction is inefficient and the sites of digestion are random.

1.1. Strategy

The target DNA must be cloned into a suitable vector such that the correct restriction enzyme sites are available for both exonuclease III digestion and protection of DNA. To create deletion mutants, the double-stranded plasmid DNA is digested with two restriction enzymes. Both restriction enzyme sites must lie at the same end of the target DNA, as shown in **Fig. 1**. The restriction enzyme that cleaves closest to or within the target (insert) DNA must leave a 3' recessed or blunt end that is susceptible to exonuclease III. The restriction enzyme that cleaves closest to the priming site or vector DNA must leave a 3' protruding end of four nucleotides or more that is resistant to exonuclease III. Exonuclease III is then added and digestion proceeds unidirectionally away from the site of cleavage into the target DNA. Aliquots are removed at timed intervals to tubes containing S1 nuclease, which removes the remaining single-stranded DNA. Klenow polymerase is then used to blunt the ends of the DNA, which is recircularized using DNA ligase *(4,5)*. A schematic diagram of the steps involved is shown in **Fig. 1**.

2. Materials

2.1. Restriction Enzyme Digestion

1. CsCl/ethidium bromide-purified plasmid DNA (*see* **Note 1**).
2. Restriction enzymes and corresponding 10X buffers suitable for generating a 3' recessed terminus or blunt end, and a four nucleotide 3' overhang (*see* **Notes 2-4**).

2.2. DNA Purification

1. TE-saturated phenol : chloroform : isoamylalcohol (25 : 24 : 1).
2. Chloroform : isoamylalcohol (24 : 1).
3. 3 M Sodium acetate (pH 5.2). Dissolve 408.3 g sodium acetate·$3H_2O$ in 800 mL H_2O. Adjust the pH to 5.2 with glacial acetic acid. Adjust the volume to 1 L with H_2O.
4. 100% Ethanol; 70% ethanol (ice cold).

2.3. Exonuclease III and S1 Nuclease Digestion

1. Exonuclease III enzyme.
2. 10X Exonuclease III buffer: 0.66 M Tris-HCl (pH 8.0), 66 mM $MgCl_2$ (*see* **Note 5**).
3. 10X S1 buffer: 5.0 mL of 5 M NaCl, 1.1 mL of 3 M potassium acetate (pH 4.5), 5.0 mL glycerol, 20 µL of 1 M $ZnSO_4$ (*see* **Note 5**).
4. S1 mix: 172 µL H_2O, 27 µL of 10X S1 buffer, 60 U S1 nuclease. Make up fresh.
5. S1 stop solution: 0.3 M Trizma base, 0.05 M EDTA (pH 8.0)

2.4. Blunt-Ending of DNA Termini

1. Klenow mix: 3 µL 0.1 M Tris-HCl (pH 8.0), 6 µL of 1 M $MgCl_2$, 20 µL H_2O, 3 U Klenow DNA polymerase. Make up fresh.
2. dNTP mix, containing 0.125 M each of dATP, dCTP, dGTP, and dTTP. Store at –20°C.

Creating Nested Deletions

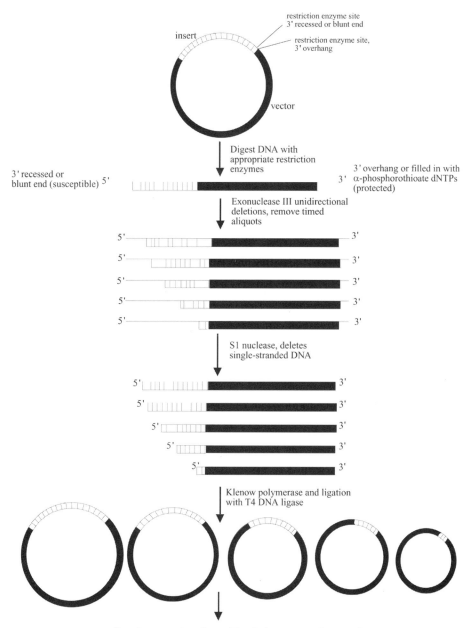

Fig. 1. Schematic diagram for the construction of unidirectional deletions using exonuclease III.

2.5. Ligation and Transformation

1. 5X Ligase buffer: 2.5 mL of 1 M Tris-HCl (pH 7.6), 0.25 mL of 1 M MgCl$_2$, 0.5 mL dithiothreitol (DTT), 0.5 mL of 0.1 M ATP, 2.5 g polyethylene glycol 6000; add water to 10 mL. Store in 0.2-mL aliquots at $-20°C$.
2. Ligase mix: 0.8 mL H$_2$O, 0.2 mL of 5X ligase buffer, 5 U T4 DNA ligase. Make up fresh.
3. *Escherichia coli* competent cells; *see* Chapter 4 or **ref. 7**.
4. SOC medium: 20 g tryptone, 5 g yeast extract, 0.5 g NaCl; add water to 980 mL and adjust to pH 7 by the addition of 5 N NaOH. Sterilize by autoclaving and add 20 mL sterile 1 M glucose.
5. Luria–Bertani (LB) agar plates containing appropriate antibiotics for selection of plasmid.

3. Methods
3.1. Restriction Enzyme Digestion

1. Digest 10 μg of the plasmid DNA with the restriction enzyme that generates the 3′ recessed terminus or blunt end according to the manufacturer's instructions (this enzyme site must lie closest to the target DNA).
2. When all the closed circular DNA has been converted to linear DNA (*see* **Note 6**), adjust the buffer and add the second restriction enzyme. This enzyme must generate a four-nucleotide 3′ overhang.
3. Alternatively, if the buffer cannot be adjusted, purify the DNA as described in **Subheading 3.2.**, resuspend the pellet in water, and digest with the second restriction enzyme.
4. Purify the DNA by extraction with phenol:chloroform or heat inactivation of the restriction enzyme, followed by precipitation of the DNA (*see* **Subheading 3.2.**).

3.2. DNA Purification

1. Add 1 volume of TE-saturated phenol:chloroform:isoamyl alcohol to the restriction digest. Mix by vortexing for 1 min and centrifuge at 12,000g for 5 min.
2. Transfer the upper aqueous phase to a fresh tube and add 1 volume of chloroform/isoamylalchol. Mix by vortexing for 1 min and centrifuge at 12,000g for 5 min.
3. Transfer the upper aqueous phase to a fresh tube and add 0.1 volume of 3 M sodium acetate and 2 volumes of 100% ethanol. Mix well and centrifuge at 12,000g for 10 min.
4. Carefully pour off the supernatant and wash the pellet with 1 mL of 70% ethanol. Draw off the ethanol with a pipet, taking care not to dislodge the pellet. Air-dry the pellet.

3.3. Exonuclease III and S1 Nuclease Digestion

1. Resuspend the digested DNA pellet in 60 μL of 1X exonuclease III buffer. Place on ice.
2. Add 7.5 μL of S1 mix to 24 numbered 1.5-mL Eppendorf tubes (or a conical bottom 96-well plate) and place on ice.
3. Incubate the DNA at 37°C for 5 min.
4. Transfer 2.5 μL of DNA to the first Eppendorf tube containing the S1 reaction mixture. Mix the solutions by pipetting up and down three to four times.
5. To the remainder of the digested DNA, add 300–500 units of exonuclease III. Mix by vortexing briefly and return the mixture to 37°C immediately. Start timing the reaction (*see* **Note 7**).
6. Remove 2.5-μL aliquots from the reaction at 30-s intervals and transfer to successive Eppendorf tubes containing S1 mix. Mix by pipetting.
7. After samples have been added to all 24 tubes, incubate at 30°C for 30 min.

8. Add 1 µL of S1 stop mixture to each Eppendorf tube and incubate at 70°C for 10 min to inactivate the S1 nuclease.
9. Centrifuge the tubes briefly to collect the liquid at the bottom of the tube and transfer onto ice. Take 2- to 3-µL samples of every other time point and analyze the extent of digestion by agarose gel electrophoresis (*see* Chapter 20 and **Notes 8** and **9**).

3.4. Blunt-Ending of DNA Termini

1. Transfer the Eppendorf tubes containing the DNA of the desired sizes to a 37°C water bath and add 1 µL of Klenow mixture. Incubate for 5 min at 37°C.
2. Add 1 µL of dNTP mix and incubate for a further 15 min at room temperature.
3. Heat inactivate the Klenow polymerase by heating at 65°C for 10 min.

3.5. Ligation and Transformation

1. Add 32 µL of ligase mix to each Eppendorf tube; mix and incubate for 1–3 h at room temperature.
2. Add 10 µL of each ligation reaction to 200-µL aliquots of *E. coli* competent cells, mix gently, and place on ice for 30 min.
3. Place in a 42°C water bath for 45 s and then place on ice for 2 min.
4. Add 800 µL of SOC medium and then shake at 37°C for 1 h. Plate the entire mixture on the appropriate selective media.
5. Extract plasmid DNA from transformants (*see* Chapters 8 and 9); analyze by restriction digestion and agarose gel electrophoresis as described in Chapter 20 (*see* **Fig. 2** and **Note 10**).

4. Notes

1. The generation of ordered sets of deletions by this method relies on the uniform digestion rate of exonuclease III. However, the enzyme also digests from nicks in double-stranded DNA molecules, creating single-stranded gaps. The effect of random nicks in the starting DNA is to randomize the resulting deletions. The greater the percentage of nicked molecules in the starting material, the more random the deletions become and the more difficult it becomes to screen for the desired (predicted) deletions among the resulting subclones. Therefore, it is important to minimize the proportion of nicked molecules in the starting DNA. This can be accomplished by (1) removing nicked (and linear) molecules from the plasmid preparation and (2) minimizing the generation of single-stranded nicks during restriction enzyme digestions. Some restriction enzyme preparations contain exonuclease activity. Before starting digestion, check an aliquot of the DNA by agarose gel electrophoresis. If the preparation contains more than 10% of relaxed circular molecules, it should be purified by conventional equilibrium sedimentation in a CsCl–ethidium bromide gradient. This step has the added advantage of removing small pieces of DNA and RNA from the closed circular DNA preparation.
2. The following restriction enzymes generate 3' recessed or blunt ends after digestion (susceptible to exonuclease III): *Eco*RI, *Sma*I, *Xma*I, *Sal*I, *Acc*I, *Hinc*II, *Hin*dIII, *Not*I, *Xba*I, *Spe*I, *Bam*HI, *Cla*I, and *Xho*I. *Bgl*II, *Bst*XI, *Apa*I, *Sac*I, *Kpn*I, *Pst*I, and *Sph*I leave 3' protruding ends after digestion (resistant to exonuclease III). Note that the restriction enzymes *Hha*I, *Pvu*I, and *Sac*II generate 3' protruding ends that are not resistant to exonuclease III digestion.
3. If no 3' protruding ends are available to block exonuclease III digestion of vector sequences, 3' recessed ends can be filled in with α-phosphorothioate dNTP mix and Klenow DNA polymerase before the second restriction digestion is performed (*8*).

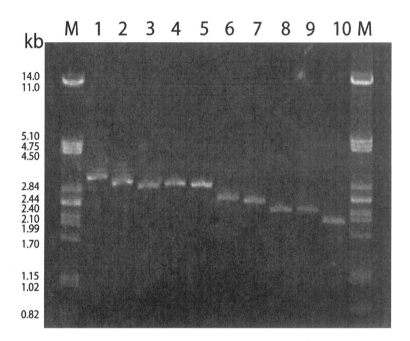

Fig. 2. Ethidium bromide-stained agarose gel of purified plasmid DNA digested with *Alw*NI. A vector containing a 600-bp insert of *Mycobacterium tuberculosis* DNA was subjected to exonuclease III deletion as described in the text. Lanes 1–10: digested plasmid DNA corresponding to 0–12 min of exonuclease III digestion at 22°C. Lane *M* indicates molecular-weight markers. Progressive deletions of approx 250 bp of insert DNA were achieved in this experiment.

4. To maximize the efficiency of cleavage, avoid using enzymes whose recognition sites are immediately adjacent to each other in a polycloning site.
5. Exonuclease III buffer and 10X S1 buffer are usually supplied with the enzyme and, therefore, do not need to be prepared.
6. To ensure that both enzymes cleave the DNA, divide the DNA into two aliquots and digest each with one of the two restriction enzymes. When all of the closed circular DNA has been converted to linear DNA, as confirmed by agarose gel electrophoresis, adjust the buffer and cleave with the second enzyme.
7. Under the described conditions, the amount of exonuclease III is saturating and approx 400–450 nucleotides are removed per minute. The rate of exonuclease III digestion can be altered by changing the incubation temperature *(4)*. Using 300–500 units of exonuclease III, the digestion rate exhibits the temperature dependence shown in the following table. There is usually a 20- to 30-s lag before the reaction begins. Different termini of DNA are degraded at different rates, so the table should only be used as a guide.

Temperature (°C)	4	22	25	30	37	45
Digestion rate (nucleotides/min)	25	80	80–125	210–230	400–450	600

8. If there does not appear to be any digestion of the experimental DNA by exonuclease III, the NaCl present after restriction digestion may be at an inhibitory concentration: A NaCl

concentration as low as 20 mM is known to affect the exonuclease III digestion rate. High EDTA concentrations may also inhibit exonuclease III by binding the Mg^{2+} ions necessary for the reaction. NaCl and EDTA can be removed by precipitating the DNA with ethanol, as described in **Subheading 3.2.**.

9. If gel electrophoresis of the digestion products indicates the presence of secondary bands, then the protecting enzyme may not have digested the DNA to completion. The resulting single-cut molecules are digested at twice the rate of doubly-cut molecules resulting in the secondary smaller species.
10. The rate of digestion by exonuclease III is dependent on the base composition of the DNA: C>>A~T>>G. As a result, different regions of DNA are degraded at different rates. When we performed this protocol with *M. tuberculosis* DNA, we found that the size of plasmids derived did not correspond to the length of time of digestion by exonuclease III. Smaller plasmids were sometimes found from earlier time-points than were found at the later time-points.

References

1. Da Silva-Tatley, F. M. and Steyn, L. M. (1993) Characterization of a replicon of the moderately promiscuous plasmid, pGSH5000, with features of both the min-replicon of pCU1 and the *ori-2* of F. *Mol. Microbiol.* **7,** 805–823.
2. Weiss, B. (1976) Endonuclease II of *Escherichia coli* is exonuclease III. *J. Biol. Chem.* **251,** 1896–1901.
3. Rogers, S. G. and Weiss, B. (1980) Exonuclease III of *Escherichia coli* K-12 an AP exonuclease. *Methods Enzymol.* **65,** 201–211.
4. Henikoff, S. (1987) Unidirectional digestion with exonuclease III in DNA sequence analysis. *Methods Enzymol.* **155,** 156–165.
5. Sambrook, J., Fritsch, E. F., and Maniatis, T. (eds.) (1989) *Molecular Cloning: A Laboratory Manual*, Cold Spring Harbor Laboratory, Cold Spring Harbor, NY.
6. Melgar, E and Goldthwait, D. A. (1968) Deoxyribonucleic acid nucleases. II. The effects of metals on the mechanism of action of deoxyribonuclease I. *J. Biol. Chem.* **243,** 4409–4416.
7. Inoue, H., Jojima, H., and Okayama, H. (1990) High efficiency transformation of *Escherichia coli* with plasmids. *Gene* **96,** 23–28.
8. Ozkaynak, E. and Putney, S. D. (1987) A unidirectional deletion technique for the generation of clones for sequencing. *BioTechniques* **5,** 770–773.

26

Transposon and Transposome Mutagenesis of Plasmids, Cosmids, and BACs

Alistair McGregor

1. Introduction

Transposons are mobile genetic elements with the capacity to "jump" to new target DNA. Although first discovered in *Zea mays* by McClintock *(1)*, they are present in DNA genomes of species from all kingdoms. Transposons fall into two major classes. Class I transposons are retroelements that transpose via an RNA intermediate that is synthesized by a reverse transcriptase (e.g., Alu elements in primates and Ty elements in yeast); class II transposons transpose directly from DNA to DNA (e.g., P elements in yeast and Tn elements in bacteria). This chapter describes the application of class II Tn elements from bacteria to mutagenize plasmid DNA by insertional mutagenesis.

Typically, class II transposons have inverted repeat sequences flanking encoded genes, such as transposase and resolvase, which are necessary for Tn transposition to target DNA. Often, these elements also encode a drug-resistance marker. The Tn elements move by excising themselves from a donor site and reinserting themselves elsewhere on target DNA, by a "cut and paste" mechanism. Insertion into target DNA is a random procedure and usually occurs as a single event over a 200-kb range because of multiple-sequence insertion immunity. Insertion immunity is controlled by the interplay among target DNA, transposon proteins, and the transposon *(2)*.

Random transposition allows the investigator to rapidly generate a library of thousands of random single-insertion mutants of a target plasmid. Mutated plasmids are selected in *Escherichia coli* by virtue of the drug-resistant marker encoded on the transposon. For each mutant plasmid, the location of the transposon insertion is defined by restriction digestion analysis of the DNA. Precise mapping of the insertion site can be obtained by DNA sequencing of the mutated plasmid using sequencing primers specific for the 5' or the 3' end of the transposon that read across the transposon–plasmid junction.

From: *Methods in Molecular Biology, Vol. 235:* E. coli *Plasmid Vectors*
Edited by: N. Casali and A. Preston © Humana Press Inc., Totowa, NJ

Transposon insertion into a gene can cause partial or complete loss of the gene function; consequently, transposon-mediated mutagenesis has tremendous potential in forward genetic studies. The simplicity and versatility of the transposon has led to its use as a molecular-biology tool in a number of different areas such as identifying essential, nonessential, and virulence genes in bacteria, parasites, and large DNA viruses *(3–6)*. Transposon-mediated mutagenesis has great potential in the rapid definition and modification of large target plasmids such as cosmids and bacterial artificial chromosomes (BACs). BAC vectors are based on the F plasmid and can accommodate >600 kb of foreign DNA sequence *(7)*. Because of their relative ease of use, a growing number of investigators are using BAC clones, particularly for the study of mammalian genomes. Random transposon mutagenesis can aid in the sequencing of an entire BAC clone using a single pair of primers. Alternatively, mutagenesis of a BAC clone can help define the function of encoded genes. This approach has been particularly successful in the study of large DNA herpesviruses involving the use of mutated, infectious BAC clones of the entire viral genome *(3,8,9)*. It should be noted that although this chapter discusses insertional transposon mutagenesis, transposons can also be used to generate deletions or inversions in plasmid clones. Investigators interested in pursuing these strategies are urged to consult Epicentre Technology, Inc about their EZ::TN plasmid-based transposon deletion kit.

The conventional approach of transposon-mediated mutagenesis involves conjugational transfer of the Tn from a donor bacterial strain to a recipient strain carrying the target plasmid. The donor plasmid is lost by an inability to propagate in the recipient strain. This can be the result of the use of a temperature-sensitive (*ts*) origin of replication (*see* below). Alternatively, the donor plasmid has an R6 gamma origin of replication that requires the *pir* gene product *in trans*, supplied only in the donor bacterial strain, to enable plasmid replication *(10)*. However, this approach is inefficient because most of the transposon insertions occur in the recipient chromosome (*see* **Fig. 1A**). A further drawback to this approach is that mutated target plasmids have to be isolated to distinguish them from colonies containing mutated bacterial chromosomes. Thus, the mutated plasmids are extracted, transformed into fresh bacteria, and colonies selected using the drug-resistance marker encoded on the transposon. This approach has been used successfully for the mutation of subgenomic fragments of murine cytomegalovirus (6) as well as full-length BAC clones of pseudorabies virus *(9)*.

Fig. 1. *(opposite page)* In vivo transposon mutagenesis of plasmid DNA. In these examples, the target plasmid encodes a Cam^R marker and the transposon encodes a Kan^R marker. **(A)** Transposition of target plasmid DNA by conjugational transfer of the transposon from a donor strain (F plus) to a recipient strain (F minus) carrying the target DNA. The transposon is encoded on a suicide plasmid, which is transferred between strains, and transposition occurs into the target plasmid DNA or bacterial chromososme. A mixed population of drug-resistant colonies is pooled to isolate mutated plasmids by transformation of a mixed DNA prep into fresh *E. coli*. **(B)** Transposon mutagenesis of target plasmid DNA using a temperature-sensitive donor plasmid introduced into bacteria via transformation of bacteria. Once bacteria are identified that carry both target and donor plasmids, growth conditions are altered to nonpermissive conditions for the donor plasmid and drug marker selection encoded on the transposon used to select for bacterial colonies carrying mutated target plasmids.

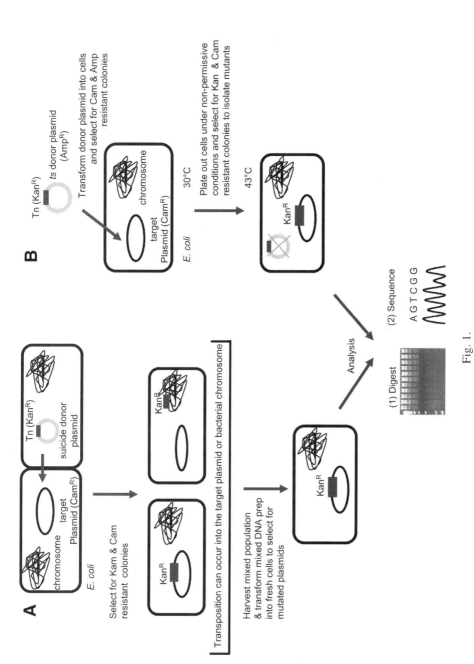

Fig. 1.

The selection of a Tn variant that preferentially targets plasmid DNA *(11)*, when coupled with the use of a *ts* donor plasmid *(12)*, has improved the efficiency of in vivo mutagenesis of plasmids (*see* **Fig. 1B**). In this approach, the Tn donor plasmid is transformed into the bacterial strain carrying the target plasmid at conditions permissive for donor plasmid replication. Once stable clones carrying both plasmids are identified, the bacteria are grown under conditions that are nonpermissive for donor plasmid replication in order to select for transposon insertion into the target DNA. To eliminate secondary transposition events, the genes required for transposition are encoded elsewhere on the donor plasmid, outside of the inverted repeats *(12)*. This approach has been successfully applied to large BAC clones *(8)*.

1.1. Tn7-Based In Vitro Mutagenesis of Plasmids and Cosmids

The simplest and most efficient approach to transposon-mediated mutagenesis of plasmid DNA is by in vitro transposition using purified and modified transposase and minimal transposon elements encoding drug-resistance markers flanked by inverted repeats. Two of the best-studied transposon systems, Tn5 and Tn7, have been successfully modified to produce in vitro transposon systems for the mutation of plasmid DNA *(13,14)*. Both systems are comparable in terms of ease of use, efficiency of mutagenesis, and commercial availability (Tn7 GPS system available from NEB, Inc; Tn5 EZ::TN system available from Epicentre Technologies, Inc). The protocol described in **Subheading 3.1.** is based on a modified in vitro Tn7 transposon mutagenesis system. To avoid redundancy, only the Tn7 system will be discussed here. However, information imparted in **Subheading 4.** will aid with all protocols. It should be noted that in vitro transposon systems have also been described for Mu, Himar1, and Ty1 transposons *(15–17)*.

The wild-type Tn7 transposon is approx 14 kb in length and encodes five transposon proteins (TnA, B, C, D, and E) that are required for the control and execution of the transposition event, and a kanamycin-resistance marker flanked by terminal inverted repeats (*see* **Fig. 2A**). However, any DNA sequence can be transposed provided that it is flanked by the transposon inverted-repeat sequences and the necessary transposition proteins are supplied *in trans*. Consequently, the transposon can be reduced to a basic element, encoding a drug-resistance marker flanked by the Tn7 inverted repeats, called a transprimer (*see* **Fig. 2B**). This transprimer can be further modified to include additional marker or reporter genes, such as *gfp* or *gpt*, without affecting the transposition efficiency *(14,18)* (*see* **Fig. 2C**). The in vitro transposition reaction requires only three transposon proteins: TnA, B, and C*. The TnC* protein is a modified version of TnC that enables transposition to occur in the absence of TnD and E provided that Mg^{2+} and ATP are also included in the reaction *(14)*.

In this method, a Tn7 transprimer is added to a reaction mixture containing target plasmid DNA and purified transposase. Transposition of target DNA occurs efficiently at 37°C in the presence of Mg^{2+} and ATP. Mutated plasmids are selected by transformation of the transposition reaction into *E. coli* and selecting for resistance to the Tn-encoded antibiotic resistance (*see* **Fig. 3**). In an in vitro reaction, it is necessary to eliminate the donor transposon. This is achieved by supplying the transposon as a poly-

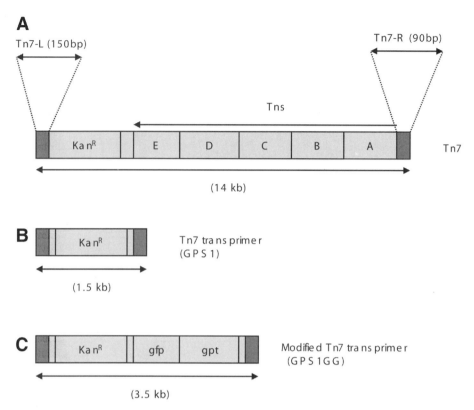

Fig. 2. The structure of the Tn7 transposon and its functional minimal derivatives based on the genome priming system (GPS) from New England Biolabs Inc. (**A**) The full-length transposon is 14 kb. It encodes five transposon proteins (TnA, B, C, D, and E) and a drug-resistance marker (Kan^R) flanked by terminal repeats. (**B**) A basic Tn7 transprimer element (Tn7GPS) encoding a drug-resistance marker flanked by inverted repeats is sufficient for in vitro transposition of target DNA provided that TnA, B, and C* are provided *in trans*. (**C**) The basic Tn7 transprimer can be modified to accommodate additional genes without affecting the ability of the Tn7 transposon to function in vitro. The example shown, pGPS1GG (available from the author), carries the additional metabolic markers *gpt* and *gfp* for selection in mammalian cells.

merase chain reaction (PCR) product or by cloning it onto a plasmid incapable of replication in normal *E. coli*, such as plasmids containing the R6 gamma origin of replication *(10,13,14)*. The latter approach is perhaps preferred, as it removes the requirement of an extra step in the gel purification of the Tn prior to mutagenesis of the target plasmid. In addition, if necessary, the Tn can be easily manipulated on the donor plasmid to introduce additional cassettes provided the plasmid is grown in a supporting bacterial strain that encodes the *pir* gene *(10)*.

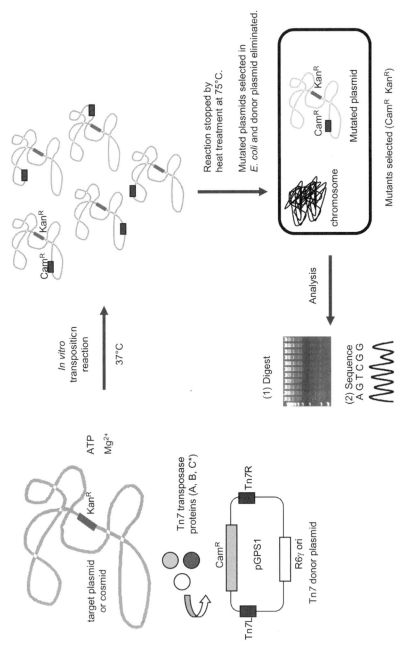

Fig. 3. Tn7-based in vitro mutagenesis of plasmids and cosmids. The T7 donor plasmid (or transprimer) is added to a reaction mixture containing target plasmid DNA and purified transposase proteins. Tn-containing plasmids are selected when the transposition reaction is transformed into bacteria by virtue of the drug-resistance marker encoded on the Tn. Mutated plasmids are subsequently isolated and analyzed.

The in vitro transposition systems work well on plasmid and cosmid target DNA molecules of up to 75 kb in size (McGregor, unpublished data), generating thousands of mutants from a single reaction. However, these systems do not work as efficiently in larger target DNA molecules such as BACs. In the author's experience, mutating large BAC clones (>100 kb) with an in vitro transposon system results in a high percentage of mutated BACs with acquired deletions as well as a transposon insertion. These deletions presumably occur as a result of transforming the transposition reaction into bacteria and the phenomenon is unique to large BAC clones. The problem can be alleviated to an extent by using DH10B-derived bacterial strains for BAC selection and maintenance (e.g., EC100; Epicentre Technology, Inc). However, this does not entirely eliminate deletion events especially when extremely large clones (>150kb) are used.

1.2. Transposome Mutagenesis of Large BAC Plasmids

As an alternative approach to the above-described in vitro transposition reaction, the use of transposome complexes are recommended for large (>100kb) BAC plasmid mutagenesis. A transposome is a stable complex formed between the Tn5 transposon or transprimer and the transposase protein *(19)*. In this protocol, the Tn5 transprimer is isolated from a donor plasmid, by restriction digestion and band purification (*see* Chapter 16) or by PCR amplification. The Tn5 transposon DNA is mixed with the transposase protein in the absence of Mg^{2+} ions to form the transposome pseudocomplex. Transposome complexes can be purchased commercially from Epicentre Technology, Inc. or, alternatively, generated by the investigator using purified components *(19)*. The stable transposome is introduced into bacteria carrying the target BAC plasmid via electroporation. Once in bacterial cells, the transposase enzyme is activated by the presence of Mg^{2+} and transposition occurs *(19)* (*see* **Fig. 4**).

2. Materials
2.1. Tn7-Based In Vitro Mutagenesis of Plasmids and Cosmids

1. Tn7 transprimer kit: Genome Priming System, GPS-1 (New England Biolabs). This kit contains the basic transprimer plasmid (with either chloramphenicol- or kanamycin-resistance markers), purified Tn proteins (A, B, C*), Tn reaction buffers, and Tn7-specific sequencing primers.
2. Target DNA (0.01–0.2 µg/µL; 3–150kb). Store at –20°C or –70°C in single-use aliquots or, alternatively, at 4°C (*see* **Note 1**).
3. Donor DNA (0.02 µg/µL).
4. Competent *E. coli* cells (DH10B or a derivative is recommended for BAC-based plasmids) (*see* **Note 2**).
5. 1 mm Gapped cuvets.
6. Electroporator.
7. Luria–Bertani (LB) medium: 10 g tryptone, 5 g yeast extract, 10 g NaCl per liter. Autoclaved.
8. LB agar: LB medium, 15 g bactoagar per liter. Autoclaved.
9. Appropriate antibiotics for selection of the Tn and target plasmid.

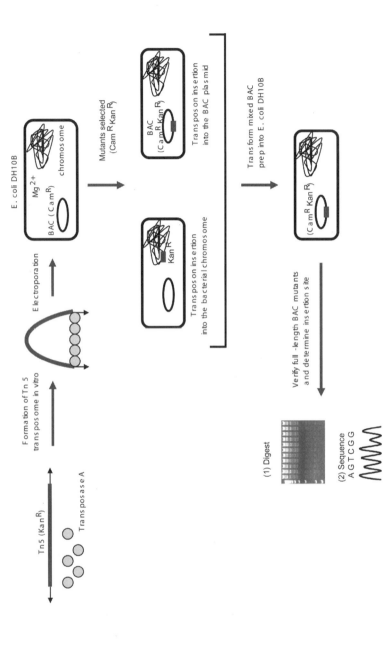

Fig. 4. Transposome mutagenesis of large BAC plasmids. The stable transposome complex, which is formed between the Tn5 transposon (encoding a Kan^R marker) and the transposase A protein in the absence of magnesium, is introduced into bacteria carrying the target BAC plasmid via electroporation. The presence of magnesium in the bacterial cell enables transposition to occur. Both bacterial genome and target plasmid DNA are transposed. To obtain BAC plasmid mutant colonies, a mixed plasmid DNA prep is generated from all kanamycin-resistant colonies and the DNA used to transform fresh DH10B cells. Mutated BACs are subsequently isolated and analyzed.

2.2. Transposome Mutagenesis of Large BAC Plasmids

1. EZ::Tn transposome (Epicentre Technology, Inc.) (*see* **Note 3**).
2. Electrocompetent *E. coli* DH10B cells carrying the target BAC (*see* **Note 2**).
3. 0.1 mm gapped cuvet.
4. Electroporator.
5. LB medium: 10 g tryptone, 5 g yeast extract, 10 g NaCl per liter. Autoclaved.
6. LB agar: LB medium, 15 g bactoagar per liter. Autoclaved.
7. Appropriate antibiotics for selection of the Tn and BAC.
8. BAC preparation kit (*see* **Note 4**).

3. Methods
3.1. Tn7-Based In Vitro Mutagenesis of Plasmids and Cosmids

1. Thaw the contents of the Tn7 GPS transprimer kit and place on wet ice.
2. Prepare the following reaction mixture in a 1.5-mL tube:

GPS 10X reaction buffer	2 µL
Donor plasmid	1 µL
Target plasmid (see Note 5)	0.1–0.5 µg
Sterile water to a volume of	18 µL.

   Add 1 µL of Tn ABC* transposase proteins, spin down the tube contents to mix, and incubate at 37°C for 10 min.
3. Add 1 µL of start solution (300 m*M* magnesium acetate) to the bottom of the tube and mix by pipetting slowly up and down.
4. Heat at 37°C for 1 h on a heating block.
5. Incubate the reaction tube at 75°C for 10 min to heat inactivate the reaction (*see* **Note 6**).
6. Remove the sample, spin briefly, and dilute to 60 µL with sterile water.
7. Use 3–5 µL of the reaction to transform *E. coli* DH10B by electroporation (1.3–2.2kV, 100 Ω, and 25 µF) (*see* Chapter 5 and **Note 7**).
8. Plate out half of the transformation reaction on LB agar plates with antibiotic selection for both the target plasmid and the transposon and incubate overnight at 37°C (*see* **Note 8**).
9. If no colonies or few colonies are obtained, refer to **Notes 9** and **10**.
10. To screen colonies, transfer colonies to 3 mL of LB medium containing antibiotics in a 14-mL snap-top tube and grow overnight in a shaker at 290 rpm and at 37°C. Prepare miniprep DNA as described in Chapters 8–10 and determine the Tn insertion site by restriction digestion (*see* Chapter 7) and DNA sequencing (*see* Chapter 22) (*see* **Note 11**).

3.2. Transposome Mutagenesis of Large BAC Plasmids

1. Thaw on ice an aliquot of electrocompetent DH10B (BAC⁺) bacteria and the Tn5 transposome complex.
2. Aliquot 1 µL of Tn5 transposome into a prechilled Eppendorf tube and mix it with 30–40 µL of electrocompetent cells.
3. Add the reaction mix to a prechilled 1-mm gapped cuvet. Transfer the cuvet to the electroporator and electroporate (1.6–2.1 kV, 100 Ω, and 25 µF) (*see* **Note 12**). Immediately place the cuvet back on ice and add 300 µL of LB medium, mix, and then transfer to a snap-top tube containing 1 mL of LB medium.
4. Shake at 200 rpm for 2 h at 37°C and then transfer to a conical flask containing 50 mL of LB medium and antibiotics to select for both the BAC plasmid and transposon. Shake

overnight at 250 rpm at 37°C. Transfer the bacterial culture to 500 mL of LB plus antibiotics in a 1-L baffled flask and shake overnight at 290 rpm at 37°C (*see* **Note 13**).
5. Harvest the cells and isolate the BAC DNA (*see* **Note 4**). Note that this DNA prep contains a mixed population and must be used to transform fresh bacterial cells to select for individual transposon-mutated BAC plasmids.
6. Carefully resuspend the mixed DNA prep in 500 μL of sterile water. Do not vortex or freeze the DNA, as this will cause strand breaks in the BAC DNA.
7. Mix 10 μL of BAC DNA with 30–40 μL of DH10B (BAC⁻) electrocompetent cells using the same procedure described in **step 3** except use a voltage setting of between 2 and 2.3 kV (*see* **Note 14**). Following electroporation, transfer the transformed cells into 1 mL of LB medium and shake at 37°C for 90 min.
8. Plate out all of the transformation reaction on LB agar plates (containing antibiotics to select for both the BAC and the transposon) and incubate overnight at 37°C (*see* **Note 8**).
9. If no colonies or few colonies are obtained, refer to **Notes 9** and **15**.
10. Isolate BAC DNA from selected colonies (*see* **Note 4**) and determine the Tn insertion site by restriction digestion (*see* Chapter 20) and DNA sequencing (*see* Chapter 22) (*see* **Note 11**).

4. Notes

1. The target DNA should be of the highest quality, and in the case of larger DNA constructs, the DNA should not be repeatedly freeze–thawed. This avoids the nicking and fragmentation of large target DNA molecules that leads to partial deletion of the target DNA when transformed into *E. coli* during the establishment of the mutated plasmid library.
2. Electrocompetent cells can be prepared as described in Chapter 5. These cells can be stored for several months in 10% glycerol at –70°C, but their efficiency will decrease with time. The highest transformation efficiency is obtained with fresh competent cells. It should be noted that high-efficiency electrocompetent cells are commercially available (e.g., DH10B Electromax cells from Invitrogen) as an alternative to preparing your own. Transformation of chemically competent cells can be used as an alternative to electroporation, but this is not recommended for cosmids or BAC clones because of poor transformation efficiencies using this technique. It is important to use high-efficiency competent cells, as this increases the yield of transposon mutants. The efficiency should be at least 10^7 transformants per microgram of transforming DNA. This is particularly important with larger target plasmids or cosmids, as these do not transform as easily as small constructs. When testing the efficiencies of the transposition reaction and the transformation reactions, it is suggested that, initially, a small target control plasmid, such as pUC19, is used first to acquaint the novice with transformation and transposition techniques.

    For stable maintenance of BACs, the *E. coli* strain DH10B (or a derivative) is recommended. This is especially the case with large plasmid constructs, which can be unstable in other RecA minus strains. The reason for the stability of BACs in DH10B is not completely understood and some BAC clones are relatively stable in other *E. coli* strains. Investigators requiring the use of other strains should investigate the stability of their individual BAC clones further.
3. Complexes are stable for up to 1 yr when stored at –20°C or –70°C (*19*), but their stability drops with repeated exposure to freeze–thawing conditions.
4. To generate high-quality DNA, for sequencing or transfection, it is recommended that the DNA be prepared by using commercially available large-scale plasmid purification kits such as the Nucleobond plasmid DNA purification kits available from Clontech. Nucleobond columns are recommended rather than Qiagen plasmid purification columns,

as the affinity of the DNA binding to the column matrix is slightly weaker and, hence, more DNA is easily eluted from these columns. This is an important factor to consider when isolating low-copy-number plasmid DNA. Alternatively, BAC DNA can be isolated using the procedure given in Chapter 12.

5. In order to obtain the maximum transposition efficiency in vitro, all reaction components must be present at optimal levels. When using one of the commercially available transposon kits, the critical factor is to have the donor transposon DNA and target DNA at the correct ratio to enable single-event transposition to target DNA to occur optimally. The donor plasmid has to be in molar excess in comparison to the target plasmid to drive the reaction (2:1 ratio of donor to target DNA). Thus, the amounts of DNA used will vary between different target DNA molecules. This is especially true with large plasmids and cosmid clones, which have significantly different molarity to small donor plasmids.

6. The transposition reaction must be heat inactivated to destroy the transposition complexes. Failure to do so severely reduces the yield of mutant plasmids obtained in a reaction. For example, heat treatment at 65°C can reduce the yield of mutants 100-fold.

7. The remainder of the reaction can be stored at 4°C for a few days, or long-term at –20°C, and used if additional colonies are required. It should be noted that large-plasmid reactions stored at –20°C should not be repeatedly freeze–thawed, as this increases the number of plasmids containing deletions.

8. Plates with colonies can be wrapped in parafilm and stored for several weeks in the refrigerator.

9. Failure to obtain antibiotic-resistant colonies after plating out on agar plates may be the result of poor competency of the cells used (*see* **Note 2**) or failure of the transposition reaction. In order to ascertain which problem has occurred, the following tests are suggested. First, plate out part of the transformation reaction, selecting for the antibiotic resistance encoded by the target plasmid in order to test the transformation efficiency of the competent cells. A large number of colonies should be obtained per plate. Second, plate out the transformation reaction using the antibiotic resistance on the transposon to determine the efficiency of the transposition reaction. Fewer colonies should be present on this plate in comparison to those obtained when selecting for the antibiotic resistance of the target plasmid, but if a very low number or no colonies are obtained, then the transposon reaction has failed (*see* **Note 10**).

10. Failure of the transposon reaction can result in no transposon mutants or a few transposon mutants with multiple transposon inserts. These results indicate that the transposon/donor plasmid is present in the reaction at quantities in excess of the target plasmid (*see* **Note 5**). Reduce the concentration of donor plasmid 10-fold and repeat the reaction. Alternatively, if possible, increase the concentration of target DNA 10-fold. Note that a larger plasmid requires more input target DNA than a smaller plasmid to achieve the same molarity.

11. Prior to sequencing the transposon insertion site on a modified plasmid, cosmid, or BAC, the initial strategy should be to map the insertion site by restriction digestion. This has an added advantage in that it enables the identification of any deletion that may have occurred. Plasmid deletion is a particular problem with larger target DNA constructs such as cosmids and BACs.

12. It should be noted that if too high a voltage is used, the transposase enzyme can be denatured. Optimal conditions for transformation should be identified in a pilot study.

13. When growing a large-scale bacterial culture for BAC DNA isolation, a limiting-growth medium such as LB medium should be used to avoid overgrowth of the culture that

results in a high amount of fragmented versions of the BAC DNA. If possible, use baffled flasks to increase the aeration of the culture and improve growth of the bacteria. In order for the bacterial strain to maintain the BAC, the appropriate antibiotic (e.g., 12.5 µg/mL chloramphenicol) must be present.

14. The voltage setting is dependent on the size of the plasmid DNA. Cosmids and BACs require a higher voltage setting (1.6–2.2 kV) than small plasmids (1.3–1.7 kV).
15. This is either the result of the loss of the BAC plasmid (*see* **Note 13**) or noncomplexed transposomes (*see* **Notes 3 and 12**).

References

1. McClintock, B. (1950) The origin and behavior of mutable loci in maize. *Proc. Natl. Acad. Sci. USA* **36,** 344–355.
2. Peters, J. E. and Craig, N. L. (2001) Tn7: smarter than we thought. *Nat. Rev. Mol. Cell. Biol.* **2,** 806–814.
3. Brune, W., Messerle, M., and Koszinowski, U. H. (2000) Forward with BACs: new tools for herpesvirus genomics. *Trends Genet.* **16,** 254–259.
4. Hensel, M., Shea, J., Gleeson, C., et al. (2000) Simultaneous identification of bacterial virulence genes by negative selection. *Science* **269,** 400–403.
5. Hutchinson, C. III, Peterson, S., Gill, S., et al. (1999) Global transposon mutagenesis and a minimal *Mycloplasma* genome. *Science* **286,** 2165–2169.
6. Zhan, X., Lee, M., Abenes, G., et al. (2000) Mutagenesis of murine cytomegalovirus using a Tn3-based transposon. *Virology* **266,** 264–274.
7. Shizuya, H., Birren, B., Kim, U.J., et al. (1992) Cloning and stable maintenance of 300-kilobase-pair fragments of human DNA in *Escherichia coli* using an F-factor-based vector. *Proc. Natl. Acad. Sci. USA* **89,** 8794–8797.
8. Brune, W., Menard, C., Hobom, U., et al. (1999) Rapid identification of essential and nonessential herpesvirus genes by direct transposon mutagenesis. *Nat. Biotechnol.* **17,** 360–364.
9. Smith, G. A. and Enquist, L. W. (1999) Construction and transposon mutagenesis in *E. coli* of a full length infectious clone of pseudorabies virus, an alphaherpesvirus. *J. Virol.* **73,** 6405–6414.
10. Metcalf, W. W., Jiang, W., Daniels, L. L., et al. (1996) Conditionally replicative and conjugative plasmids carrying *lacZ* alpha for cloning, mutagenis and allele replacement in bacteria. *Plasmid* **35,** 1–13.
11. Haas, R., Kahrs, A.F., Facius, D., et al. (1993) Tn*Max*—a versatile mini-transposon for the analysis of cloned genes and shuttle mutagenesis. *Gene* **130,** 23–31.
12. Posfai, G., Koob, M. D., Kirkpatrick, H. A., et al. (1997) Versatile insertion plasmids for targeted genome manipulations in bacteria: isolation, deletion, and rescue of the pathogenicity island LEE of the *Escherichia coli* O157:H7 genome. *J. Bacteriol.* **179,** 4426–4428.
13. Goryshin, I. Y. and Reznikoff, W. S. (1998) Tn5 in vitro transposition. *J. Biol. Chem.* **273,** 7367–7374.
14. Biery, M.C., Stewart, F.J., Stellwagen, A.E., et al. (2000) A simple in vitro Tn7-based transposition system with low target site selectivity for genome and gene analysis. *Nucleic Acids Res.* **28,** 1067–1077.
15. Happa, S., Taira, S., Heikkinen, E., et al. (1999) An efficient and accurate integration of mini Mu transposons in vitro: a general methodology for functional genetic analysis and molecular biology applications. *Nucleic Acids Res.* **27,** 2777–2784.

16. Akerley, B. J., Rubin, E. J., Camilli, A., et al. (1998) Systematic identification of essential genes by in vitro Mariner mutagenesis. *Proc. Natl. Acad. Sci. USA* **95,** 8927–8932.
17. Devine, S. E. and Boeke, J. D. (1994) Efficient integration of artificial transposons into plasmid targets in vitro: a useful tool for DNA mapping, sequencing and genetic analysis. *Nucleic Acid Res.* **22,** 3765–3772.
18. McGregor, A. and Schliess, M. R. (1999) Generation of GPCMV mutants via the use of a novel transposon based in vitro insertion system, *24th International Herpesvirus Workshop* (abstract 7.016).
19. Goryshin, I. Y., Jendrisak, J., Hoffman, L.M., et al. (2000) Insertional transposon mutagenesis by electroporation of released Tn5 transposition complexes. *Nat. Biotechnol.* **18,** 97–100.

27

In Vitro Transcription and Translation

Farahnaz Movahedzadeh, Susana González Rico, and Robert A. Cox

1. Introduction

In this chapter, we describe the use of plasmid vectors in transcription and translation systems in vitro to investigate aspects of the biology of the gene and the protein for which it codes. An in vitro, or cell-free, assay reproduces a relatively complex physiological process by mixing the essential purified components of the system under controlled conditions outside of the cell. Such systems allow the basic steps of transcription and translation to be studied individually, and the products obtained at each step to be altered in different ways according to the needs of the research . Thus, an in vitro system is convenient when it is necessary to modify a product, for example, by introducing mutations, labels, tags, or fusions.

1.1. In Vitro Transcription

Typically, the gene or sequence of interest is cloned under the influence of a strong promoter, such as those from phages SP6, T7, or T3, and is transcribed by the cognate phage RNA polymerase (RNAP). There are several commercially available plasmids that contain these promoters in the flanking regions of the multiple cloning sites (*see* Chapter 2). Alternatively, the target gene can be cloned with its own promoter and transcribed by its cognate RNAP. The decision of which promoter to use depends on the type of experiment being designed and the questions being addressed. In those cases, where the only requirement is obtaining the correct RNA transcript, it is not important which promoter and RNAP is responsible for transcribing the gene. However, investigation of promoter strength or of the control of transcription initiation requires that the endogenous promoter and cognate RNAP be used.

When cloning the target gene, it is important to consider the nature of the gene sequence to be transcribed, especially when the expression of genes that form part of an operon or when regions controlling gene expression are being investigated *(1–6)*. The secondary structure of the RNA transcript can be of key importance to transcription and translational control and it is sometimes difficult to produce in vitro an RNA molecule with the same secondary structure as occurs in vivo. In some cases, distant sequences can affect the secondary structure of a transcript and, therefore, its interaction with other regulatory proteins or RNAP. Introducing mutations into the sequence of the RNA can have an effect on both the secondary structure of the transcript and its activity. Some changes in secondary structure may be transient; for example, the secondary structure of the nascent transcript will change as new nucleotides are added, and, therefore, regulatory sequences can become available or unavailable at different stages of the transcription process.

If all of the necessary elements are present and nucleotides and energy are provided, an appropriate RNAP can be used to transcribe the cloned gene in vitro. A protocol for the synthesis of RNA by in vitro transcription of linear templates is given in **Subheading 3.1.**. If desired, a tagged nucleotide can be used to label the RNA transcript. The RNA transcript can be purified and used in a variety of different experiments, including the study of transcriptional elements such as terminators, antiterminators, and regulatory motifs. The labeled RNA transcript can also be used in studies of the regulatory functions of proteins, by means of analysis of RNA–protein interactions using assays such as band-shift and filter-binding. Newer techniques, such as BIACORE®, measure real-time interactions between molecules by monitoring changes in the surface plasmon resonance of a biosensor chip. By systematically changing the sequence of an RNA transcript, it is possible to identify motifs that interact with proteins.

1.2. In Vitro Translation

Not only can RNA transcripts be synthesized in vitro, but they can also be translated in a cell-free protein synthesizing system containing the appropriate components. In order to do this, it is necessary to clone the gene so that it can be transcribed as a single RNA transcript containing a suitable ribosome-binding site (Shine-Dalgarno sequence) *(7)* and a correct open-reading-frame. Use of in vitro translation systems makes it possible to synthesize proteins in a rapid and controlled way; up to 300 ng of protein can be obtained in a 50 µL reaction volume. The reaction substrate can be RNA, or DNA if a "coupled" transcription–translation system is used. The RNA template can be total RNA, mRNA, or an in vitro generated RNA transcript. The stability of the RNA is important when using this approach and it is recommended that a ribonuclease inhibitor be included in the reaction mix.

An advantage of using in vitro translation systems, instead of overexpression of a protein in vivo, is that although the yield is usually lower, the protein can be labeled very easily with modified amino acids. The most commonly used label is [^{35}S]-methionine, but it is also possible to use other radiotracers and amino acids, as well as commercially available biotinylated or fluorescently labeled lysine residues or other nonradioactive labels *(8)*. When a single mRNA species is used as the template, the resultant labeled product should be radiochemically pure, and for many applications, it will be possible to use the in vitro synthesized protein directly without any further purification. However, as the components of the in vitro system are much less complex than the whole cell, the purification of the protein of interest is greatly simplified, and it is possible to obtain a biochemically pure product relatively quickly. It is also possible to produce fusion proteins containing parts from different proteins or "tags" that facilitate its identification and/or purification (*see* Chapter 28). The in vitro translation system is also useful when the protein of interest is toxic to the cell or is expressed at very low levels in the in vivo system. This is often the case when expressing proteins involved in the regulation of gene expression.

The applications of the in vitro translational systems are summarized as follows:

1. *Genetic verification and detection*: An in vitro transcription–translation assay is commonly used to verify that a particular sequence of DNA (or cDNA) codes for a protein.
2. *Functional analysis*: In many cases, proteins expressed in an in vitro system display their natural activity and can be assayed directly without further protein purification. The system is also useful for mutation analysis. In addition, it is possible to study posttranslational modifications of the protein such as phosphorylation, adenylation, autoproteolytic activities, and, by using microsomal membranes, glycosylation, methylation, and signal sequence cleavage.
3. *Detection of molecular interactions*: The technique allows the preparation of RNA or proteins, labeled either radioactively or with a tag. The labeled product can be used in a variety of systems to study protein–protein, protein–DNA, or protein–RNA interactions. The cell-free system can also be used to test the effects of antisense oligonucleotides in transcription–translation processes, as well as any factor that could interfere with the translational process.
4. *Molecular structure and localization analysis*: The technique can be used to monitor the macromolecular assembly and folding of proteins (chaperonins, membrane association, etc.). The in vitro synthesis of milligram amounts of proteins is sufficient for structural studies such as X-ray crystallography and magnetic resonance spectroscopy.
5. *Preparative synthesis*: Finally, the in vitro transcription–translation analysis can be used for large-scale synthesis of proteins. For proteins that cannot be expressed in vivo because they are toxic or unstable, this may be the only avenue to obtaining protein.

1.2.1. Commercial Kits

The cell-free systems used to translate RNA in vitro are basically cell lysates that contain ribosomes, tRNAs, factors for initiation, elongation, and termination, aminoacyl-tRNA synthases, amino acids, and ATP/GTP. Most of these systems can be prepared in the laboratory using relatively simple but time-consuming procedures. A variety of systems for in vitro translation are available commercially, and although they are expensive, they are recommended because they give highly reproducible results. These systems use either a DNA template (coupled transcription–translation systems) or an RNA template. The principal commercially available systems used for in vitro translation are based either on *Escherichia coli* or rabbit reticulocyte lysates. The translational apparatus is sufficiently conserved between prokaryotes and eukaryotes that heterologous cell-free systems (such as an *E. coli* cell-free extract and human mRNA or a rabbit cell-free extract and bacterial mRNA) function with high fidelity, irrespective of the origin of the cell-free system.

1.2.1.1. *E. coli* System

This extract is obtained by disrupting cells of *E. coli* (*9,10*). A strain of *E. coli* deficient in OmpT endoproteinase and Lon protease activity is used, in order to increase the stability of the expressed protein. Large particles are removed from the lysate by centrifugation. The supernatant (S30 extract) is treated with microccocal nuclease, which inactivates the endogenous mRNA, but not tRNA or ribosomes *(11)*. The nuclease can then be heat inactivated and target mRNA added to the system. This procedure significantly reduces the background synthesis of proteins, retaining the full activity of all of the other components of the system required for translation. This system, although it does not support post-translational processing, allows rapid screening for translation of prokaryotic or eukaryotic DNAs cloned in *E. coli* high-expression vectors. It is also a good model for studing the control of gene expression in prokaryotes and to screen chemicals that may interfere with bacterial translation. The mRNA species requires a ribosome-binding site to function in protein biosynthesis. The yields of protein obtained with mRNA template are 1–10% of yields obtained in coupled transcription–translation systems (*see* **Subheading 1.2.1.4.**).

1.2.1.2. Rabbit Reticulocyte Lysate

This eukaryotic system is widely used for the in vitro translation of mRNA. The reticulocytes are prepared from New Zealand white rabbits using a standard protocol *(12,13)* that allows the isolation and purification of reticulocytes from other cellular components. The purified reticulocytes are then lysed and treated with micrococcal nuclease, similarly to *E. coli* systems, to reduce background endogenous protein synthesis. Most of the commercially available reticulocyte systems have been optimized by the inclusion of an additional source of energy, additional tRNA species to expand the range of mRNA that can be translated, salts, and hemin to prevent inhibition of initiation. In some cases, the system can be further optimized by altering parameters such as the concentration of Mg^{2+} and K^+ or by the addition of dithiothreitol (DTT).

The rabbit reticulocyte lysate system offers several advantages over *E. coli* extracts. The utilization of inappropriate ribosome-binding sites by *E. coli* ribosomes can lead to abbreviated products resulting from initiation of translation at internal ATG codons that are preceded by a spurious ribosome-binding site. Because different mechanisms are used by prokaryotic and eukaryotic ribosomes to bind mRNA, the use of a eukaryotic in vitro translation system reduces the synthesis of abbreviated products. Furthermore, the commercially available eukaryotic systems translate mRNA more efficiently than the available *E. coli* systems. A protocol for the translation of mRNA in an in vitro reticulocyte system is given in **Subheading 3.2.**

1.2.1.3. Canine Pancreatic Microsomal Membranes

To study cotranslational processing of in vitro translated eukaryotic proteins, canine pancreatic microsomal membranes can be used in conjunction with the rabbit reticulocyte lysate. In the presence of these microsomal vesicles, processes such as signal peptide cleavage, membrane insertion, membrane translocation, and protein glycosylation can be studied. Microsomes are prepared using a standard procedure *(14)* and isolated from the membrane fraction and membrane-bound ribosomes, as well as mRNA. This assures a consistent performance with minimal translational inhibition and background translation.

1.2.1.4. Coupled Transcription–Translation Systems

These systems utilize a DNA template; they are called coupled because both transcription and translation reactions occur in the same tube. The principle is very simple: In the intact bacterial cell, translation of the nascent mRNA chain is initiated before transcription is complete; thus, both processes occur simultaneously. In vitro coupled systems work very similarly to transcribe and translate a particular DNA sequence to protein in a single-reaction tube. As transcription and translation are coupled, the problems of RNA stability and secondary structure are diminished. The DNA template can be a plasmid (circular or linearized), polymerase chain reaction (PCR) product, or a reverse transcription (RT) PCR product containing the uninterrupted (no introns) coding sequence of interest, downstream from, and in the correct orientation with, a promoter and translation start site.

Promega provides an *E. coli* S30 extract for use with linear templates (mRNA or DNA) and another for use with circular DNA. The gene to be expressed requires a promoter (such as *lac*UV5, *tac*, λP_L, and λP_R) and transcripts require a ribosome-binding site. Hybrid transcription–translation systems are also available commercially. The more recent products comprise rabbit or, less frequently, wheat germ *(15)* ribosomes combined with *E. coli* transcriptional components. Promega recommends their TNT® Quick series, which is based on rabbit ribosomes, for use with plasmid DNA and SP6, T3, or T7 promoters. The TNT Quick for PCR DNA kit is optimized for PCR products with an incorporated T7 promoter.

Systems designed for large-scale synthesis of proteins have been described, based on flow methodology ("continuous-exchange cell-free technology") *(16,17)* to provide a constant supply of energy and to remove low-molecular-weight inhibitors. Roche has refined this approach and offers kits based on a modified *E. coli* system. Two versions are available: one for circular and the other for linear templates including PCR products with incorporated T7 promoters and ribosome-binding sites. Roche claims that 5 mg of protein have been synthesized in a 1-mL reaction volume within 24 h.

2. Materials
2.1. In Vitro Transcription from Phage Promoters (see Note 1)

1. Linear template DNA (0.2–1 µg/µL) (*see* **Notes 2–4**).
2. 5X transcription buffer: 200 m*M* Tris-HCl (pH 7.9), 30 m*M* MgCl$_2$, 10 m*M* spermidine, 50 m*M* NaCl.
3. 15–20 U RNA polymerase (SP6, T7, or T3).
4. 100 m*M* dithiothreitol (DTT).
5. 20–40 U Ribonuclease inhibitor (e.g., RNasin, Promega).
6. 2.5 m*M* rATP, rGTP, and rUTP mixture.
7. 100 µ*M* rCTP solution.
8. 50 µCi/µL [α-^{32}P] rCTP (specific activity of 400 Ci/mmol).
9. Nuclease-free water.
10. 1 U/µL RNase-free DNase (e.g., RQ1 DNase, Promega).
11. Loading buffer: 98% formamide, 10 m*M* EDTA, 0.1% xylene cyanol, 0.1% bromophenol blue.

2.1.1. Purification of Labeled RNA Transcripts

1. 6% Denaturing polyacrylamide gel containing 8 M urea (*see* **Note 5**).
2. 10X TBE gel electrophoresis buffer: 890 mM Tris, 890 mM boric acid, 25 mM EDTA.
3. Elution buffer: 0.5 M ammonium acetate, 1 mM EDTA, 0.2% (w/v) sodium dodecyl sulfate (SDS).
4. Phenol : chloroform : isoamyl alcohol (25 : 24 : 1).
5. 3 M Sodium acetate (pH 5.2): Prepare a 3 M sodium acetate solution and adjust to pH 5.2 by the addition of glacial acetic acid.
6. 100% Ethanol.

2.2. In Vitro Translation of RNA Using Rabbit Reticulocyte Lysate

1. Rabbit reticulocyte lysate (Promega).
2. 1 mM Amino acid mix minus methionine (provided with the rabbit reticulocyte lysate kit, Promega).
3. 10 mCi/mL [^{35}S]-methionine (specific activity of 1200 Ci/mmol).
4. Ribonuclease inhibitor (e.g., RNasin, Promega).
5. RNA substrate in nuclease-free water (1 mg/mL) (*see* **Notes 6** and **7**).
6. Nuclease-free water.

3. Methods
3.1. In Vitro Transcription from Phage Promoters

1. Prepare the reaction mixture at room temperature (*see* **Note 8**), as follows:

5X Transcription buffer	4 µL
100 mM DTT	2 µL
Ribonuclease inhibitor	20–40 U
2.5 mM rATP, rGTP and rUTP mixture	4 µL
100 µM rCTP	2.4 µL
Template DNA (0.2–1 µg/µL)	1–2 µL
[a-^{32}P] rCTP	5 µL
RNA polymerase (15–20 U)	1 µL
Add nuclease-free water to a final volume of	20 µL.

2. Incubate for 30 min at 37°C.
3. Degrade the DNA template with 1 µL of RNase-free DNase for 15 min at 37°C.
4. Stop the reaction by adding 20 µL of loading buffer.

3.1.1. Purification of Labeled RNA Transcripts

1. Incubate the sample for 3 min at 90–100°C to denature the RNA and place immediately on ice.
2. Run the total volume on a 6% polyacrylamide–urea gel in 1X TBE electrophoresis buffer.
3. Following electrophoresis, cover the gel with plastic film and expose it to X-ray film for 5–10 min at room temperature to locate the full-length transcript. After aligning the developed film on the gel, excise a gel slice that contains the transcript from the gel using a sterile razor blade. Re-expose the cut gel to ensure that the correct region of the gel has been removed.
4. Elute the RNA from the gel slice by incubating it in 0.5–1 mL of elution buffer for 1–2 h (or overnight) at 37°C. Transfer the elution buffer to a fresh tube (leaving the gel slice behind).

5. Extract with 1 volume of phenol : chloroform : isoamyl alcohol. Vortex for 1 min and centrifuge in a microcentrifuge for 2 min. Transfer the upper, aqueous phase to a fresh tube.
6. Precipitate the RNA by the addition of 1/10 volume of 3 M sodium acetate (pH 5.2) and 2.5 volumes of 100% ethanol. Pellet in a microcentrifuge for 20 min and carefully remove the supernatant.
7. Resuspend the probe RNA in 20 µL of nuclease-free water, the concentration should be approximately 1 mg/mL. The RNA should be used within 3 d of preparation.
8. The mRNA transcript can be subsequently be used as a template for the in vitro translation system described in **Subheading 3.2.** (*see* **Note 9**).

3.2. In Vitro Translation of RNA Using Rabbit Reticulocyte Lysate

1. Thaw the components of the reaction mix on ice.
2. Denature the RNA transcript at 67°C for 10 min and place immediately on ice.
3. Set up a reaction mix as follows (*see* **Note 10**):

5X Transcription buffer	4 µL
Rabbit reticulocyte lysate	35 µL
Amino acid mix minus methionine	1 µL
[^{35}S] Methionine	2 µL
Ribonuclease inhibitor	40 U
RNA substrate (1 mg/mL, *see* **Note 7**)	2 µL

    Add nuclease-free water to a final volume of 50 µL.
4. Incubate at 30°C for 90–100 min.
5. Analyze the results of translation by an appropriate method; for example, SDS-PAGE (*see* Chapter 29) or measurement of the incorporation of radioactive amino acids into protein or other assays for the protein of interest (enzymatic activity, antibody reactivity, etc.). Alternatively, purify a tagged protein by appropriate methods.

4. Notes

1. All reagents, except **items 1, 8**, and **11**, can be purchased either as separate items or as a Riboprobe kit from Promega. Store at –20°C.
2. The gene of interest must be cloned under the control of a strong promoter such as T3, T7, SP6, or an appropriate *E. coli* promoter. It is important to obtain highly pure and clean template DNA. The template must be CsCl gradient or gel purified and must not contain glycerol or salts. If a PCR product is used as a template (*see* **Note 4**), it must be purified by conventional methods to avoid contamination from primers and aberrant PCR products. Residual ethanol and phenol : chloroform : isoamyl alcohol will also interfere with the transcription reaction.
3. If the DNA template is a circular plasmid, it must be linearized by digestion with an appropriate restriction enzyme prior to transcription. This enables the generation of a runoff transcript and prevents the transcription of downstream vector sequences. A restriction enzyme with a single recognition site, near to the 3' end of the gene is used. Avoid the use of enzymes that generate 3' overhangs, as extraneous transcripts are produced when templates with 3' overhangs are transcribed in vitro with SP6 or T7 RNAPs *(18)*. However, if there is no other option, the 3' overhangs can be blunt ended by the 3' to 5' exonuclease activity of DNA polymerase I large (Klenow) fragment (*see* Chapter 15).
4. It is sometimes convenient to avoid cloning steps by amplifying the gene of interest by PCR, using an upstream primer that contains the appropriate bacteriophage promoter

sequence in addition to the priming sequence. Bacteriophage promoters such as T7, T3, or SP6 are small enough (around 20 nucleotides) to be included in a PCR primer without causing too much interference. We have successfully used a 24-nucleotide T7 promoter sequence: 5'-AAT TCT AAT ACG ACT CAC TAT AGG -3' *(19)*. The T7 promoter sequence should be located at the 5' end of the primer. To generate transcripts of the sense strand, incorporate the promoter into the 5' primer. Conversely, utilize a promoter on the 3' primer to generate an antisense probe.

5. Assemble the apparatus according to the manufacturer's instructions and prepare a denaturing acrylamide gel mix as described in Chapter 22.
6. The template must contain a ribosomal binding site (RBS), located approximately seven bases upstream of the start codon. It is important to verify that the sequence is correct by sequencing of the DNA template (*see* Chapter 22). If a tag, such as 6×His, GST, or an epitope, is to be fused to the protein, ensure that it is in frame with the rest of the open reading frame and that the fusion does not create a stop codon.
7. When using whole-cell RNA, it is important to note that most preparations are composed mainly of rRNA. Therefore a much higher concentration of RNA is needed (final concentration 100–200 µg/mL) than when using an in vitro-generated transcript.
8. Do not set up the reaction on ice because the DNA template will condense and precipitate in the presence of spermidine at 0°C.
9. Unlabeled RNA transcripts, to be used as substrates for in vitro translation, can be similarly purified, as well as checked for degradation and size. Perform two in vitro transcription reactions, one incorporating a radioactively labeled nucleotide and the other using nonradioactive nucleotides. Run the two samples on a polyacrylamide gel. Visualize the radioactive transcript by autoradiography and use it to identify the position of the nonradioactive RNA of interest. Excise the appropriate gel slice and purify the RNA transcript.
10. It is recommended to use a positive control (usually provided with the commercial system) and a negative control (without RNA template). Use the negative control to assess the background incorporation of radioactive amino acids into proteins resulting from translation of residual endogenous mRNA in the lysate.

Acknowledgments

We thank Simon A. Cox for help in the preparation of this manuscript. FM is funded by the Wellcome Trust grant no. 051880/Z/97/2 and EEU grant no. ITPMRB44. SGR received financial support from a Research Development Award in Tropical Medicine, The Wellcome Trust grant no. 056016/Z/98/Z.

References

1. Arnstein, H. R. V and Cox, R. A. (1992) *Protein Biosynthesis: In Focus*, Oxford University Press, Oxford.
2. Grunberg-Manago, M. (1999) Messenger RNA stability and its role in the control of gene expression in bacteria and phages. *Annu. Rev. Genetics* **33,** 193–227.
3. Weissmann, C., Billeter, N. A., Goodman, H. M., et al. (1973) Structure and function of phage RNA. *Annu. Rev. Biochem.* **42,** 303–308.
4. Zengel, J. M. and Lindahl, L. (1992) Ribosomal protein L4 and transcription factor NusA have separable roles in mediating termination of transcription within the leader region of the S10 operon of *E. coli*. *Genes Dev.* **6,** 655–662.
5. Zengel, J. M., and Lindahl, L. (1993) Domain I of 23S rRNA competes with a paused transcription complex for ribosomal protein L4 of *E. coli*. *Nucleic Acids Res.* **21,** 2429–2435.

6. Zengel, J. M., Mueckl, D., and Lindahl, L. (1980) Protein L4 of the *E. coli* ribosome regulates an eleven gene r - protein operon. *Cell* **21,** 523–525.
7. Shine, J. and Dalgarno, L. (1974) The 3'-terminal sequence of *Escherichia coli* 16S ribosomal RNA: complementarity to nonsense triplets and ribosome binding sites. *Proc. Natl. Acad. Sci. USA* **71,** 1342–1346.
8. Noren, C. J., Anthony-Cahill, S. J., Griffith, M. C., et al. (1989) A general method for site-specific incorporation of unnatural amino acids into proteins. *Science* **244,** 182–188.
9. Zubay, G. (1980) The isolation and properties of CAP, the catabolite gene activator. *Methods Enzymol.* **65,** 856–877.
10. Lesley, S. A., Brow, M. A. D., and Burgess, R. R. (1991) Use of in vitro protein synthesis from polymerase chain reaction generated templates to study interaction of *Escherichia coli* transcription factors with core RNA polymerase and for epitope mapping of monoclonal anitbodies. *J. Biol. Chem.* **266,** 2632–2638.
11. Jackson, R. I. and Hunt, T. (1983) Preparation and use of nuclease-treated lysates for the translation of eukaryotic messenger RNA. *Methods Enzymol.* **96,** 50–74.
12. Darnbrough, C., Legon, S., Hunt, T., et al. (1973) Initiation of protein synthesis: evidence for messenger RNA-independent binding of methionyl-transfer RNA to the 40 S ribosomal subunit. *J. Mol. Biol.* **76,** 379–403.
13. Pelham, H. R. and Jackson, R. J. (1976) An efficient mRNA-dependent translation system from reticulocyte lysates. *Eur. J. Biochem.* **67,** 247–256.
14. Walter, P. and Blobel, G. (1983) Preparation of microsomal membranes for cotranslational protein translocation. *Methods Enzymol.* **96,** 84–93.
15. Erickson, A. H. and Blobel, G. (1983) Cell-free translation of messenger RNA in a wheat germ system. *Methods Enzymol.* **96,** 38–50.
16. Spirin, A. S., Boranov, V. I., Ryabova, L. A., et al. (1988) A continuous cell-free translation system capable of producing polypeptides in high yield. *Science* **242,** 1162–1164.
17. Boranov, V. I. and Spirin, A. S. (1993) Gene expression in cell-free system on preparative scale. *Methods Enzymol.* **217,** 123–142.
18. Schenborn, E. T. and Nierendorf R. C. (1985) A novel transcription property of SP6 and T7 RNA polymerases: Dependence on template structure. *Nucleic Acids. Res.* **13,** 6223–6236.
19. Movahedzadeh, H., Gonzalez-y-Merchand, J. A., and Cox, R. A. (2001) Transcription start site mapping, in Mycobacterium tuberculosis *Protocols: Methods in Molecular Medicine*, (Parish, T. and Stoker, N. G., eds.), Humana, Totowa, NJ, vol. 54, pp. 105–124.

28

Vectors for the Expression of Recombinant Proteins in *E. coli*

Sally A. Cantrell

1. Introduction

Escherichia coli is the most commonly used and best characterized organism for overexpressing foreign and nonforeign proteins. The use of *E. coli* confers several immediate advantages to the user: rapid and high-level expression as a result of the speed of cell growth to high density; low complexity and low cost of growth media; and the ability to target proteins to the desired subcellular location (*1*). In addition, a number of mutant host strains are available that can improve recombinant protein expression. For example, the use of host strains with mutations in cytoplasmic protease genes decreases protein degradation. Hosts that encode tRNAs for codons seldom found in *E. coli* can successfully express some otherwise difficult to express proteins that contain these rare codons (*see* Chapter 3).

The researcher is faced with a myriad of vector options when designing recombinant proteins. Each year, new plasmid vectors and mutant host strains with characteristic advantages are made commercially available. How to choose the correct vector and where to find it is the subject of this chapter. Additionally, because each protein is unique, the scientist must know how to modify the chosen system to optimize expression of the gene of interest (*see* Chapter 29). The fundamental questions to ask when determining the system to use are as follows: Which type of promoter is required? Whether the protein should be produced as a fusion to direct its subcellular localization, increase its solubility, or facilitate its purification? The choice will often depend on the downstream use of the expressed protein.

2. Cloning Into Expression Vectors
2.1. Transcription Vs Translation Vectors

There are two types of expression vector: transcription vectors and translation vectors. Transcription vectors are utilized when the DNA to be cloned has an ATG start codon and a prokaryotic ribosome-binding site. Translation vectors contain an efficient ribosome-binding site and, therefore, it is not necessary for the target DNA to contain one. This is particularly useful in cases where the initial portion of the gene must be cleaved in an effort to improve solubility. Another consideration when choosing a transcription or translation vector is the source of the DNA to be expressed. Eukaryotic genes are usually cloned into translation vectors, whereas prokarytic genes are cloned into transcription vectors (2). The reason for this is that prokaryotic genes usually have a ribosome-binding site that is compatible with the host *E. coli* translation machinery, whereas eukaryotic genes do not. Prokaryotic gene expression may be improved by use of an engineered promoter and ribosome-binding site.

2.2. Cloning Site

2.2.1. Polylinkers

Cloning sites in the first vectors were located in antibiotic-resistance genes (*see* Chapter 2). More modern plasmids also contain an antibiotic-resistance gene for selection of plasmid-bearing *E. coli*. However, most modern cloning vectors have a "polylinker" or "polycloning" site that provides restriction enzyme sites for subcloning DNA fragments (*see* Chapters 2 and 16). Many vectors are constructed such that the polylinker lies within the coding sequence of the α-peptide of the β-galactosidase gene (*lacZ*) (3). In a suitable host, the intact vector produces the LacZ protein that can be detected using colorimetric assays. When cloned DNA interrupts the *lacZ* reading frame and prevents the production of the LacZ protein, colonies appear white rather than blue when plated on the chromogenic substrate 5-bromo-4-chloro-3-indolyl-β-D-galactoside (X-Gal) (*see* Chapter 19). PETBlue™ vectors (*see* **Table 1**), which are derived from pET vectors, have the added advantage of supporting blue–white screening for recombinant clones.

2.2.2. TA Cloning

Many vectors are designed to be compatible with the TA cloning method that allows direct cloning of polymerase chain reaction (PCR) products (4,5) (*see* Chapter 17). This technique eliminates the need for restriction enzyme sites to be incorporated into the primers to facilitate cloning of the PCR product. The terminal transferase activity of thermostable DNA polymerases, such as *Taq*, adds a single deoxyadenosine (A) to the 3' end of PCR products. As a result, any PCR-amplified product can be cloned into a vector with a single overhanging deoxythymidine (T). Proof-reading DNA polymerases do not have this activity. PCR products generated by these enzymes can be incubated with *Taq* polymerase in the presence of dATP in order to add 3' A overhangs. The pBAD-TOPO, pCR-TOPO, and pTrc-TOPO vectors (*see* **Table 1**) all make use of the TA cloning methodology.

2.3. Plasmid Copy Number

Cloning vectors are usually only 3–4 kb in size because all extraneous DNA that is not necessary for plasmid maintenance or replication is removed so that the vector has only a limited number of restriction enzyme sites. Most *E. coli* cloning vectors have a

Table 1
Expression Vectors

Vector	Resistance marker[a]	Promoter	Origin	Fusion tag (location)[b]	Cleavage molecule	Localization	Company
pGEX	ampR	*tac*	pBR322	GST (N)	Thrombin	N/A	Amersham Biosciences
pEZZ[c,d]	ampR	*lac*	pUC	ZZ peptide (N)	N/A	Extracellular	Amersham Biosciences
pPRO series	cmR	Ltet0-1	ColE1	6×HN (N)	Enterokinase	N/A	BD Biosciences Clontech
pHAT series	ampR	*lac*	pUC	HAT (N)[e]	Enterokinase	N/A	BD Biosciences Clontech
pBAD series[g]	ampR	*araBAD*	pBR322	His (C)[g]	Enterokinase	N/A	Invitrogen
pBAD/gIII A, B, C	ampR	*araBAD*	pBR322	His (C)[g]	N/A	Periplasm	Invitrogen
pBAD/Thio-TOPO[f]	ampR	*araBAD*	pUC	His (C), thioredoxin (N)[g]	Enterokinase	Cytoplasmic side of inner membrane	Invitrogen
pThioHis A, B, C	ampR	*trc*	ColE1	His-thioredoxin (N)	Enterokinase	Cytoplasmic side of inner membrane	Invitrogen
pLEX	ampR	*pL*	pUC	N/A	N/A	N/A	Invitrogen
Gateway pDEST	ampR	T7	pBR322	GST (N, C), 6×His (N)[g]	N/A	N/A	Invitrogen
pRSET A, B, C	ampR	T7	ColE1	6×His (N)[g]	Enterokinase	N/A	Invitrogen
pCR-T7-TOPO[f]	ampR	T7	ColE1	6×His (N, C)[g]	Enterokinase	N/A	Invitrogen
pTrcHis TOPO[f,h]	ampR	*trc*	pBR322	6×His (N, C)[g]	Enterokinase	N/A	Invitrogen
pTrcHis A, B, C	ampR	*trc*	pBR322	6×His (N)[g]	Enterokinase	N/A	Invitrogen
Impact™ CN series (pTYB vectors)	ampR	T7 *lacO*	ColE1	Intein/chitin binding domain (N, C)	Thiols	N/A	New England BioLabs
Impact™ pTWIN	ampR	T7 *lacO*	ColE1	Intein/chitin (N, C)	Temp/pH, Thiols	N/A	New England BioLabs
pMAL™ series	ampR	*tac*	pBR322	Maltose-binding protein (MBP) (N)	Xa, Enterokinase, Genenase I	Periplasmic, cytoplasmic	New England BioLabs

(*continued*)

Table 1 (continued)

Vector	Resistance marker[a]	Promoter	Origin	Fusion tag (location)[b]	Cleavage molecule	Localization	Company
pET series[c]	ampR or kanR	T7 or T7/lac	pBR322	His (N, C, I), T7 (N, I), S (N, I), Trx (N), CBD (N, C), KSI (N), HSV (C), PKA (N), Dsb (N), GST (N), Nus (N)	Thrombin, Enterokinase, Xa	Periplasmic, cytoplasmic	Novagen
PinPoint™	ampR	tac	pUC	Biotin (N)	Xa	N/A	Promega
pGEMEX	ampR	T7	pUC	T7 Gene 10 (N)	N/A	N/A	Promega
pQE series	ampR	T5 lacO	ColE1	His (N, C)	Xa	N/A	Qiagen
pHB6	ampR	trc	ColE1	HA (N), 6×His (C)	N/A	N/A	Roche
pVB6	ampR	trc	ColE1	VSV-G (N), 6×His (C)	N/A	N/A	Roche
pXB, pBX	ampR	trc	ColE1	Protein C (N, C)	N/A	N/A	Roche
pBH	ampR	trc	ColE1	HA (C)	N/A	N/A	Roche
pBV	ampR	trc	ColE1	VSV-G (C)	N/A	N/A	Roche
pBX	ampR	trc	ColE1	Protein C (C)	N/A	N/A	Roche
pFLAG	ampR	tac	pBR322	FLAG (N, C), ompA (N)	Enterokinase	Periplasmic, cytoplasmic	Sigma
pCAL series	ampR	T7 lacO	pBR322	Calmodulin binding (N, C). kemptide (N), FLAG™ (N)	Thrombin	N/A	Stratagene

[a] ampR, ampicillin resistance; cmR, chloramphenicol resistance; kanR, kanamycin resistance.
[b] (N) amino terminus; (C) carboxyl terminus; (I) internal tag.
[c] Blue–white screening.
[d] Noninducible system.
[e] Modified His tag.
[f] TA cloning vector.
[g] Epitope for antibody detection.
[h] Suitable for eukaryotic protein expression.

ColE1 origin of replication that allows for the high (15–60 copies) accumulation of plasmid DNA in the cell *(6,7)*. Some examples of plasmids in this category are the pPRO series, the Impact™ series, and the pQE series (*see* **Table 1** for more options). Expression levels may be raised further by increasing the plasmid copy number by using origins of replications from pUC plasmid-derived vectors that are present at a few hundred copies per cell. The pEZZ, pHAT, pBAD/Thio-Topo, pLEX, PinPoint™, and pGEMEX plasmids are all maintained at high-copy-number and can be found listed in **Table 1**. The pET vectors, which are derived from pBR322, are medium-copy-number plasmids (15–20 copies). Other medium-copy-number plasmids include the Gateway™ pDEST, pTrcHis, pMAL™, pFLAG, and pCAL vectors. Multicopy plasmids can pose a problem when the target gene generates a product that is toxic to the cell. As multicopy plasmids are distributed randomly during cell division, cells that do not contain the toxic gene product are more likely to survive. pSC101 is similar to the pET vectors, but its low copy number (5–10 copies per cell) enables many proteins that are toxic to cells when in the pET system to be tolerated *(8)*.

3. Expression System

3.1. Promoters

Proper promoter selection is of the utmost importance when designing an expression system. In fact, expression vectors were originally classified by the nature of their promoters because a strong promoter was considered the most important asset of these vectors *(9)*. A strong promoter has a high frequency of transcription initiation and generates the target protein as 10–30% of the total cellular protein production *(10)*. An expression vector's promoter is located upstream of the ribosome-binding site and is under the control of a regulatory gene that is either located on a plasmid or within the host chromosome. Because many recombinant protein products are toxic to the host, the promoter must be tightly regulated so that the target protein is expressed only at the appropriate time and, thus, puts minimal stress on the host *(11–13)*. When expression is incompletely repressed, plasmid instability may occur, and the cell growth rate may decrease, resulting in reduced target protein production *(14,15)*. In order to accomplish transient induction, the promoter must initially be completely repressed and then easily induced with the addition of an inducer at the time that expression is desired.

Historically, the most commonly used promoters have been the lactose (*lac*) *(6)* and tryptophan (*trp*) *(16)* promoters. These two promoters were combined to create the hybrid promoters *tac* *(17)* and *trc* *(18)* that are also commonly used. Other common promoters are the phage lambda promoters *(19)*, the phage T7 promoter (T7) *(20)*, and the alkaline phosphatase promoter (*phoA*) *(21)*. Properties of promoters commonly used in *E. coli* expression plasmids are outlined in **Table 2**.

3.1.1. The lac Promoter

The *lac* system is based on the promoter of the *E. coli lac* operon and was first utilized by Jacques Monod in his research on inducible enzyme synthesis *(24,25)*. In the presence of lactose or the gratuitous inducer isopropyl-β-D-thiogalactoside (IPTG) the *lac* repressor (LacI) changes conformation and can no longer bind to the *lac* opera-

Table 2
Promoters Commonly Used in *E. coli* Expression Vectors

Promoter	Regulator	Inducer
λ P_L	λcIts857	Thermal
λP_R, P_L	λcIts857	Thermal
araBAD	AraC	L-Arabinose
cspA	None	Thermal
lac	lacI, lacI^q lacI(ts), lacI^q(ts)	Isopropyl-β-D-thiogalactopyranoside(IPTG) Thermal
*lac*UV5	lacI, lacI^q lacI(ts), lacI^q(ts)	IPTG Thermal
Ltet0-1	tet repressor protein	Tetracycline or anhydro-tetracycline
phoA	phoB (positive) phoR (negative)	Phosphate starvation
T7	(RNAP)[a]	IPTG
T3/*lac* operator	lacI, lacI^q lacI(ts), lacI^q(ts)	IPTG Thermal
T5/*lac* operator	lacI, lacI^q lacI(ts), lacI^q(ts)	IPTG Thermal
T7/*lac* operator	lacI, lacI^q lacI(ts), lacI^q(ts)	IPTG Thermal
tac	lacI, lacI^q lacI(ts), lacI^q(ts)	IPTG Thermal
tetA	None	Tetracycline or anhydro-tetracycline
trc	lacI, lacI^qlacI(ts) lacI^q(ts)	IPTG Thermal
trp	None	Tryptophan starvation

[a] *See* **Subheading 3.1.5.**
Source: Data compiled from **refs. *10*, *22*,** and ***23*.**

tor *(26)*. The removal of LacI from the operator allows RNA polymerase to bind to the *lac* promoter and initiate transcription *(27)*. The *lac* operon has been engineered to optimize *E. coli* protein expression. For example, the *lac*UV5 promoter, a derivative of the *lac* promoter, has a mutation that renders it insensitive to repression in response to high glucose levels and allows rich media to be used for the growth of the expression strain. The *lac* promoters are valuable tools for achieving graded expression of proteins, particularly if *lacY* mutant hosts are used. For a further discussion of the variety of mutant hosts available, *see* Chapter 3. Vectors containing the *lac* promoter include pEZZ and pHAT (*see* **Table 1**). Protocols for the use of *lac*-derived promoters are described in detail in Chapter 29.

3.1.2. The tac and trc Promoters

The *tac* and *trc* promoters were born out of a desire to create stronger, more tightly regulated promoters. Although both are highly active promoters, they have incomplete repression in the uninduced state. This is not problematic unless the expressed protein is toxic to the cell. The *tac* promoter was created by combining the tightly regulated *lac*UV5 operator with the *trp* promoter, which is three times as strong as the *lac* promoter *(28)*. This hybrid promoter gives high-level expression utilizing the same induction and repression system as the *lac* promoter. The *tac* promoter contains only 16 base pairs (bp) between the consensus –35 and –10 sequences, which is 1 bp short of the consensus distance. Therefore, 1 bp was inserted to create the *trc* promoter. As expected, this proved to be a very efficient promoter in vitro *(29)*. However, when tested with two different reporter genes, it was less efficient in *E. coli* than the *tac* promoter *(30)*. It was ultimately determined that the strongest promoters in *E. coli* are not necessarily those that conform to the consensus promoter sequences *(31,32)*. Examples of commercially available vectors containing *tac* promoters are PinPoint, pMAL, and pGEX (*see* **Table 1**). Some commercially available *trc* vectors are pThioHisABC, pTrcHis TOPO, pTrcHis ABC, and pHB6.

3.1.3. The ara Promoter

The arabinose promoter is part of a regulated expression system that is more complex than the lactose system. The arabinose operon contains three genes (*araB*, *araA*, and *araD*) that encode enzymes that catalyze the conversion of arabinose to xyulose-5-phosphate, an intermediate that can be utilized to generate energy *(33)*. These three genes are transcribed from the P_{BAD} promoter. The *araC* gene encodes a transcriptional repressor that is transcribed from a second promoter (P_C) that is divergent from P_{BAD} *(34)*. The repressor protein, AraC, can act in two radically different manners depending on whether or not it is bound by arabinose *(35)*. In the absence of arabinose, the AraC repressor protein binds the O_2 operator and the I_1 site (*see* **Fig. 1**), causing the DNA in this region to adopt a looped conformation *(36)*. This prevents transcription from the P_{BAD} promoter because the RNA polymerase is sterically blocked from binding to the promoter. However, when arabinose is present, the AraC repressor changes shape and, subsequently, binds preferentially to the adjacent I_1 and I_2 sites. When this happens, the loop configuration is released and transcription may occur. A second layer of complexity is added to the system by its response to the presence or absence of glucose. Low levels of glucose cause an increase in the levels of cAMP in the cell, which, in turn, causes the cAMP receptor protein (CRP) to become active. Binding of the CAP site by the activated CRP leads to full induction of expression of the arabinose genes *(37)*. The expression of the *araBAD* genes is induced to approx 300 times their uninduced level *(38)*. One final nuance of this system is that the level of the repressor protein, AraC, is autoregulated *(39)*. Thus, when the level of AraC increases, AraC binds the O_1 operator and prevents *araC* mRNA transcription. The major advantages of this promoter system are that it can be partially induced depending on the concentration of inducer used and it is tightly repressed in the absence of inducer *(40)*.

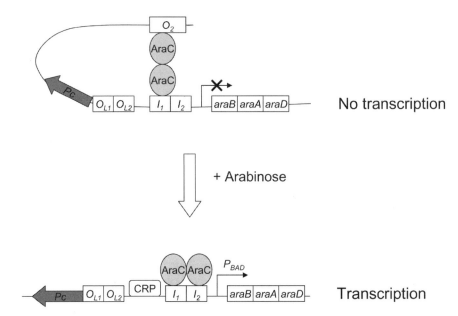

Fig. 1. Regulation of the arabinose operon. The arabinose operon consists of three structural genes, *araB*, *araA*, and *araD*. The *araC* gene encodes a transcriptional repressor that is transcribed from the promoter (P_C). In the absence of arabinose, the AraC repressor protein binds the operator site, O_2, and the I_1 site, causing the DNA in this region to adopt a looped conformation preventing transcription. In the presence of arabinose, the AraC repressor changes shape and binds the adjacent I_1 and I_2 sites, causing the loop to be released so that transcription may occur. Binding of the CAP site by cAMP receptor protein (CRP) leads to full induction of expression.

The pBAD vector series contains the arabinose promoter (*see* **Table 1**). The versatility within this system is substantial. One of the vectors available from this series includes a specific signal sequence for secreted expression to the periplasm (*see* **Subheading 4.1.2.**). The target protein can also be N-terminal or C-terminal tagged with a variety of epitopes.

3.1.4. The trp Promoter

The *trp* promoter regulates transcription of the tryptophan operon that encodes the genes required for tryptophan synthesis. Expression of the operon is inhibited by the *trp* repressor and induced by tryptophan starvation and by 3-indolylacrylic acid (IAA) *(25)*. In the uninduced state, a single copy of the repressor gene is sufficient to repress multiple copies of the operon *(25)*. One of the advantages of this promoter is its strength, but this is a concomitant disadvantage because, without an appropriately placed transcriptional terminator, it can result in a high level of read-through transcription of plasmid DNA.

3.1.5. The T7 Promoter

The bacteriophage T7 DNA-dependent RNA polymerase (T7 RNAP) is increasingly used to express a wide range of proteins in bacterial hosts. In this system, target genes are cloned downstream of the bacteriophage T7 promoter on medium-copy-number plasmids. Because the target gene does not utilize the host's polymerase, basal expression of toxic gene products cannot occur in the absence of T7 RNAP *(41)*.

There are two ways to induce expression in T7 RNAP-dependent systems. First, RNAP can be delivered by infection of the host cell with the T7 phage *(42)*. The phage polymerase acts *in trans* on its cognate promoter located on the expression plasmid. This method provides for complete repression in the absence of induction. Second, RNAP can be supplied by incorporating the T7 RNAP gene in the host strain's genome (e.g., *E. coli* BL21). T7 RNAP expression in this strain is from the IPTG-inducible *lac*UV5 promoter. However, the use of a *lac*-based promoter results in some basal expression in the absence of induction, which is problematic when dealing with toxic gene products. One strategy to address this issue is the co-overexpression of phage T7 lysozyme that degrades T7 RNA polymerase. The T7 lysozyme is supplied from the compatible pLysS and pLysE plasmids that are available from Novagen. Using these plasmids causes a lag time between induction of the RNA polymerase and maximum expression of the protein because the concentration of the polymerase must be raised high enough to overcome the inhibition by the lysozyme. The entire bacteriophage T7 promoter system, originally developed by Studier and colleagues *(43)*, is now available from Novagen as the pET (plasmids for expression by T7 RNAP) series.

3.1.6. The Lambda Promoters

The lambda promoters, λP_L and λP_R, exhibit up to 10-fold greater efficiency than the *lac* promoter *(25)*. Transcription from these promoters is repressed by the λcI gene product. A mutation in the repressor, *c*I857, renders it temperature sensitive so that expression of target genes from these promoters can be turned on and off through temperature change. At 28°C, the repressor is active and transcription is inhibited, but when the temperature is raised to 42°C, the repressor is inactivated and the promoter is induced. Temperature-sensitive promoters are advantageous when doing large-scale protein production because chemical inducers such as IPTG are toxic and expensive. One drawback to expression systems that are induced by a temperature rise is that proteolytic activity may be increased during the heat-shock response. Additionally, proteins are more likely to fold into their native forms at lower temperatures.

3.1.7. Cold-Shock Promoters

To circumvent the problems associated with the heat-shock response, promoters that are activated by a temperature downshift may be used. One such example is the promoter for the major *E. coli* cold-shock protein (CspA) *(44)*, which is repressed at and above 37°C and active at 10°C *(45)*. A disadvantage to this system is that it becomes repressed 1–2 h after temperature downshift so high-level protein accumulation is not possible *(46)*.

3.2. Nonpromoter Regulatory Elements

Factors other than the promoter also play an important role in obtaining high-level expression. Studies examining the sequences surrounding the transcriptional and translational start sites have revealed their importance in determining expression levels *(47–49)*. Upstream (UP) elements located 5' of certain bacterial promoters are A+T-rich sequences that increase transcription by interacting with the α-subunit of RNA polymerase *(50)*. A portable UP element was shown to increase the transcription of one gene more than 300-fold, highlighting the role that specific sequences may play in increasing the strength of a given promoter. It is important to remember that the same gene cloned into two different vectors can vary widely in expression levels. This may have more to do with how sequences from the vector interact with sequences from the insert to form secondary structures than the choice of promoter *(51)*.

4. Gene Fusions
4.1. Subcellular Localization

The localization of a protein in the host cell may affect its production and tertiary structure. Recombinant proteins can be directed to one of three compartments: cytoplasm, periplasm, or extracellular medium (secreted).

4.1.1. Cytoplasmic Localization

The cytoplasm, where most proteins are expressed, is a reducing environment. Proteins that are expressed in the cytoplasm may be soluble and can be extracted directly from the supernatant of lysed cells. Alternatively, some proteins, such as those containing hydrophobic or membrane-associated domains, colocalize to the insoluble cellular fraction. Insoluble proteins often aggregate to form inclusion bodies. There are three reasons why inclusion bodies may be desirable *(2)*. First, they are easily isolated by centrifugation during purification to yield concentrated and pure protein. Second, inclusion bodies protect the protein from proteolytic degradation. Third, toxic proteins are more readily tolerated during cell growth when found in this inactive state. Depending on downstream uses, proteins found in inclusion bodies may need to be refolded. These proteins can be solubilized with salt solutions, detergents, or denaturing agents. Finding a reagent that solubilizes and retains proper enzyme activity is generally an empirical process. One common choice of reagent is BugBuster™ Protein Extraction Reagent (Novagen). Detailed discussion of this topic is beyond the scope of this chapter but can be found elsewhere *(52,53)*.

4.1.2. Periplasmic Localization

The periplasm is an oxidizing environment that contains enzymes that catalyze the formation of disulfide bonds. Proteins may be directed to the periplasm in order to isolate active, folded proteins containing disulfide bonds and to avoid cytoplasmic insolublility or degradation *(54–56)*. If the protein in its native host is secreted, transmembrane, or contains disulfide bonds that are necessary for correct folding, then the periplasm is most likely the best compartment for expression *(57)*. Targeting a protein to the periplasm has the disadvantage that, generally, lower expression levels are obtained in the periplasm than in the cytoplasm. In addition, proteins must change conformation when passing through the cytoplasmic membrane into the periplasm. Some proteins are incompatible with passage through the membrane and can get stuck during this translocation process and are degraded *(58)*. Protein extraction from the periplasm involves bursting the bacterial outer membrane by osmotic shock.

A signal sequence, fused to a target gene, is used to direct heterologous proteins to the periplasm. The vectors, pET-39b(+) and pET-40b(+) encode the Dsb tag (*see* **Table 1**). This tag contains a signal sequence that directs the target protein to the periplasm and catalyzes the formation and isomerization of disulfide bonds *(59,60)*. The pMAL™ vectors also target proteins to the periplasm due to the presence of the *malE* signal sequence.

4.1.3. Extracellular Localization

In theory, the secretion of proteins into the extracellular milieu provides several advantages over localizing the protein to the cytoplasm or periplasm. First, there is less proteolytic activity here because *E. coli* secretes very few proteins. Second, purification is simplified because the protein does not need to be extracted from the bacterial cell and there are fewer other proteins present. Third, the protein is likely to be soluble and found in its native conformation. However, it is not yet possible to reliably target specific proteins across the inner and outer membranes in substantial quantities. Proteins can be secreted in small quantities using endogenous *E. coli* protein secretion pathways. The hemolysin gene is used to create hybrids for secretion *(61–63)*. Other secretory proteins containing leader sequences that can be used to target heterologous proteins for secretion are OmpA and PelB. The pEZZ vector contains a "ZZ" peptide derived from OmpA to facilitate protein secretion (*see* **Table 1**). **Table 3** contains additional information on tags that may be used to target proteins to the cytoplasm or to the periplasm, or for secretion.

4.2. Protein Purification and Detection

Fusion partners serve several purposes in addition to helping to determine the subcellular location of a target protein. Tags with specific binding sites were developed to facilitate common purification and detection procedures for different proteins that had otherwise unique structures and physicochemical properties. Tags are incorporated into the primary amino acid sequence of a target protein and may be located at the N-terminus or C-terminus. There are a number of considerations to take into account when deciding where to fuse the tag. Some proteins are expressed well as C-terminal fusions, whereas others express better as N-terminal fusions; this is likely the result of structural constraints that allow for proper folding in one orientation or another. Likewise, the proximity of the active site to the N-terminus or C-terminus is important. In most instances, biological activity is more likely to be maintained if the active site is furthest away from a potentially interfering fusion tag. Also, proteins that contain N-terminal signal sequences are posttranslationally processed; thus, fusions to these proteins should be placed at the C-terminus. The pTYB vectors (*see* **Table 1**) have been designed such that a gene can be cloned with either a N-terminal or a C-terminal

Table 3
Fusion Tags and Signal Peptides

Tag	Size (aa)	Basis for detection and/or purification	Uses and comments	Localization signal
Biotin purification tag	386	Avidin or streptavidin	Western blot, peptide tag biotinylation in vivo by *E. coli*	None
CBD-Tag (calmodulin-binding domain)	36	Calmodulin resin	Purification (mild elution conditions), small tag less likely to affect protein of interest	None
CBD-Tag (Cellulose-binding domain)	156, 114, 107	Polyclonal antibody, cellulose	Western blot, purification	None
Chitin-binding domain	495	Anti-chitin-binding domain antibody, chitin beads	Purification, Western blot	None
c-*myc* epitope	10	Monoclonal antibody	Western blot, immunoprecipitation, purification	None
DHFR epitope (dihydrofolate reductase)	243	N/A	Stabilizes small peptides, poorly immunogenic (fusion protein will be recognized when making antibody)	None
Dsb-Tag™	208 or 236	N/A	Soluble protein, periplasmic disulfide bond formation, isomerization	Periplasmic
FLAG-Tag™	8, 24	Monoclonal antibody, polyclonal antibody, small antibody recognition site	Western blot, immunoprecipitation, purification	None
GST-Tag™ (Glutathione-S-transferase)	234	Monoclonal antibody, enzymatic activity, glutathione affinity	Purification, Western blot, quantitative assay	None
HA-Tag (Hemaglutinin)	9	Monoclonal antibody	Western blot, immunoprecipitation, purification	None
His-Tag®	6, 8, or 10	Monoclonal antibody, metal chelation chromatograpy (i.e., His-Bind resin)	Purification; native or denaturing conditions	None
HSV-Tag® (Herpes Simplex Virus)	11	Monoclonal antibody	Western blot, immunofluorescence	None

Tag	Size	Detection/Purification	Purpose	Localization
KSI (Keto-steroid Isomerase)	125	N/A	Highly expressed hydrophobic domain, small protein production/purification	None
MBP (Maltose Binding Protein)	378	Polyclonal antibody, amylose/maltose resin	Small protein/peptide production/purification; enhances solubility	Periplasmic, cytoplasmic (if signal sequence deleted)
Nus-Tag™ (Nut Utilization Substance)	491	N/A	Promotes cytoplasmic solubility	None
OmpA epitope	21	N/A	Signal sequence for secretion of fusion protein	Periplasmic
PelB/ompT	20 or 22	N/A	Protein export/folding	Periplasmic
PKA (Protein Kinase A)	5	Use PKA site for labeling with ^{32}P	In vitro phosphorylation	None
Protein C tag	12	Monoclonal antibody	Western blot, immunoprecipitation, purification	None
S-Tag™	15	S-protein	Western blot, quantitative assay, purification	None
T7-Tag®	11 or 260	Monoclonal antibody	Western blot, immunoprecipitation, purification	None
Trx-Tag (Thioredoxin)	109	N/A	Soluble protein, thioredoxin promotes cytoplasmic disulfide bond formation in specific reductase deficient *E. coli* host strains	None
V-5 epitope	14	Monoclonal antibody	Western blot, immunoprecipitation, purification	None
VSV-G tag (Vesicular stomatitis virus glycoprotein)	11	Monoclonal antibody	Western blot, immunoprecipitation, purification	None
"ZZ" peptide (protein A signal sequence)	126	IgG coupled to Sepharose 6 Fast Flow Agarose Matrix	Gene expression not inducible	Culture medium

Source: Data compiled from manufacturer's literature.

tag using the same restriction sites. When a C-terminal tag is desired, the target gene must not contain a stop codon in order to allow translation to continue to the end of the adjacent tag. Conversely, C-terminal fusions encoded by vectors can be avoided by including a stop codon in the insert. Also, the tag will not function unless cloned in frame with the target protein open reading frame. Many commercially available expression vectors are available with cloning sites for cloning in all three reading frames to facilitate proper construction of the translational fusion.

A widely used fusion tag involves the fusion of 6–10 histidine amino acid residues onto the target protein at either the N-terminus or C-terminus. The histidine tag can subsequently be used for purification, because it will bind nickel ions immobilized on a column *(64)*. The interaction between the amino acid side chains and the metal ions is reversible; thus, proteins fused to the histidine motif can be separated from other proteins that do not contain polyhistidine. Many vectors include the histidine tag as one of their features: pPRO, pBAD, pThioHis, pDEST, pRSET, pET, and pQE (*see* **Table 1** for more options). The histidine fusion protein will bind to the nickel ions under native or denaturing conditions and either can be used for purification. Cysteine and tryptophan side chains can also be used to bind proteins to immobilized transition metal ions and zinc *(65–67)*. If the purification procedure results in a small protein yield, it may be the result of the low affinity of the tag for the column. This may be caused by an interaction between the target protein and the affinity tag that blocks or distorts the binding site. Shortening or lengthening the fused polypeptide can sometimes alleviate this problem. Occasionally, low yields result from the target protein being truncated because of proteolytic degradation or the ribosome initiating translation at a site downstream of the true start codon *(68)*. When fusion affinity tags are attached to both ends of a target protein, they can then be used successively to ensure purification of the full-length protein.

PinPoint encodes a short peptide that is biotinylated in vivo and allows for purification of the target protein on a strepatavidin column. Many systems purify proteins based on the use of antibodies through immunoprecipitation (*see* **Table 3**). These include T7-Tag™, HSV-Tag™, FLAG-Tag™, and "ZZ" tag. These fusions are reported to have little influence on the native conformation of the expressed gene. Larger tags, such as GST (pGEX and pET vectors) and MBP (pMAL vector), employed to improve solubility of hydrophobic proteins, can also be used for purification by immunoprecipitation.

Other tags are useful for quantifying purified protein by Western blots or spectrophotometrically. For example, the S-Tag™ (pET vector, Novagen) has high affinity for the ribonuclease–S-protein and can be quantified by incubating lysate containing the fusion protein with purified S-protein in the FRETWORKS™ assay. This interaction reconstitutes RNase activity that cleaves the FRET ArUAA substrate to produce a fluorescent product *(69)*. The GST tag (pGEX and pET vectors) can also be used to quantitatively assay for fusion proteins in crude extracts.

4.3. Protease Cleavage

In many cases, the presence of a tag makes little difference in the subsequent usage of the protein; however, a protein in its native state without a tag is necessary for

E. coli Expression Vectors

certain applications such as crystallography. Tags can be removed by including a specific protease cleavage site between the tag and the cloned gene. The site is recognized by a highly specific protease, providing a mechanism for removing the fusion partner and producing a final product with native termini. Factor Xa, thrombin, and enterokinase are commercially available products that have well-defined target sites. Almost all vectors available include cleavage sites.

The IMPACT™ system from New England BioLabs offers a tag that combines the benefits of the chitin-binding domain with a self-splicing protein element called an intein. With the intein tag, the target protein can be purified in a single chromatographic step without the use of a protease. Affinity purification takes place on a chitin column and then the addition of thiols such as DTT, β-mercaptoethanol, or cystein causes intein-mediated cleavage between the intein and the target protein.

5. Troubleshooting Tips for Insoluble Proteins

Certain vectors and host strains enhance the likelihood of expressing a soluble protein. One approach to increasing the soluble yields of aggregated proteins is to improve folding of newly synthesized proteins through the co-overexpression of cytoplasmic molecular chaperones *(70)*. The ATP-dependent DnaK-DnaJ-GrpE and GroEL-GroES systems of *E. coli* increase soluble protein yields and are available on cloning vectors from the institutions that developed them *(71,72)*. Thioredoxins are necessary for the formation of disulfide bonds in the cytoplasm *(73)*. The use of vectors that encode a thioredoxin fusion protein (Trx), such as the pET-32 series (*see* **Table 1**), increases the yield of soluble product in the cytoplasm *(74)*. Host strains such as BL21*trx*B, Origami, Origami B, and AD494 are deficient in certain reductases and promote the formation of disulfide bonds *(49)* (*see* Chapter 3). The use of one of these strains in combination with Trx may improve the solubility and activity of a given target protein. Large fusion tags sometimes improve the solubility of hydrophobic proteins. The maltose-binding protein (MBP) and chitin-binding domains, protein A, dihydrofolate reductase (DHFR), Nut Utilization Substance (Nus), Trx, and glutathione-*S*-transferase (GST) are all promoted as solubility-enhancing tags. One study found that MBP was a far better solubilizing partner than either Trx or GST *(75)*. The MBP fusion tag is encoded on the pMAL vector. The Trx and GST fusions are also commonly utilized and can be found on the pET and pGEX vectors, respectively. Although these tags can help with solubility, large fusions may affect the conformation of the protein of interest so it may be necessary to cleave off the purification tag (*see* **Subheading 4.3.**). Further information regarding other proteins that can be used to enhance the solubility of fusion proteins may be found in **Table 3**. Finally, protein solubility may be improved by decreasing the rate of protein synthesis by using low induction temperatures and growing cells in minimal media.

6. Conclusions

The *E. coli* plasmid vectors available to researchers are continually fine-tuned, making it easier to express a wide variety of proteins in any given expression system. A list of commercially available expression vectors is included in **Table 1**. Because of the unique qualities of any given protein, there is no guarantee that any one vector will work for expression. Recent innovations include donor vectors that serve to rapidly transfer the gene of interest into multiple different expression vectors (Creator™, Clontech). A similar approach uses plasmids with the potential to easily exchange essential expression components such as the promoter, origin of replication, antibiotic resistance genes, and purification tags (ProTet 6XHN tags, Clontech).

There are still some limitations to the prokaryotic expression system. One of the most important is the inability to accomplish posttranslational modifications such as glycosylation, acetylation, or amidation that are necessary for the correct expression of some eukaryotic proteins. When proteins cannot be expressed in *E. coli*, useful vectors are commerially available. In spite of this drawback, *E. coli* remains one of the most attractive hosts for heterologous protein expression. Vectors are commercially available that allow expression of target proteins in vertebrate cells and insect cells in addition to *E. coli* from a single expression vector containing unique promoters for each organism (pTriEx, Novagen).

References

1. Fernandez, J. M. and Hoeffler, J. P. (eds.) (1999) *Gene Expression Systems*, Academic, San Diego, CA.
2. pET System Manual 9th ed. (2000) Novagen, Inc., Madison, WI.
3. Gronenborn, B. and Messing, J. (1978) Methylation of single-stranded DNA in vitro introduces new restriction endonuclease cleavage sites. *Nature* **272**, 375–377.
4. Clark, J. M. (1988) Novel non-templated nucleotide addition reactions catalyzed by prokaryotic and eucaryotic DNA polymerases. *Nucleic Acids Res.* **16**, 9677–9686.
5. Mead, D. A., Pey, N. K., Herrnstadt, C., et al. (1991) A universal method for the direct cloning of PCR amplified nucleic acid. *BioTechnology* **9**, 657–663.
6. Yanisch-Perron, C., Vieirira, J., and Messing, J. (1985) Improved M13 phage cloning vectors and host strains: Nucleotide sequences of the M13mp18 and pUC19 vectors. *Gene* **33**, 103–119.
7. Vieira, J. and Messing, J. (1982) The pUC plasmids: an M13mp7-derived system for insertion mutagenesis and sequencing with synthetic universal primers. *Gene* **19**, 259–268.
8. Dersch, P., Fsihi, H., and Bremer, E. (1994) Low-copy-number T7 vectors for selective gene expression and efficient protein overproduction in *Escherichia coli*. *FEMS Microbiol. Lett.* **123**, 19–26.
9. Brosius, J. (1999) Expression Vectors Employing the *trc* promoter, in *Gene Expression Systems* (Fernandez, J. M. and Hoeffler, J. P., eds.), Academic, San Diego, CA, pp. 45–94.
10. Hannig, G. and Makrides, S. C. (1998) Strategies for optimizing heterologous protein expression in *Escherichia coli*. *Trends Biotechnol.* **16**, 54–60.
11. Brown, W. C. and Campbell, J. L. (1993) A new cloning vector and expression strategy for genes encoding proteins toxic to *Escherichia coli*. *Gene* **127**, 99–103.
12. Doherty, A. J., Connolly, B. A., and Worrall, A. F. (1993) Overproduction of the toxic protein, bovine pancreatic DNaseI in *Escherichia coli* using a tightly controlled T7-promoter-based vector. *Gene* **136**, 337–340.
13. Suter-Crazzolara, C. and Unsicker, K. (1995) Improved expression of toxic proteins in *E. coli*. *BioTechniques* **19**, 202–204.

14. Mertens, N., Remaut, E., and Fiers, W. (1995) Tight transcriptional control mechanism ensures stable high-level expression from T7 promoter-based expression plasmids. *BioTechnology* **13**, 175–179.
15. Bentley, W. E., Mirjalili, N., Anderson, D. C., et al. (1990) Plasmid-encoded protein: the principal factor in the "metabolic burden" associated with recombinant bacteria. *Biotechnol. Bioeng.* **35**, 668–681.
16. Goeddel, D. V., Yelverton, E., Ulrich, A., et al. (1980) Human leukocyte interferon produced by *E. coli* is biologically active. *Nature* **287**, 411–416.
17. Brosius, J. (1984) Plasmid vectors for the selection of promoters. *Gene* **27**, 161–172.
18. Amann, E. and Brosius, J. (1985) 'ATG vectors' for regulated high-level expression of cloned genes in *Escherichia coli*. *Gene* **40**, 183–190.
19. Elvin, C. M., Thompson, P. R., Argall, M. E., et al. (1990) Modified bacteriophage lambda promoter vectors for overproduction of proteins in *Escherichia coli*. **Gene** *87*, 123–126.
20. Tabor, S. and Richardson, C. C. (1985) A bacteriophage T7 RNA polymerase/promoter system for controlled exclusive expression of specific genes. *Proc. Natl. Acad. Sci. USA* **82**, 1074–1078.
21. Chang, C. N., Kang, W. J., and Chen, E. Y. (1986) Nucleotide sequence of the alkaline phosphatase gene of *Escherichia coli*. *Gene* **44**, 121–125.
22. Makrides, S. C. (1996) Strategies for achieving high-level expression of genes in *Escherichia coli*. *Microbiol. Rev.* **60**, 512–538.
23. Stevens, R. C. (2000) Design of high-throughput methods of protein production for structural biology. *Structure Fold. Des.* **8**, R177–R185.
24. Denhardt, D. T. and Colasanti, J. (1988) A survey of vectors for regulating expression of cloned DNA in *E. coli*, in *Vectors: A Survey of Molecular Cloning Vectors and Their Uses* (Rodriguez, R. L. and Denhardt, D. T., eds.), Butterworths, Stoneham, MA, pp. 179–203.
25. Hu, M. C. and Davidson, N. (1987) The inducible *lac* operator-repressor system is functional in mammalian cells. *Cell* **48**, 555–566.
26. Kamashev, D. E., Esipova, N. G., Ebralidse, D. D., et al. (1995) Mechanism of Lac repressor switch-off: orientation of the Lac repressor DNA binding domain is reversed upon inducer binding. *FEBS Lett.* **375**, 27–30.
27. Bourgeois, S. and Pfahl, M. (1976) Repressors. *Adv. Protein Chem.* **30**, 1–99.
28. de Boer, H. A., Comstock, L. J., and Vasser, M. (1983) The *tac* promoter: a functional hybrid derived from the *trp* and *lac* promoters. *Proc. Natl. Acad. Sci. USA* **80**, 21–25.
29. Mulligan, M. E., Brosius, J., and McClure, W. R. (1985) Characterization *in vitro* of the effect of spacer length on the activity of *Escherichia coli* RNA polymerase at the TAC promoter. *J. Biol. Chem.* **260**, 3529–3538.
30. Brosius, J., Erfle, M., and Storella, J. (1985) Spacing of the –10 and –35 regions in the *tac* promoter: effect on its in vivo activity. *J. Biol. Chem.* **260**, 3539–3541.
31. Deuschle, U., Kammerer, W., Gentz, R., et al. (1986). Promoters of *Escherichia coli*: a hierarchy of *in vitro* strength indicates alternate structures. *EMBO J.* **5**, 2987–2994.
32. Kammerer, W., Deuschle, U., Gentz, R., et al. (1986) Functional dissection of *Escherichia coli* promoters: Information in the transcribed region is involved in late steps of the overall process. *EMBO J.* **5**, 2995–3000.
33. Miyada, G. C., Stoltzfus, L., and Wilcox, G. (1984) Regulation of the *araC* gene of *Escherichia coli*: catabolite repression, autoregulation, and effect on *araB*AD expression. *Proc. Natl. Acad. Sci. USA* **81**, 4120–4124.
34. Wilcox, G., Boulter, J., and Lee, N. (1974) Direction of transcription of the regulatory gene *araC* in *Escherichia coli* B-r. *Proc. Natl. Acad. Sci. USA* **71**, 3635–3639.

35. Englesberg, E. and Wilcox, G. (1974) Regulation: Positive control. *Annu. Rev. Genet.* **8,** 219–242.
36. *pBAD TOPO TA Cloning Kit, Version G,* Invitrogen Corp., Carlsbad, CA.
37. Casadaban, M. (1976) Regulation of the regulatory gene for the arabinose pathway, *araC.* *J. Mol. Biol.* **104,** 557–566.
38. Schleif, R. (2000) Regulation of the L-arabinose operon of *Escherichia coli. Trends Genet.* **16,** 559–565.
39. Lee, N. L, Gielow, W. O., and Wallace, R. G. (1981) Mechanism of *araC* autoregulation and the domains of two overlapping promoters, P_C and P_{BAD}, in the L-arabinose regulatory region of *Escherichia coli. Proc. Natl. Acad. Sci. USA* **78,** 752–756.
40. Carson, M. J., Barondess, J. J., and Beckwith, J. (1991) The FtsQ protein of *Escherichia coli*: membrane topology abundance, and cell division phenotypes due to overproduction and insertion mutations. *J. Bacteriol.* **173,** 2187–2195.
41. Durbin, R. (1999) Gene expression systems based on bacteriophage T7 RNA polymerase, in *Gene Expression Systems* (Fernandez, J. M. and Hoeffler, J. P., eds.), Academic, San Diego, CA, pp. 9–44.
42. McAllister, W. T., Morris, C., Tosenberg, A. H., et al. (1981) Utilization of bacteriophage T7 late promoters in recombinant plasmids during infection. *J. Mol. Biol.* **153,** 527–544.
43. Studier, F. W. and Moffatt, B. A. (1986) Use of bacteriopage T7 RNA polymerase to direct selective high-level expression of cloned genes. *J. Mol. Biol.* **189,** 113–130.
44. Phadtare, S. Alsina, J., and Inouye, M. (1999) Cold-shock response and cold-shock proteins. *Curr. Opin. Microbiol.* **2,** 175–180.
45. Vasina, J. A., Peterson, M. S., and Baneyx, F. (1998) Expression of aggregation-prone recombinant proteins at low temperatures: a comparative study of the *Escherichia coli cspA* and *tac* promoter system. *Biotechnol. Prog.* **9,** 211–218.
46. Baneyx, F. (1999) Recombinant protein expression in *Escherichia coli. Curr. Opin. Biotechnol.* **10,** 411–421.
47. Gaal, T., Barkei, J., Dickson, R. R., et al. (1989) Saturation mutagenesis of an *Escherichia coli* rRNA promoter and initial characterization of promoter variants. *J. Bacteriol.* **171,** 4852–4861.
48. Hsu, L. M., Giannini, J. K., Leung, T. W., et al. (1991) Upstream sequence activation of *Escherichia coli argT* promoter in vivo and in vitro. *Biochemistry* **30,** 813–822.
49. Josaitis, C. A., Gaal, T., Ross, W., et al. (1990) Sequences upstream of the –35 hexamer of rrnB P1 affect promoter strength and upstream activation. *Biochim. Biophys. Acta* **1050,** 307–311.
50. Aiyar, S. E., Gourse, R. L., and Ross, W. (1998) Upstream A-tracts increase bacterial promoter activity through interactions with the RNA polymerase α-subunit. *Proc. Natl. Acad. Sci. USA* **95,** 14,652–14,657.
51. Shatzman, A. R. and Rosenberg, M. (1987) Expression, identification, and characterization of recombinant gene products in *Escherichia coli*, in *Methods in Enzymology* (Berkger, S. L and Kimmel, A. R., eds.), Academic, Orlando, FL, Vol. 152, pp. 661–673.
52. Rudolph, R. and Lilie, H. (1996) *In vitro* folding of inclusion body proteins. *FASEB J.* **10,** 49–56.
53. Mukhopadhyay, A. (1997) Inclusion bodies and purification of proteins in biologically active forms. *Adv. Biochem. Eng. Biotechnol.* **56,** 61–109.
54. Rietsch, A., Belin, D., Martin, N., et al. (1996) An in vivo pathway for disulfide bond isomerization in *Escherichia coli. Proc. Natl. Acad. Sci. USA* **93,** 13,048–13,053.
55. Raina, S. and Missiakas, D. (1997) Making and breaking disulfide bonds. *Annu. Rev. Microbiol.* **51,** 179–202.

56. Sone, M., Akiyama, Y., and Ito, K. (1997) Differential in vivo roles played by DsbA and DsbC in the formation of protein disulfide bonds. *J. Biol. Chem.* **272**, 10,349–10,352.
57. Jones, P. (ed.) (1998) *Vectors: Expression Systems*, Wiley, Chichester.
58. Gentz, R., Kuys, Y., Zwieb, C., et al. (1988) Association of degradation and secretion of three chimeric polypeptides in *Escherichia coli*. *J. Bact.* **170**, 2122–2120.
59. Missiakas, D., Georgeopoulos, C., and Raina, S. (1994) The *Escherichia coli dsbC* (*xprA*) gene encodes a periplasmic protein involved in disulfide bond formation. *EMBO J.* **13**, 2013–2020.
60. Zapun, A., Missiakas, D., Raina, S., et al. (1995) Structural and functional characterization of DsbC, a protein involved in disulfide bond formation in *Escherichia coli*. *Biochemistry* **34**, 5075–5089.
61. Blight, M. A., Chervaux, C., and Holland, I. B. (1994) Protein secretion pathways in *Escherichia coli*. *Curr. Opin. Biotechnol.* **5**, 468–474.
62. Kern, I. and Ceglowski, P. (1995) Secretion of streptokinase fusion proteins from *Escherichia coli* cells through the hemolysin transporter. *Gene* **163**, 53–57.
63. Mackman, N., Baker, K., Gray, L., et al. (1987) Release of a chimeric protein into the medium from *Escherichia coli* using the C-terminal secretion signal of haemolysin. *EMBO J.* **6**, 2835–2841.
64. Setlow, J.D. (ed.) (1990) *Genetic Engineering: Principle and Methods*, Plenum, New York.
65. Porath, J. (1985) Immobilized metal ion affinity chromatography—a powerful method for protein purification, in *Modern Methods in Protein Chemistry* (Tschelsche, H., ed.), Walter de Gruyter, Berlin, pp. 85–95.
66. Sulkowski, E. (1985) Purification of proteins by IMAC. *Trends Biotechnol.* **3**, 1–7.
67. Hemdan, E. S. and Porath, J. (1985) Development of immobilized metal affinity chromatography II: interaction of amino acids with immobilized nickelimionodiacetate. *J. Chromatogr.* **323**, 255–264.
68. Preibisch, G., Ishihara, H., Trippier, D., et al. (1988) Unexpected translation initiation within the coding region of eukaryotic genes expressed in *Escherichia coli*. *Gene* **72**, 179–186.
69. *pTriEx System Manual*, Novagen, Madison, WI.
70. Thomas, J. G., Ayling, A., and Baneyx, F. (1997) Molecular chaperones, folding catalysts and the recovery of active recombinant proteins from *E. coli*: to fold or to refold. *Appl. Biochem. Biotechnol.* **66**, 197–238.
71. Nishihara, K., Kanemori, M., Kitagawa, M., et al. (1998) Chaperone coexpression plasmids: differential and synergistic roles of DnaK-DnaJ-GrpE and GroEL-GroES in assisting folding of an allergen of Japanese cedar pollen, Cryj2, in *Escherichia coli*. *Appl. Environ. Microbiol.* **64**, 1694–1699.
72. Castanie, M. P., Berges, H., Oreglia, J., et al. (1997) A set of pBR322-compatible plasmids allowing the testing of chaperone-assisted folding of proteins overexpressed in *Escherichia coli*. *Anal. Biochem.* **254**, 150–152.
73. Stewart, E. J., Aslund, F., and Beckwith, J. (1998) Disulfide bond formation in the *Escherichia coli* cytoplasm: an in vivo role reversal for the thioredoxins. *EMBO J.* **17**, 5543–5550.
74. LaVallie, E.R., DiBlasio, E. A., Kovacic, S., et al. (1993) A thioredoxin gene fusion expression system that circumvents inclusion body formation in the *E. coli* cytoplasm. *BioTechnology* **11**, 187–193.
75. Kapust, R. B. and Waugh, D. S. (1999) *Escherichia coli* maltose-binding protein is uncommonly effective at promoting the solubility of polypeptides to which it is fused. *Protein Sci.* **8**, 1668–1674.

29

Expression of Recombinant Proteins From *lac* Promoters

Charles R. Sweet

1. Introduction

The Gram-negative bacterium *Escherichia coli* enjoys widespread use in modern biology as both a model organism and a microbiological tool. One of the keys to its popularity lies in the functionality of the *lac* operon. This regulated *E. coli* operon has been manipulated and utilized in diverse and creative ways since it was first studied by Jacob and Monod in 1961 *(1)*. Excellent in-depth reviews of the *lac* operon and its uses in classical genetics are provided in **refs. 2** and **3**. Its enduring use can be attributed to the development of a wide range of engineered mutants and plasmid constructs designed to take advantage of specific features of the operon, particularly its inducibility by β-galactosides. This chapter will focus on the use of the *lac* operon in *E. coli* protein expression, although it has also been utilized for protein expression in mammalian systems *(4)*. The application of *lacZ* as a reporter gene is discussed in Chapter 30.

1.1. The lac *Operon*

In the bacterium, the *lac* operon is necessary for the utilization of lactose as a carbon source. The operon (*see* **Fig. 1**) consists of three genes: *lacZ*, *lacY*, and *lacA*. Upstream of these genes is a promoter and operator sequence and the *lacI* gene. The operator is the binding site for the LacI transcriptional repressor, which is transcribed independently of the other *lac* genes.

The first gene of the operon encodes β-galactosidase (LacZ) that cleaves the glycosidic linkage of lactose to produce the sugars glucose and galactose *(5)*. The second gene encodes the lactose permease (LacY). This membrane protein is involved in transport of β-galactosides into the cell *(5)*. LacA is encoded by the final gene of the operon. It is unclear what role this protein plays in lactose metabolism; however it has been characterized as a thiogalactoside transacetylase *(6)*.

These proteins are necessary only when the bacterium is required to use lactose as a carbon source; thus, their expression is regulated in response to the concentration of

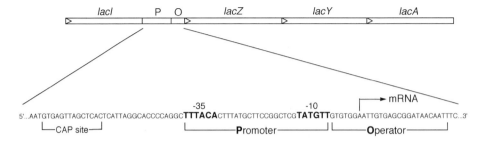

Fig. 1. The *lac* genetic locus and binding sequences. The wild-type *E. coli lac* operon and control elements are shown. When CAP is bound to the promoter (P) and LacI is not bound to the operator (O), RNA polymerase binds to the promoter and initiates transcription of *lacZ*, *Y*, and *A* as a polycistronic message. The sequence of the regulatory region is shown, from the CAP-binding site to the end of the operator region. The –10 and –35 regions of the promoter are shown in **bold**.

lactose and glucose. In the absence of lactose, the LacI repressor binds to the operator sequence of the *lac* promoter and prevents RNA polymerase from binding to the promoter *(7)*. LacI also binds lactose and other β-galactosides, causing LacI to lose affinity for the operator. This relieves repression of transcription of the operon when these molecules are present *(7)*. Expression of the *lac* operon is also controlled by glucose, via the catabolite gene activator protein (CAP). This protein, encoded by *crp*, binds cAMP and the *lac* promoter to increase transcription initiation by RNA polymerase. The absence of glucose stimulates the production of cAMP, which increases the binding of CAP to the *lac* promoter and induces *lac* expression when lactose metabolism is necessary *(8,9)*.

1.2. The Tools of lac *Expression*

1.2.1. IPTG Induction of lac *Expression*

The essential element involved in *lac*-mediated protein expression is not the *lac* genes, but rather the *lac* regulatory apparatus. Generally, a gene selected for overexpression is cloned into a plasmid vector behind the *lac* promoter. Expression of this gene is then under *lac* control, either as an independent gene or fused to a fragment of *lacZ* or other genes (*see* **Fig. 2A**). This promoter is then induced (selectively activated) by the addition of a β-galactoside, most commonly isopropyl-β-D-thiogalactoside (IPTG; *see* **Fig. 3**). LacI binds IPTG, preventing LacI binding of the operator and triggering transcription of the operon. As IPTG is not cleaved by β-galactosidase, it persists in the cell and greatly stimulates the expression of *lac*-controlled genes even at low concentrations.

Fig. 2. *(opposite)* Schematic of *lac* expression strategies. Genetic elements from the *E. coli lac* operon are shown in white, those from bacteriophage T7 are shown in gray, and the hashed element represents the cloned gene encoding the protein to be expressed. (**A**) *lac*UV5 and *trp/lac* expression. LacI binds to the operator (O), preventing RNA polymerase from binding to the promoter (P). IPTG serves as a coactivator to relieve LacI-mediated repression of transcription.

Recombinant Protein Expression

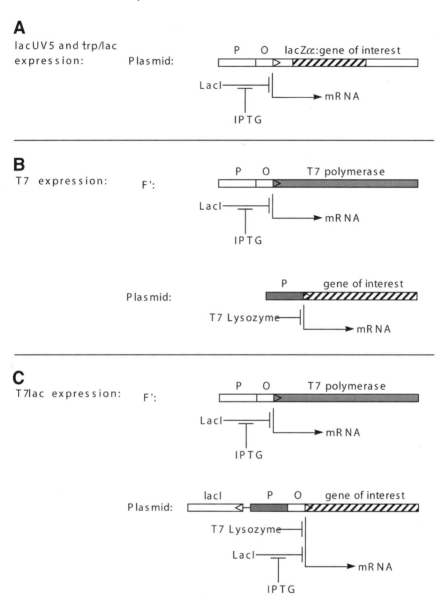

Fig. 2. *(continued)* The gene of interest is shown as an insertion into the sequence of *lacZα*. **(B)** T7 expression. The *lac*UV5 promoter, subject to the regulation shown in **(A)**, drives the expression of T7 RNA polymerase, carried on the F' genetic element. T7 polymerase then trancribes the gene of interest, from the T7 promoter (T7P) that is carried on the plasmid. T7 polymerase is inhibited by T7 lysozyme. **(C)** T7*lac* expression. In addition to the control shown in **(B)**, T7*lac* plasmids carry the *lac* operator downstream of the T7 promoter, preventing binding of T7 polymerase to the promoter in the absence of IPTG, as described above. In order to fully occupy the plasmid copies of the operator, LacI is also expressed on the plasmid.

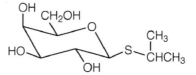

Fig. 3. Structure of IPTG.

1.2.2. Engineered lac Genes and Promoter Systems: The Molecular Toolbox

Over the last few decades, a major effort has been made to engineer the *lac* promoter, *lac* proteins, and several accessory proteins, in order to optimize *lac*-dependent expression and increase the flexibility and functionality of the system. The basic molecular tools created in this effort are discussed in the following subsections. Understanding of these concepts is crucial to successful experimental design.

1.2.2.1. THE *LAC*UV5 PROMOTER: REMOVAL OF CAMP REGULATION

To facilitate the exploitation of the *lac* promoter system as a "protein factory," the *lac*UV5 mutant promoter is commonly used (*see* **Fig. 2A**). This promoter is relatively insensitive to CAP-mediated regulation compared to the wild-type promoter and, thus, the cAMP–CAP–promoter complex is no longer necessary for RNA polymerase to bind to the promoter and initiate transcription *(10)*. Therefore, protein expression can be driven by *lac*UV5 in rich media such as Luria–Bertani (LB) media.

1.2.2.2. LACI^q: A TIGHTER INHIBITOR

Both the wild-type and *lac*UV5 mutant promoters give rise to a significant level of basal expression in the absence of induction. This can cause problems if the target protein is toxic in *E. coli*, as even a small amount of toxic protein can cause poor bacterial growth and a reduction in protein expression over time. LacI^q, a mutant *lac* repressor, is often used to modulate *lac* expression. This repressor binds the *lac* operator more tightly than wild-type LacI and, thus, decreases basal expression by around 10-fold *(11)*. LacI^q is usually expressed from the F' element or, alternatively, from a derivative of the pLysS/E or Rosetta™ helper plasmids.

1.2.2.3. THE *TRP/LAC* PROMOTERS: MAXIMIZED PROMOTER ACTIVITY

One disadvantage of the *lac* promoter for mediating protein expression is that its promoter sequence TTTACA (–35) and TATGTT (–10) differs in two bases from the canonical *E. coli* promoter sequence TTGACA (–35) and TATATT (–10). In particular, the G to T change in the –35 region is deleterious to the activity of the promoter *(2)*. To alleviate this problem, a promoter has been engineered that bears the –35 region from the *trp* (tryptophan) promoter and the –10 region and the operator from *lac*UV5. There are two forms of the *trp/lac* promoter, *trc* and *tac*, which give rise to similar levels of protein expression *(12,13)*. The increased activity of *trc* and *tac* requires that LacI^q or overexpressed LacI be present to suppress transcription in the absence of induction.

1.2.2.4. The T7 System: High-Level Overexpression

One of the most widely used recombinant protein expression systems is the T7 system, in which the gene of interest is removed from direct *lac*-mediated control (*see* **Fig. 2B**). Instead, the gene of interest is cloned on a plasmid (such as one of the pET series of vectors) behind a promoter that is recognized by the bacteriophage T7 RNA polymerase, which, in turn, is inserted into the host strain under *lac*UV5 control *(14)*. Therefore, the polymerase, controlled by *lac*, is inducible by IPTG. Once induced, this highly active polymerase transcribes the target gene, and protein yields in excess of 50% of total cellular protein can be obtained *(15)*. This activity is so efficient, however, that it interferes with the production of essential cellular proteins and is eventually toxic to the cell. For this reason, T7 strains should not be passaged or stored after induction.

1.2.2.5. The T7*lac* System: Complete Inhibition of "Leaky" Expression

Perhaps the best mechanism for prevention of basal expression is the modified T7 regulatory scheme known as T7*lac* *(16)*. In this system, the plasmid bearing the gene of interest contains the *lac* operator sequence just downstream of the T7 promoter (*see* **Fig. 2C**). The binding of the LacI repressor to this operator blocks transcription from the promoter. This plasmid also bears *lacI*, because without it, the operator on the multicopy plasmid would titrate out the endogenous LacI, resulting in expression of both the polymerase and the target gene.

1.3. Optimization of lac *Expression*

The abundance of commercial expression vectors, host strains, and helper plasmids that are now available allows great flexibility in attempts to express and purify proteins in *E. coli* and has made the *lac* genetic locus one of the foundations of modern molecular biology. In **Subheading 3.1.**, the basic protocol for IPTG-induced recombinant protein overexpression is described. An overview of prokaryotic protein expression strategies and procedures can be found in **refs. 17** and **18**.

The success or failure of *lac*-based expression efforts is often rooted in the design phase of the experiment. With so many vectors and strains available, the investigator may fall into the habit of choosing a system already on hand or one that has worked in the past. However, the capricious nature of protein expression, especially heterologous expression, dictates that one must carefully examine and optimize all aspects of the expression system for each individual protein.

When designing the expression construct, ensure that there is a Shine–Dalgarno sequence (ribosome binding site, or RBS) four to eight bases upstream of the target gene's start codon *(19)*. If a fusion/epitope-tagged construct is used, make certain that the fusion is to the end of the protein that is furthest from the active site of the protein and that it will not be cleaved by posttranslational processing. Also, ensure that the antibiotic selection marker is compatible with the host strain and helper plasmids.

For some applications, it is desirable to use *lac*-based protein expression in a constitutive manner, such as complementation of an *E. coli* mutant with a heterologous protein. In this case non-T7 vectors that bear the *lac* or *trp/lac* promoters, such as pBluescript or pTrc99a, can be used. Without IPTG induction, these constructs produce significant levels of basal transcription. IPTG induction can be used in this situation to increase the expression of the target genes.

If an acceptable level of expression is not attained at the first attempt or if the expressed protein lacks a required activity, empirical methods of optimization must be employed. The common problems of toxicity and solubility are addressed in **Subheadings 1.3.1.** and **1.3.2.**, respectively. If the difficulty with expression is not the result of one of these factors, various substitutions can be made in the expression system, starting with variation of the expression strain and helper plasmids (*see* Chapter 3). Target proteins will often express better in one of the various expression strains than in others. Often switching between an *E. coli* B strain, such as BL21, and an *E. coli* K12 strain, such as Novablue, can have a profound effect on the solubility of the target protein. The reasons for these differences are still poorly understood.

1.3.1. Strategies for Dealing with Toxic Proteins

Overexpression of proteins in *E. coli* often has the potential of interfering with essential cellular processes. Even when expressing only moderately toxic proteins, the expression plasmid should be transformed into the expression host strain to generate a new expression clone prior to each expression. Methods of reducing or eliminating basal *lac* expression may also be employed to reduce toxicity. Expression of LacI^q and overexpression of LacI are often utilized to limit basal *lac* expression (*see* **Subheading 1.2.2.2.**). Alternatively, the hybrid promoter/operator T7*lac* utilizes the *lac* operator to prevent basal transcription by T7 polymerase (*see* **Subheading 1.2.2.5.**). Basal expression can be further reduced in T7-based systems by using plasmids (pLysS/E) bearing T7 lysozyme, which inhibits the action of T7 polymerase *(20)*. The plasmid pLysE expresses more lysozyme than pLysS, because of the orientation of the gene on the pACYC-derived plasmid. The "E" variant is somewhat toxic to *E. coli*, but it provides the most stringent level of control. It is important to note that T7 lysozyme also cleaves the bacterial cell wall *(21)*. In these strains, any membrane disruption event, such as freeze–thaw, will provide the lysozyme access to the periplasmic space, resulting in rapid lysis of the cell. If this effect is undesirable, add glycerol to the cracking buffer before freezing harvested cultures bearing pLysS/E (*see* **Subheading 3.1.**). If the protein of interest is highly toxic to *E. coli*, basal expression can be completely eliminated by constructing the expression clone such that it does not carry the T7 polymerase, but rather induction is accomplished by infecting the strain with phage CE6 bearing the gene for T7 polymerase (*see* Chapter 28).

1.3.2. Strategies for Dealing with Insoluble Proteins

Another problem encountered during the overexpression and purification of proteins using *lac*-mediated expression is target protein insolubility. This occurs when misfolded target proteins aggregate as inclusion bodies. Misfolding is the result of a large number of factors, including accelerated protein synthesis, incompatibility of the protein with the *E. coli* cytosolic environment, the absence of necessary chaperones, the miscoding of low-usage codons by the *E. coli* protein synthesis machinery, or the absence of required cofactors. Many of the same empirical methods for maximizing protein expression (*see* **Subheading 4.**), such as changes in growth medium, temperature, length of induction, and addition of exogenous cofactors, are also advantageous in maximizing solubility and stability. There are also several strains and procedures specifically designed to address this problem (*see* Chapters 3 and 28).

1.4. Purification of His-Tagged Proteins

Several vectors are commercially available for use in purifying proteins through fusion "tags," including glutathione-*S*-transferase (GST), cellulose-binding domain (CBD), maltose-binding protein (MBP), and the 6×His epitope tag (*see* Chapter 28). The histidine tag is perhaps the most popular, because of its small size and the availability of many different vectors that encode this epitope. The most commonly used column resins for affinity purification of 6×His-tagged proteins use nickel as the ligand (e.g., Qiagen Ni-NTA resin [nickel-bound nitrolo-tri-acetic acid]). Resins utilizing cobalt instead of nickel, such as the Clontech Talon™ resin, are also used. Cobalt has less affinity for the histidine tag than nickel, which results in less nonspecific binding. This can increase the resolution of the column. The most common method of protein elution is to compete the histidine tag off the resin using imidazole, a structural analog of histidine. A basic procedure for Ni-NTA purification of an overexpressed 6×His-tagged protein is described in **Subheading 3.3**. For more detailed and alternative procedures, refer to **ref. 22** and the manufacturer's literature *(23)*.

2. Materials
2.1. Expression of Recombinant Protein

1. Plasmid-bearing recombinant gene of interest (*see* **Note 1**).
2. *E. coli* host strain (*see* **Subheading 1.3.**).
3. LB growth medium, prepared according to manufacturer's instructions (*see* **Notes 2** and **3**).
4. 1 *M* IPTG: Prepare in purified water, filter sterilize, and store in small aliquots at –20°C (*see* **Note 4**).
5. Cracking buffer: 26.04 g HEPES [4-(2-hydroxy-ethyl)-piperazine-1-ethane-sulfonic acid] sodium salt, 100 mL glycerol (optional, often aids protein stability), 17.54 g NaCl; bring up to 1 L with purified water and mix thoroughly. Adjust to desired pH (usually 7.5) using concentrated HCl and store at 4°C (*see* **Note 5**).
6. French press or alternative equipment for cell lysis.

2.2. SDS-PAGE Visualization of the Target Protein

1. Precast sodium dodecyl sulfate–polyacrylamide gel electrophoresis (SDS-PAGE) gel (*see* **Note 6**).
2. 5X SDS-PAGE running buffer: 15.1 g Tris-HCl, 94 g glycine, 5 g SDS; bring to 1 L with purified water. Dilute to 1X for experimental use.
3. 2X SDS-PAGE sample buffer: 1 mL of 1 *M* Tris-HCl (pH 6.8), 2 mL glycerol, 0.4 g SDS, 0.31 g dithiothreitol (DTT), 1 mL of 1 mg/mL bromophenol blue dye; bring to 10 mL with purified water. Store in aliquots at –20°C.
4. Coomassie blue stain solution: 100 mL acetic acid, 100 mL methanol, 800 mL purified water, 0.25 g Coomassie blue G-250 dye. Filter using Whatman no. 1 paper and store at room temperature.
5. Destain solution: 10% acetic acid, 10% methanol in purified water.

2.3. Purification of 6×His-Tagged Recombinant Proteins

1. Chromatography apparatus.
2. 50% (v/v) Ni-NTA slurry.
3. Running buffer: Tris-HCl, HEPES, and others can be used; a pH of at least 7.0 and a concentration of at least 50 mM is recommended (*see* **Note 7**).
4. 1 M Imidazole.

3. Methods
3.1. Expression of Recombinant Protein

1. Grow a 5-mL overnight culture of the expression construct at 37°C with vigorous shaking. Use the same growth medium as will be used for expression. For controls, grow a culture of the expression strain bearing an empty vector (the expression plasmid with no target gene inserted) and a culture of the expression strain alone. If the target strain does not grow properly, these controls will allow discrimination among problems with the target gene, the expression plasmid, and the medium used to grow the expression strain.
2. Inoculate an expression culture using a small amount of the overnight culture, to an optical density at 600 nm (OD_{600}) of 0.02–0.04 (1 : 100 dilution of an overnight culture of *E. coli* grown in LB under standard conditions). Pilot expression experiments during the optimization process should be performed at the 5-mL scale; 100 mL or more of expression culture is usually needed if purified protein is desired.
3. Follow the growth of the culture using a spectrophotometer. When the culture reaches an OD_{600} of 0.6 (mid log-phase growth), induce the culture by the addition of IPTG to a final concentration of 1 mM. Be sure to grow and harvest an uninduced control under the same conditions, for comparison and troubleshooting purposes.
4. Grow the cultures for at least 2 h postinduction (*see* **Note 8**).
5. Harvest the culture by centrifugation at 4000g for 15 min at 4°C (for large cultures, increase the spin time).
6. Resuspend the cell pellet in 200 mL per L of culture of ice-cold cracking buffer to wash the cell pellet, and then centrifuge again at 4000g for 15 min at 4°C.
7. Resuspend the washed pellet in cracking buffer—this time in 40 mL/L of culture. If the target protein is to be purified via affinity chromatography, this volume can be increased, up to fivefold. This reduces the viscosity of the load that is added to the column, improving the performance (and speed) of the column.
8. Lyse the resuspended cultures by passing through a French pressure cell (use at least 14,000 psi) at 4°C (*see* **Note 9**).
9. Centrifuge the lysed cell suspension at 10,000g for 20 min to remove cellular debris and inclusion bodies from the lysate.
10. Aliquot and freeze the cell-free lysate extract at –80°C until needed for experiments or freeze in bulk if slated for purification.

3.2. SDS-PAGE Visualization of the Target Protein

1. Prepare an appropriate volume (*see* **Note 10**) of each extract (including uninduced and empty vector controls) by mixing with an equal volume of 2X SDS-PAGE sample buffer and then boil the samples for 5 min.
2. Assemble the gel apparatus and fill with 1X SDS-PAGE running buffer.

3. Run the gel in an electrophoresis apparatus, according to the manufacturer's instructions for the particular electrophoresis system and electrode. When the dye front of the sample buffer reaches the bottom of the plate, remove the gel.
4. Stain the gel with Coomassie blue stain for 2–16 h with slow shaking.
5. Incubate the gel in destain solution with slow shaking, until the protein bands are clearly visible. To accelerate the destaining process, place a paper towel or sponge in the destain solution with the gel to absorb excess dye.
6. If a visible protein band cannot be determined to be the target by comparison between the expression extract and uninduced and empty vector controls, additional methods will be necessary to verify expression. Such methods include Western blot immunodetection (*see* **Notes 11** and **12**) and enzymatic activity assays specific to the target protein.

3.3. Purification of 6×His-Tagged Recombinant Proteins (see Note 12)

All steps in this protocol are performed at 4°C.

1. Prepare a Ni-NTA column, using a volume of resin appropriate to the scale of purification (Ni-NTA has a binding capacity of roughly 5–10 mg of protein per milliliter). Choose the width of the column so that there is at least 1 in. of column depth. For all steps except column loading, use a maximal flow rate of 1 mL/min (*see* **Note 13**).
2. Wash the column with 10 column volumes (CV) of the preferred running buffer (*see* **Note 7**).
3. Load the previously prepared cell lysate at a slow flow rate (no more than 0.5 mL/min, ideally 2–3 CV/h), or incubate the resin with the lysate for several hours with gentle agitation and then prepare the column, as in **step 1**, with the protein-loaded resin (*see* **Note 14**).
4. Wash the column with 10 CV of running buffer.
5. Wash the column with 5 CV of running buffer containing 25 mM imidazole.
6. Elute proteins from the column in a stepwise manner, with 5 CV each of running buffer containing 50 mM, 100 mM, 200 mM, and 400 mM imidazole (*see* **Note 15**). Collect 0.5 CV elution fractions. On Ni-NTA columns, most 6×His-tagged proteins elute between 100 and 200 mM imidazole. If a gradient maker is available, elution using a 25- to 500-mM imidazole gradient can increase resolution.
7. Determine the location and purity of fractions containing the purified protein by SDS-PAGE (*see* **Note 16**). Often the target protein band will be visible in the elution fractions, even if it was not visible in the cell-free soluble lysate.

4. Notes

1. This construct should be designed, generated, and verified through the methods described elsewhere in this volume.
2. LB is the most commonly used medium for the growth and expression of *E. coli*. However, either an ultrarich broth such as Terrific Broth (TB), or a minimal medium such as Minimal A, may create better culture conditions for expression of the target protein. Altering the growth medium in which the expression is performed can also have an effect on the level of protein expression. For soluble, nontoxic proteins, an ultrarich growth media such as TB is likely to yield the greatest protein production per liter of culture, because of the high bacterial growth that can be obtained. Conversely, a difficult protein may express better in minimal medium, such as Minimal A.
3. If the target protein requires any cofactors (i.e., metals, vitamin-derived small molecules), supplement the expression medium with appropriate (extracellular) forms and

concentrations of these cofactors. If the cofactor is not present, the target protein may either misfold (leading to inclusion bodies) or bind all of the available cellular cofactors (leading to toxicity).
4. IPTG is unstable at room temperature, so only thaw the working stock immediately prior to induction of the culture.
5. The composition of the cracking buffer is important, as it becomes either the load buffer for the first step of a purification effort or it becomes the final extract buffer if crude cell lysates are used in further experiments. Cracking buffers made using HEPES are effective and generally compatible with a wide range of downstream applications. Alternatively, phosphate-buffered saline (PBS) is a common and easily prepared cracking buffer. Tris-based buffers are also popular, but suffer the drawback that the pH of Tris-buffered solutions can vary substantially with temperature.
6. Precast gels are available for a number of common electrophoresis systems. Select an SDS-PAGE gel with a polyacrylamide concentration appropriate to your target protein's size: For large proteins, use 4–10%; for small proteins, use 12–20%.
7. Do not use buffers with reducing agents or EDTA, as they will cause the nickel to dissociate from the resin.
8. Protein expression is dependent on both length of the induction period and temperature, and the optimal condition must be determined through trial and error. To optimize the expression of a protein, examine the level of expression at various postinduction timepoints and during growth at various temperatures, in order to find the time and temperature combination that maximizes expression. Standard induction temperatures include 37°C, 30°C, 25°C, and 18°C, although any temperature within the growth range of the host strain can be used. The length of induction should be increased as the temperature is decreased. An induction time of 2–6 h is usually optimal at 37°C, whereas an induction time of 16 h (overnight) is recommended at 18°C. Heterologous recombinant proteins are often expressed well at low temperature.
9. Alternatively, lysozyme/EDTA treatment or sonication can be used to disrupt the cells.
10. On minigels such as the commonly used Bio-Rad Mini-protean™ system, a recommended volume for adequate visualization of protein is 5 µL of cell-free lysate extracts (generally 10–30 µg of total protein) and 10 µL of nickel column elution fractions.
11. 6×His-Tagged proteins can be visualized by Western blot using an anti-His antibody *(24)*.
12. If the target protein is highly hydrophobic (e.g., membrane proteins), special steps may be required during purification and visualization *(25)*.
13. The use of a low-pressure liquid chromatography (LC) apparatus and fraction collector, if available, will increase the ease and reproducibility of the purification.
14. Nonspecific protein binding to nickel columns can be reduced by loading the lysate on the column in a low concentration (10 mM) of imidazole or by the addition of up to 300 mM NaCl to the running buffer. In addition, the elution gradient can be made shallower or steeper to optimize purification
15. If the presence of imidazole is detrimental to downstream applications of the purified protein, a change in pH can be used to elute the protein, or the pure protein can be dialyzed to remove the imidazole. If further purification is needed, another chromatography step will also remove the imidazole along with contaminant proteins.

Acknowledgments

The author would like to thank Dr. Heather C. Prince and Dr. Mara K. Vorachek-Warren for their critical reading of this work.

References

1. Jacob, F. and Monod, J. (1961) Genetic regulatory mechanisms in the synthesis of proteins. *J. Mol. Biol.* **3**, 318–356.
2. Miller, J. H. (1992) *A Short Course in Bacterial Genetics*, Cold Spring Harbor Laboratory, Cold Spring Harbor, NY.
3. Neidhardt, F. C. (ed.) (1996) Escherichia coli *and* Salmonella typhimurium: *Cellular and Molecular Biology*, American Society for Microbiology, Washington, DC.
4. Hu, M. C. T. and Davidson, M. (1987) The inducible *lac* operator-repressor system is functional in mammalian cells. *Cell* **48**, 555–566.
5. Zabin, I. and Fowler, A. V. (1980) β-galactosidase, the lactose permease protein, and thiogalactoside transacetylase, in *The Operon*, 2nd ed. (Miller, J. H. and Reznikoff, W. S., eds.), Cold Spring Harbor Laboratory, Cold Spring Harbor, NY, pp. 89–122.
6. Lewendon, A., Ellis, J., and Shaw, W. V. (1995) Structural and mechanistic studies of galactoside acetyltransferase, the *Escherichia coli lacA* gene product. *J. Biol. Chem.* **270**, 26,326–26,331.
7. Miller, J. H. (1980) The *lacI* gene: its role in *lac* operon control and its uses as a genetic system, in *The Operon*, 2nd ed. (Miller, J. H. and Reznikoff, W. S., eds.), Cold Spring Harbor Laboratory, Cold Spring Harbor, NY, pp. 31–88.
8. Emmer, M., deCrombrugghe, B., Pastan, I., et al. (1970) Cyclic AMP receptor protein of *E. coli*: its role in the synthesis of inducible enzymes. *Proc. Natl. Acad. Sci. USA* **66**, 480–487.
9. Zubay, G., Schwartz, D., and Beckwith, J. (1970) Mechanism of activation of catabolite-sensitive genes: a positive control system. *Proc. Natl. Acad. Sci. USA* **66**, 104–110.
10. Eron, L. and Block, R. (1971) Mechanism of initiation and repression of in vitro transcription of the *lac* operon. *Proc. Natl. Acad. Sci. USA* **68**, 1828–1832.
11. Calos, M. P. (1978) DNA sequence for a low-level promoter of the *lac* repressor gene and an 'up' promoter mutation. *Nature* **274**, 762–769.
12. Amann, E. and Brosius, J. (1985) 'ATG vectors' for regulated high-level expression of cloned genes in *Escherichia coli*. *Gene* **40**, 183–190.
13. Amann, E., Ochs, B., and Abel, K. (1988) Tightly regulated *tac* promoter vectors useful for the expression of unfused and fused proteins in *Escherichia coli*. *Gene* **69**, 301–315.
14. Studier, F. W., Rosenberg, A. H., Dunn, J. J., et al. (1990) Use of T7 polymerase to direct expression of cloned genes. *Methods Enzymol.* **185**, 60–89.
15. *pET System Manual*, Novagen, Madison, WI, available online from www.novagen.com.
16. Dubendorff, J. W. and Studier, F. W. (1991) Controlling basal expression in an inducible T7 expression system by blocking the target T7 promoter with *lac* repressor. *J. Mol. Biol.* **219**, 45–59.
17. Sambrook, J. and Russell, D. W. (2001) *Molecular Cloning: A Laboratory Manual*, 3rd ed., Cold Spring Harbor Laboratory, Cold Spring Harbor, NY.
18. Ausubel, F. M., Brent, R., Kingston, R. E., Moore, D. D., Seidman, J. G., Smith, J. A., and Struhl, K. (ed.) (1989) *Current Protocols in Molecular Biology*. Wiley, NY.
19. Shine, J. and Dalgarno, L. (1974) The 3'-terminal sequence of *Escherichia coli* 16S ribosomal RNA: complementarity to nonsense triplets and ribosome binding sites. *Proc. Natl. Acad. Sci. USA* **71**, 1342–1346.
20. Studier, F. W. (1991) Use of bacteriophage T7 lysozyme to improve an inducible T7 expression system. *J. Mol. Biol.* **219**, 37–44.
21. Inouye, M., Armheim, N., and Sternglanz, R. (1973) Bacteriophage T7 lysozyme is an N-acetylmuramyl-L-alanine amidase. *J. Biol. Chem.* **248**, 7247–7252.

22. Deutscher, M. P. (ed.) (1990) *Guide to Protein Purification, Methods in Enzymology* Vol. 182, Academic Press, San Diego, CA.
23. *The QIAexpressionist™: An Ni-NTA Purification Guide*, Qiagen, Valencia, CA, available online from www.qiagen.com.
24. *QIAexpress™: Detection and Assay Handbook*, Qiagen, Valencia, CA, available online from www.qiagen.com.
25. von Jagow, G. and Schägger, H. (1994) *A Practical Guide to Membrane Protein Purification: Separation, Detection, and Characterization of Biological Macromolecules.* Academic, San Diego, CA.

30

Plasmid-Based Reporter Genes

Assays for β-Galactosidase and Alkaline Phosphatase Activities

Minghsun Liu

1. Introduction

Reporter genes encode easily measurable traits. Most commonly, they are used to investigate the expression of other genes for which functional assays are not available or for which measurement of expressed product is difficult. This is often done through constructing a genetic fusion in which expression of the target gene also drives expression of the reporter gene product. Measurement of reporter gene product gives a direct readout for expression of the target gene *(1)*. Even with the advent of technologies such as DNA microarrays, reporter genes continue to provide a powerful tool for studying gene expression *(2–4)*.

Table 1 lists some of the reporter genes commonly used in bacterial systems. They have been developed for a wide variety of applications. **Table 2** lists some factors to be considered in choosing a suitable reporter gene system. A broad range of assay formats (colorimetric, radioactive, fluorescent, luminescent, enzyme-linked immunosorbent assay [ELISA], or *in situ* staining) are available for many of the reporter genes listed in **Table 1**. An increasing number of vendors are also providing commercial preparations that minimize the use of hazardous reagents while increasing assay sensitivity and ease of use. In addition to reagents, different assay formats also require different types of instrument for detection, and the performance of many assays will depend on the particular detection instrument used. Unfortunately, there does not appear to be a centralized listing of the latest offerings except in the advertisement or product information sections of several major scientific journals.

1.1. Choosing a Reporter Gene Vector

Many of the reporter genes listed in **Table 1** are available in different types of vectors that have been tailored to specific applications. In choosing a vector system, considerations should be given to several factors.

From: *Methods in Molecular Biology, Vol. 235:* E. coli *Plasmid Vectors*
Edited by: N. Casali and A. Preston © Humana Press Inc., Totowa, NJ

Table 1
Some Reporter Genes Commonly Used in Bacterial Systems

Reporter gene	Advantages	Limitations
β-Galactosidase (LacZ)	Simple assays in a wide variety of formats; high dynamic range (substrate dependent)	Endogenous activity in some bacterial species
Chloramphenicol acetyltransferase (CAT)	Lack of background activity in most bacterial species	Radioactive substrate required for high sensitivity; narrow dynamic range; radioactive assay can take hours to days
Alkaline phosphatase (PhoA)	Simple assays in a wide variety of formats; high sensitivity (substrate dependent)	*In situ* detection requires proper localization of reporter
Green fluorescent protein (GFP)	*In situ* measurement of activity; single-cell sensitivity possible	Wild-type GFP has a long half-life
Bacterial luciferase (LuxAB)	*In situ* measurement of activity; rapid assay possible	Dependent on cellular generation of energy
Kanamycin resistance (NPT-II)	May be used for studying protein targeting	Limited assay format
β-Glucuronidase (GUS)	Wide variety of assay formats; no background activity in plants	Endogenous activity in mammalian cells

1.1.1. Copy Number

For investigating the promoter activity of a gene of interest, placing a promoter region in a high- or medium-copy-number reporter gene vector may titrate out regulatory proteins (which are typically expressed from a single copy on the chromosome) that bind the fragment and allow some of the promoter fragments to escape regulatory control. This may result in artificially high or low reporter activity depending on the nature of the regulatory proteins involved. The copy number of a particular plasmid may also vary depending on bacterial growth conditions, leading to misinterpretation of results. A more realistic condition is created by using low-copy-number plasmids or by integrating a single copy of the reporter gene vector into the chromosome.

One option for creating chromosomally integrated, single-copy reporter genes in *Escherichia coli* is to use λ bacteriophage-based constructs *(5,6)*. For bacterial strains unable to support replication of ColE1-based plasmids (e.g., pBR322, pUC19), several ColE1-based reporter gene vectors are available for use as "suicide" vectors *(7,8)*. They allow convenient molecular manipulations in *E. coli* before being transferred and integrated into the target bacterial strain (*see* **Fig. 1**). For *E. coli* and related bacterial species, plasmids based on the R6K replicon have also been used as "suicide" vectors *(9,10)*. DNA replication from the R6K origin requires the π protein encoded by the *pir* gene. Without the π protein, R6K-based plasmids can integrate into the

Table 2
Factors to Consider When Choosing a Reporter Gene

Characteristics	Comments
Stabililty	An extremely long half-life is undesirable, but products must be stable long enough for assays to be completed
Sensitivity	Must be able to detect signals using available reagents and equipment
Dynamic range	The ability to detect reporter activity over a wide range is ideal
Codon usage	Ideally should be optimized for the host bacteria
Energy/substrate requirement	Some reporters require cellular energy reserve (e.g., LuxAB), O_2 (e.g., GFP), or specific substrates (e.g., LacZ)
Background expression	A low background activity allows specific detection of reporter activity

chromosome if a region of homology exists between the plasmid and chromosome. Plasmid integration into the chromosome can be selected for using antibiotic resistance encoded by the plasmid.

1.1.2. Extraneous Control Elements

The presence of unexpected regulatory elements on the vector or read-through transcription that originates upstream of the reporter gene may lead to artificial results. Some reporter gene vectors have tandem copies of a strong transcriptional terminator (such as the one from the *rrnB* rRNA operon) to insulate fusions from exogenous transcription *(5)*. Inclusion of proper vector controls is needed to exclude these extraneous effects.

1.1.3. Fusion Junction

In many reporter gene vectors, multiple cloning sites (MCSs) are provided 5' to the reporter gene for cloning of the target sequence. For transcriptional fusions, an appropriate translational start site is included on the vector. An unintended translational fusion, in addition to the intended transcriptional fusion, may be created when the fusion is in-frame and when another appropriately placed translational start signal is present. Polymerase chain reaction (PCR) incorporating synthetic oligonucleotides can be used to easily create precisely engineered fusion junctions *(11)* (*see* Chapters 15 and 17). Designing fusion junctions in a consistent manner allows results from different constructs to be compared.

1.2. Two Commonly Used Reporter Genes
1.2.1. β-Galactosidase

β-Galactosidase is encoded by the *lacZ* gene as part of the *lac* operon. The *lac* operon is one of the earliest regulatory networks to be studied *(12,13)* (*see* Chapter 29).

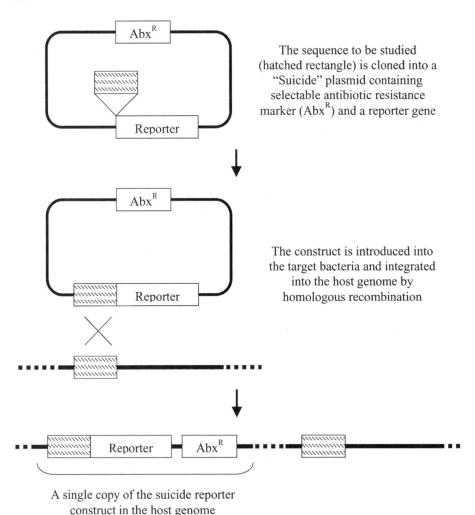

Fig. 1. Using a suicide plasmid to make a single-copy reporter construct.

As a result, a wide variety of assays have been developed to measure LacZ activity. Its activity can be easily monitored using 5-bromo-4-chloro-3-indolyl-β-D-galactoside (X-gal), methylumbelliferyl-β-D-galactoside (MUG), fluorescein digalactoside (FDG), or o-nitrophenyl-β-D-galactoside (ONPG). These are artificial LacZ substrates that are metabolized to produce measurable color changes. Newer, commercial formulations offering various improvements over these substrates are available from several vendors *(14–16)*.

A comprehensive discussion of *lacZ*-based vectors is beyond the scope of this chapter; however, several good reviews are available *(3,13)*. Some of the *lacZ*-based reporter gene systems also contain *lacY* that encodes a β-galactoside permease. *E. coli* without a functional LacY is phenotypically Lac⁻. However, for in vitro assays using chromogenic substrates, a functional LacY is not required. It should be noted that the *lacZ* gene in many common cloning vectors, such as pUC19 and pBluescript, is present as a truncated *lacZ* allele called *lacZα*. This fragment complements the "remaining portion" of *lacZ* that is encoded by the partially deleted genomic allele, *lacZΔM15* (*see* Chapter 3). However, *lacZα* is not usually used as a reporter gene.

The β-galactosidase assay presented in **Subheading 3.1.** is based on the protocol described by Miller but uses different buffers *(13)*. The substrate ONPG is colorless but is hydrolyzed by β-galactosidase to produce the yellow-colored product, nitrophenol, which is quantified spectrophotometrically by measuring the optical density (OD_{420}).

1.2.2. Alkaline Phosphatase

In *E. coli*, alkaline phosphatase (AP) is encoded by the *phoA* gene *(17)*. The wild-type *phoA* gene has a signal sequence directing export of alkaline phosphatase into the periplasm, where it is active. Alkaline phosphatase is not active in the highly reducing environment of the cytoplasm. Genetic fusions to *phoA* have been used extensively for detecting extracytoplasmic proteins or protein domains *(18)*. Several reviews have detailed discussions on some of the available *phoA*-based reporter vectors *(13,18)*. Alkaline phosphatase activity is detected using a series of procedures analogous to the ones available for β-galactosidase.

The alkaline phosphatase assay presented in **Subheading 3.2.** is identical to the original protocol except for some minor modifications *(17)*. In this protocol, the colorless substrate *p*-nitrophenyl phosphate (PNPP) is hydrolyzed by alkaline phosphatase, producing nitrophenol, which is quantified as described for β-galactosidase.

2. Materials
2.1. β-Galactosidase Assay

1. PM2 buffer: 36 mM NaH_2PO_4, 67 mM Na_2HPO_4, 0.1 mM $MgCl_2$, 2 mM $MgSO_4$ (*see* **Note 1**).
2. PM2SH: Add 135 µL of β-mercaptoethanol to 50 mL of PM2 buffer (*see* **Note 2**).
3. 1 M Na_2CO_3 (*see* **Note 1**).
4. 4 mg/mL ONPG: Add 16 mg ONPG to 4 mL PM2SH (sufficient for up to 25 samples). (*see* **Note 2**).
5. $CHCl_3$.
6. 0.1% (w/v) sodium dodecyl sulfate (SDS).

2.2. Alkaline Phosphatase Assay

1. 1 M Tris-HCl, pH 8.0.
2. Resuspension buffer: 10 mM Tris-HCl (pH 8.0), 0.1 M NaCl.
3. 4 mg/mL PNPP: Add 16 mg PNPP to 4 mL Tris buffer (*see* **Note 2**).
4. 1 M K_2HPO_4 (*see* **Note 3**).
5. $CHCl_3$.
6. 0.1% (w/v) SDS.

3. Methods

3.1. β-Galactosidase Assay

1. Grow cultures of the strains to be assayed. The assay should be performed using mid-log cultures (*see* **Note 4**).
2. Chill the cultures on ice for at least 10 min, to prevent further growth, and measure the cell density at OD_{600}.
3. Transfer 1 mL of cells to a tube and permeabilize the cells by adding 60 µL of $CHCl_3$ and 30 µL of 0.1% SDS. Vortex for 10 s; then place on ice (*see* **Note 5**).
4. Add 15–30 µL of permeabilized cells to PM2SH buffer to a final volume of 800 µL. Equilibrate the tubes at 30°C for 1–2 min.
5. Add 160 µL of ONPG to each tube and record the time as T_1. Incubate the samples at 30°C until a pale yellow color develops (*see* **Note 6**).
6. Stop the reaction by adding 400 µL of 1 M Na_2CO_3, mix by inverting the tubes once or twice, and record the time as T_2 (*see* **Note 7**).
7. Spin the tubes for 3 min at >15,000g and read the OD_{420} (blank is 800 µL of PM2SH) (*see* **Note 8**).
8. Calculate β-galactosidase activity as follows (*see* **Notes 9–11**):

$$\text{β-gal units} = (1000\, OD_{420}) / (tv\, OD_{600})$$

where $t = T_1 - T_2$ (in min) and v is the volume of cells assayed (in mL).

3.2. Alkaline Phosphatase Assay

1. Grow cultures of the strains to be assayed. The assay should be performed using mid-log cultures (*see* **Note 4**).
2. Pellet cells from 1 mL of culture, wash the bacterial pellet once with 1 mL of resuspension buffer, and then resuspend the pellet in 1 mL of resuspension buffer. Determine the cell density by measuring the OD_{600}.
3. Add 50–100 µL of washed cells to Tris buffer to a final volume of 1 mL. Permeabilize the cells by adding 60 µL of $CHCl_3$ and 30 µL of 0.1% SDS. Vortex for 10 s and incubate at 37°C for 5 min.
4. Add 100 µL of PNPP to each tube and record the time as T_1. Incubate the samples at 37°C until a pale yellow color develops (*see* **Note 6**).
5. Stop the reaction by adding 100 µL of cold 1 M KH_2PO_4 to each tube, mix by inverting the tubes once or twice, and record the time as T_2.
6. Spin the tubes for 3 min at >15,000g and read the OD_{420} (blank is Tris buffer) (*see* **Note 8**).
7. Calculate alkaline phosphatase activity as follows (*see* **Notes 9–11**):

$$\text{AP units} = (1000\, OD_{420}) / (tv\, OD_{600})$$

where $t = T_1 - T_2$ (in min) and v is the volume of cells assayed (in mL).

4. Notes

1. These solutions can be autoclaved or filter sterilized. Their performances are not affected by storage at ambient temperature for up to 6 mo.
2. Prepared fresh as needed.
3. Keep cold at approx 4°C.
4. Non-log-phase cells may be used, but exponentially growing cells give more consistent and reproducible results.

5. Use tubes that are $CHCl_3$ resistant. Cultures should turn slightly milky white and then segregate into two phases.
6. Short incubation times (< 5 min) can result in artificially high numbers. A reaction time of 15–30 min appears to work the best. Stagger the starting time of each tube by a specific amount of time (15–30 s allows sufficient time for adding reagents and vortexing) to allow exact timing.
7. Na_2CO_3 terminates the reaction by increasing the pH.
8. The centrifugation step pellets cell debris and eliminates the light scattering that it causes during spectrophotometry. An alternative to this is to read the OD_{550} and correct the measurement by subtracting $1.75 \times OD_{550}$ from the denominator of the equation in **step 8**.
9. These units are arbitrary units. Conversion to specific activity is possible but is not usually reported in the literature. Typical values range between 500 and 30000 units for β-galactosidase, and between 100 and 5000 units for AP. The background should be below 30 units for β-galactosidase and 10 units for AP. If the OD_{420} is greater than 1.0, dilute the sample in PM2SH (for β-galactosidase) or Tris buffer (for AP) or repeat the assay using fewer cells (this is probably a more accurate method).
10. Assays performed using the same set of reagents on the same bacterial strain should be highly reproducible with less than 5% variation between reactions. However, greater variation (up to 20%) may be observed with assays performed on different days using the same strain. This can be the result of variability in reagent preparations (because some of the reagents are prepared fresh as needed). If results of assays performed on different days are to be combined, each measurement may first be expressed as a percentage of a positive control level.
11. When no activity is detected, the use of positive controls can rule out problems with reagents. For β-galactosidase, one of the common *E. coli* cloning strains (e.g., DH5α or XL1Blue) carrying a cloning vector suitable for blue–white selection (e.g., pUC19, pBluescript) may be used as a positive control (*see* Chapter 2). Many of the common *E. coli* strains used in the lab carry the wild-type *phoA* allele and can be used as controls for alkaline phosphatase (*see* Chapter 3).

References

1. Slauch, J. M. and Silhavy, T. J. (1991) Genetic fusions as experimental tools. *Methods Enzymol.* **204**, 213–248.
2. Merrell, D. S. and Camilli, A. (2000) Detection and analysis of gene expression during infection by in vivo expression technology. *Phil. Trans. R. Soc. Lond. B: Biol. Sci.* **355**, 587–599.
3. Slauch, J. M., Mahan, M. J., and Mekalanos, J. J. (1994) In vivo expression technology for selection of bacterial genes specifically induced in host tissues. *Methods Enzymol.* **235**, 481–492.
4. Schenborn, E. and Groskreutz, D. (1999) Reporter gene vectors and assays. *Mol. Biotechnol.* **13**, 29–44.
5. Simons, R. W., Houman, F., and Kleckner, N. (1987) Improved single and multicopy *lac*-based cloning vectors for protein and operon fusions. *Gene* **53**, 85–96.
6. Ostrow, K. S., Silhavy, T. J., and Garrett, S. (1986). *Cis*-acting sites required for osmoregulation of *ompF* expression in *Escherichia coli* K-12. *J. Bacteriol.* **168**, 1165–1171.
7. Merriman, T. R. and Lamont, I. L. (1993) Construction and use of a self-cloning promoter probe vector for gram-negative bacteria. *Gene* **126**, 17–23.

8. Martinez de Tejada, G., Miller, J. F., and Cotter, P. A. (1996) Comparative analysis of the virulence control systems of *Bordetella pertussis* and *Bordetella bronchiseptica*. *Mol. Microbiol.* **22**, 895–908.
9. Platt, R., Drescher, C., Park, S. K., and Phillips, G. J. (2000) Genetic system for reversible integration of DNA constructs and *lacZ* gene fusions into the *Escherichia coli* chromosome. *Plasmid* **43**, 12–23.
10. Schmoll, T., Ott, M., Oudega, B., et al. (1990) Use of a wild-type gene fusion to determine the influence of environmental conditions on expression of the S fimbrial adhesin in an *Escherichia coli* pathogen. *J. Bacteriol.* **172**, 5103–5111.
11. Landt, O., Grunert, H. P., and Hahn, U. (1990) A general method for rapid site-directed mutagenesis using the polymerase chain reaction. *Gene* **96**, 125–128.
12. Muller-Hill, B. (1996) *The lac Operon: A Short History of a Genetic Paradigm*, Walter de Gruyter, New York.
13. Miller, J. H. (1992) *A Short Course in Bacterial Genetics*, Cold Spring Habor Laboratory, Cold Spring Harbor, NY.
14. Heuermann, K. and Cosgrove, J. (2001) S-Gal: an autoclavable dye for color selection of cloned DNA inserts. Biotechniques 30, 1142-1147.
15. Cheng, G., Thompson, R. P., and Gourdie, R. G. (1999) Improved detection reliability of beta-galactosidase in histological preparations. *BioTechniques* **27**, 438–440.
16. Beale, E. G., Deeb, E. A., Handley, R. S., et al. (1992) A rapid and simple chemiluminescent assay for *Escherichia coli* beta-galactosidase. *BioTechniques* **12**, 320–323.
17. Brickman, E. and Beckwith, J. (1975) Analysis of the regulation of *Escherichia coli* alkaline phosphatase synthesis using deletions and phi80 transducing phages. *J. Mol. Biol.* **96**, 307–316.
18. Manoil, C., Mekalanos, J. J., and Beckwith, J. (1990) Alkaline phosphatase fusions: sensors of subcellular location. *J. Bacteriol.* **172**, 515–518.

31

Plasmid-Based Reporter Genes

Assays for Green Fluorescent Protein

Sergei R. Doulatov

1. Introduction

Green fluorescent protein (GFP) of the jellyfish *Aqueorea victoria* is a 238-amino-acid, 28-kDa protein that absorbs light with an excitation maximum of 395 nm and fluoresces with an emission maximum of 509 nm *(1)*. GFP owes its unique spectral properties to its chromophore *(2)* that consists of a Ser65, Tyr66, and Gly67 tripeptide *(3)*. Autocatalytic cyclization of this tripeptide, induced by oxidation of Tyr66, is a necessary posttranslational step for proper fluorescence (strong reducing agents reversibly convert GFP into a nonfluorescent form) *(4)*. This can occur in the absence of any cofactors, making GFP an extremely useful tool for a wide range of applications in a variety of heterologous systems *(5–7)*. GFP activity can be assayed both qualitatively and quantitatively using a variety of techniques, including simple plate counting, fluorescence and confocal microscopy, flow cytometry, and fluorometry. Transcriptional and translational fusions of GFP to a gene or protein of interest can be used as gene expression reporters and subcellular localization tags. GFP is a small protein (28 kDa) compared to other reporters (e.g., β-galactosidase is 465 kDa) and GFP fusions often retain the native protein function *(8–10)*. This makes GFP useful as a generic tag for studying protein synthesis, translocation, and other protein–protein interactions. GFP is also widely used as a reporter in many genetic techniques, including transposon mutagenesis, promoter/enhancer traps, and one-component hybrid systems. GFP can be visualized using microscopy in both live and fixed cells, making it an excellent tool for studying dynamic changes in living cells. The generation of transgenic organisms carrying GFP fusion constructs has been instrumental in studying in vivo changes in protein expression and localization under a variety of conditions (e.g., during embryonic development).

In addition, the recent generation of multiple GFP derivatives with altered fluorescent properties allows multicolor imaging using laser-scanning microscopy, and the generation of GFP derivatives with altered chromophore stability *(11–14)* facilitates studies involving rapid changes in gene expression *(15)*. GFP fluorescence can be visualized using standard fluorescein isothiocyanate (FITC) excitation–emission filter sets due to a minor absorption peak at 470 nm; however multiple red-shifted GFP derivatives have been generated, for which the main absorption peak is around 490 nm, giving greater sensitivity in fluorescence microscopy and FACS. These derivatives include second-generation *GFPmut*1–3 derivatives, commercially available third-generation enhanced GFP (EGFP) and its yellow (EYFP), cyan (ECFP), and red (DsRed2) derivatives (available from Clontech), as well as red-shifted GFP (rsGFP) and blue fluorescent protein (BFP) (available from Quantum). From these, additional derivatives have been generated with altered splicing, thermal, and folding properties and codon usages optimized for particular organisms *(16)*. *GFPmut*1–3 and *GFPuv* (Clontech) are suited for use in bacteria. For use in yeast, *GFPmut*3.1 expressed from a yeast promoter is the best choice. EGFP and its derivatives work well in plant cells, although plant-optimized derivatives are also available (e.g., *mgfp5-ER*) *(16)*. Any third-generation GFP derivative (Clontech, Quantum) will work well in mammalian cells and many of them are optimized for human codon usage. Many GFP derivatives show marked thermosensitivity. Optimum fluorescence occurs at 25°C and decreases at higher temperatures *(17)*. However, mutations that increase the efficiency of protein folding and chromophore formation, as occurs in EGFP, for example, markedly reduce this thermosensitivity.

Techniques to study protein–protein interactions using fluorescence resonance energy transfer (FRET) and protein mobility using fluorescence recovery after photobleaching (FRAP) have been recently adapted for use with GFP derivatives. FRET occurs when a donor and acceptor molecule that have overlapping excitation and emission spectra approach within 10–50 Å of each other. For example, when using BFP as a donor and GFP as an acceptor, ultraviolet (UV) excitation of BFP results in energy transfer to a proximal GFP, inducing green fluorescence that is not normally observed upon excitation with UV light *(18,19)*. FRAP involves irreversible photobleaching of GFP-tagged molecules in a small area of the cell by a focused laser beam. Subsequent diffusion of surrounding nonbleached molecules into this area leads to a recovery of fluorescence from this region, which is recorded at low laser power *(20)*.

In some organisms, GFP fails to fluoresce and detection requires that the GFP be expressed from a strong promoter. For example, the author has found this to be the case with the genus *Bordetella*. In many other cell types, GFP does fluoresce, but either the fluorescence diminishes with time, only appears in a subpopulation of cells, or appears in the whole population but shows variability between cells. These phenomena might be caused by selection against GFP-expressing cells, as GFP expression might be energetically demanding. This effect might only become apparent under additional selection pressures such as antibiotic selection or in vivo growth. To counteract this, stable transfectants are selected in order to reduce variability in copy number and to limit GFP expression to those copies of the gene that have been stably integrated, or cell populations are sorted to select for those cells that stably express GFP.

Assays for GFP

The following sections of this chapter describe some of the protocols that are utilized for the visualization, detection, and quantification of GFP in both bacterial and eukaryotic cells. The construction and verification of plasmids bearing GFP fusions involves standard cloning techniques that are presented elsewhere in this volume. Protocols for the culture of mammalian cells and their transfection with GFP-expressing reporter constructs are beyond the scope of this volume and the reader should refer to an appropriate source for this information.

2. Materials

2.1. Direct Colony Examination

Short-wavelength UV lamp or appropriate imaging device.

2.2. Fluorescence Microscopy

1. Dulbecco's phosphate-buffered saline (DPBS, pH 7.4).
2. 4% Paraformaldehyde in DPBS: Add 4 g of paraformaldehyde to 80 mL of DPBS and heat to dissolve. Cool and readjust pH if necessary; then dilute to a final volume of 100 mL.
3. Mounting medium: 90% glycerol (v/v) in 100 mM, Tris-HCl (pH 7.5); or commercial preparation (e.g., Molecular Probes).
4. 22 × 22 mm Cover slips.
5. Microscopy cover slides.
6. Purified recombinant GFP protein (Clontech).
7. 10 mM Tris-HCl (pH 8.0).

2.3. Fluorometry

1. PBS (pH 7.4).
2. 10 mM Tris-HCl (pH 8.0).
3. Purified recombinant GFP protein (Clontech).

2.4. Flow Cytometry

1. PBS (Ca^{2+} and Mg^{2+} free).
2. 2% Paraformaldehyde in PBS: Add 2 g of paraformaldehyde to 80 mL of PBS and heat to dissolve. Cool and adjust pH to 7.4. Dilute to a final volume of 100 mL.
3. Fluorescent microbeads (Clontech).

3. Methods

3.1. Colony Examination

This is the quickest and easiest way of visualizing fluorescent bacterial colonies. In the case of questionable fluorescence, use the fluorescence microscopy protocol in **Subheading 3.2.** for a more sensitive assessment of fluorescent cells.

1. Plate bacteria onto an appropriate medium and grow until single colonies are distinguishable.
2. In a dark-room, illuminate the plates using a short-wavelength UV lamp or an appropriate imaging device.
3. Examine colonies for green fluorescence by eye. An image can be taken for a permanent record.

3.2. Fluorescence Microscopy

Individual cells may be visualized using fluorescent, epifluorescent, or inverted microscopes. Simultaneous detection of multiple GFP variants can be performed by confocal microscopy *(17,21)*. Purified recombinant GFP can be used to optimize the lamp and filter set conditions and to approximately quantitate the GFP fluorescence from the cells *(22)*. Described here is a basic protocol for visualizing the distribution of GFP in fixed tissue culture cells grown on cover slips, which has been adapted from various sources *(5,6,21,23–26)*. GFP in tissue sections can also be visualized by fluorescence microscopy following embedding and sectioning of tissue samples. For the preparation of bacteria, yeast, or nonadherent mammalian cells for fluorescence microscopy, refer to **Notes 1** and **2**.

1. Remove the culture medium from the tissue culture dish and wash the cells once with DPBS.
2. Cover the cells with freshly prepared 4% paraformaldehyde in DPBS and incubate for 30 min at room temperature (*see* **Note 3**).
3. Remove the paraformaldehyde solution and then wash the cells twice with DPBS allowing 10 min per wash (*see* **Note 4**).
4. Remove the cover slip from the tissue culture dish using forceps.
5. If purified GFP is available, dilute it to 0.1 mg/mL and 0.01 mg/mL in 10 mM Tris-HCl (pH 8.0) (*see* **Note 5**). Spot 1–2 µL of the diluted protein onto the microscope slide. Allow it to air-dry for a few seconds and mark its position on the other side of the slide to aid in focusing.
6. Place a small drop of mounting medium on the cover slide and slowly place the cover slip on top, with the cell side down.
7. Remove excess mounting medium by holding two pieces of filter paper against opposing edges of the cover slip until no more medium is drawn from underneath. Allow the mounting medium to dry completely.
8. Examine the slides by fluorescence microscopy (*see* **Note 6**) under high magnification using general FITC or variant-optimized filter sets (*see* **Note 7**). View the purified GFP spots under low magnification to evaluate the lamp and filter conditions.

3.3. Fluorometry

This is perhaps the easiest and most sensitive method for comparing and quantifying fluorescence from a large number of samples *(5,6,25,27–31)*. However, the instrument error is relatively high and greatly increases at low- and high-fluorescence intensities. For high precision, assay three independently grown cultures of each sample and several dilutions of each culture. To assay a large number of samples, a single dilution is usually sufficient for an approximate comparison of fluorescence intensities. A positive control comprising GFP expressed from a constitutive promoter is recommended, as well as a negative control comprising a clone carrying the vector alone. Purified GFP is required for construction of a standard curve if quantitative measurement is required.

1. a. For liquid cultures (*see* **Note 1**): Determine the cell density by measuring the OD_{600} or by counting in a hemacytometer, as appropriate. Pellet the cells by centrifugation and wash them once with PBS. Resuspend the samples in PBS (*see* **Note 8**).

b. For adherent cell lines: Harvest using an appropriate method (*see* **Note 9**) and count cells using a hemacytometer. Resuspend the cells in PBS (*see* **Note 8**).
2. Prepare several twofold dilutions of each culture in PBS.
3. For quantitative measurements, construct a standard curve of fluorescence versus amount of GFP. Prepare a serial dilution series of purified recombinant GFP in 10 m*M* Tris-HCl (pH 8.0) (*see* **Note 5**). Place the appropriate filters in the fluorometer (*see* **Note 10**). Set the range of fluorescence values from 0 to 1000. Set the fluorescence of the undiluted sample to a value of 1000. Plot the relative fluorescence units (RFU) as a function of the dilution factor of each sample on a log scale.
4. Measure the fluorescence of the experimental samples. Allow at least 10 s for the fluorometer to report an accurate value, but do not leave samples too long in order to avoid photobleaching of the fluorescence.
5. Normalize the fluorescence according to the optical density of the cultures. For quantitative measurement, use the standard curve to estimate the amount of GFP present in the sample (*see* **Notes 11** and **12**).

3.4. Flow Cytometry

The distribution of fluorescence intensities in a population of GFP-expressing cells can provide information on the population dynamics of gene expression. Flow cytometry allows analysis of this population distribution and subsequent sorting of particular subpopulations *(5,6,8,11,17,25,31–33)*. One disadvantage of flow cytometry for detecting GFP expression is that it has a relatively low sensitivity compared to other methods and autofluorescence can be a problem. Some cell types, particularly those with high concentrations of flavin coenzymes or aromatic compounds, show marked autofluorescence that interferes with visualization of GFP. Mock-transfected cells can be used to gauge the extent of autofluorescence of the cells to be used. The use of GFP variants with absorption maxima at longer wavelengths (e.g., DsRed) can minimize the effects of autofluorescence as cells absorb less energy of longer wavelengths *(27)*. Flow cytometry can be made a quantitative method by the use of commercially available fluorescent microbeads as a standard for calibrating the flow cytometer.

1. a. For liquid cultures (*see* **Note 1**): Determine the cell density by measuring the OD_{600} or by counting in a hemacytometer, as appropriate. Pellet the cells by centrifugation and wash them once with PBS. Resuspend samples in PBS to a cell density of 1×10^6/mL.
   b. For adherent cell lines: Harvest using an appropriate method (*see* **Note 9**). Count the number of cells using a hemacytometer and resuspend the cells in PBS to a density of 2×10^6 cells/mL.
2. Add an equal volume of freshly prepared 2% paraformaldehyde in PBS and incubate for 10 min (*see* **Notes 13** and **14**).
3. Run a negative control sample through the flow cytometer. Adjust the flow rate so that the cell count is optimal for the instrument.
4. If flow cytometry microbead standards are available, follow the manufacturers' instructions to generate a calibration plot.
5. Assay the experimental samples. Use the FL1 emission channel that is normally used to detect FITC fluorescence to monitor GFP fluorescence.
6. Bacteria can be detected by side scatter collected with logarithmic amplifiers. Mammalian cells can be detected with forward scatter using linear amplifiers.

7. Collect data for at least 50,000 cells. Determine the integral mean fluorescence of the population using the flow cytometry software according to the manufacturers' instructions.
8. Wash the flow cytometer tubing with PBS in between assaying different samples. If bacteria are assayed, perform several washes and then assay a buffer-only sample to ensure that no bacteria remain is the tubing (*see* **Note 13**).

4. Notes

1. Grow cells in the appropriate liquid medium to exponential phase: 1×10^8 to 1×10^9 cells/mL for bacteria, 1×10^6 to 1×10^7 for yeast, and 1×10^4 and 1×10^5 for mammalian.
2. Mix 5 µL of liquid culture with 5 µL of molten 1% low-melting-point agarose at 37°C. Spot the mixture directly onto a prewarmed microscope slide and quickly place a cover slip on top. Allow the agarose to solidify at room temperature and examine GFP fluorescence.
3. Alternatively, PBS and 2% paraformaldehyde in PBS may be used in place of DPBS solutions. Fixation of cells with organic reagents results in the partial loss of fluorescent properties of GFP. Fixation with ethanol is not recommended unless GFP is localized to the membrane or a subcellular compartment.
4. Washed cells can be stored at 4°C. The rate of photobleaching is higher in fixed specimen preparations than in nonfixed specimens; thus, fixed cells should be stored protected from light.
5. Diluted samples can be aliquoted and stored at –20°C, protected from light, for up to 1 yr without loss of fluorescence intensity. Avoid repeated freeze–thawing of aliquots.
6. Most fluorescent microscopes are equipped with mercury arc light sources, which are suitable for most applications. However, the light emitted by mercury lamps is not spectrally uniform and this may pose problems when trying to compare the relative intensity of fluorescence when using multiple fluorescent proteins. Mercury lamps also emit relatively high-energy light that may result in faster rates of fluorescence photobleaching, although commercially available GFP derivatives are generally more resistant to this effect (*27*). Xenon arc lamps produce less light that is more spectrally uniform than that emitted from mercury lamps and, thus, are well suited to visualizing all GFP variants.
7. Generally, standard FITC filter sets work well with most second- and third-generation GFP derivatives, although filter sets that are optimized for GFP fluorescence (e.g., Chroma filter set 41017, Chroma Technology, Brattleboro, VT) often give much better results. The optimal choice of filter sets is dependent on both the particular GFP derivative used and the light source. User manuals for commercially available GFP variants usually contain detailed instructions for the choice of filter sets.
8. The number of cells to be assayed varies depending on the level of GFP present in the sample. This will, in turn, be dependent on a number of factors, including the copy number of the GFP gene in the cells assayed and the level of GFP expression in the cells. For each GFP reporter, the number of cells required to produce fluorescence in the detectable range must be determined empirically. Relative fluorescence can then be normalized to a per cell basis in order to compare between samples.
9. The method of harvest is dependent on the cell type used and the reader should refer to an appropriate source for information.
10. Use an excitation filter of 365 nm and an emission filter of 510 nm for assaying wild-type GFP or *GFPuv*. Use an excitation filter of 450–490 nm and an emission filter of 510 nm for assaying EGFP, EYFP, ECFP, rsGFP, and their derivatives. Use an excitation filter of 350–380 nm and an emission filter of 420–470 nm when assaying BFP.
11. If the fluorescence of the sample is less than 10, set the fluorescence of the first dilution of the standard curve series to 1000, adjust the standard curve accordingly, and reassay

the sample. If the fluorescence of the sample is greater than 1000, prepare and assay dilutions of the sample.

12. To increase the sensitivity of the assay, measure the fluorescence of cell lysates. The protein concentration of the cell lysates can be measured by an appropriate method and then the amount of GFP present in each sample can be expressed as a fraction of the total cellular protein.

13. Bacteria are readily deposited in the flow cytometer tubing. Fixing cells prior to flow cytometry is recommended to prevent contamination of the flow cytometer. However, if cell sorting of viable subpopulations is desired, then decontamination of the flow cytometer can be achieved by washing the tubing of the flow cytometer once with 1% sodium dodecyl sulfate followed by thorough washing with PBS.

14. Eukaryotic cells may be stained with 0.5 µg/mL propidium iodide (PI) in order to differentiate between live and dead cells (do not use in conjunction with fixation in which case all cells are dead). PI is sensitive to light; thus, cells stained with PI should be kept on ice and in the dark until FACS analysis. Use the FL3 emission channel to detect PI fluorescence.

References

1. Shimomura, O., Johnson, F. H., and Saiga, Y. (1962) Extraction, purification and properties of aequorin, a bioluminescent protein from the luminous hydromedusan, *Aequorea. J. Cell Comp. Physiol.* **59,** 223–227.
2. Ormö, M., Cubitt, A. B., Kallio, K., et al. (1996) Crystal structure of the *Aequorea victoria* green fluorescent protein. *Science* **273,** 1392–1395.
3. Cody, C. W., Prasher, D. C., Westler, W. M., et al. (1993) Chemical structure of the hexapeptide chromophore of *Aequorea* green-fluorescent protein. *Biochemistry* **32,** 1212–1218.
4. Heim, R., Prasher, D. C., and Tsien, R. Y. (1994) Wavelength mutations and posttranslational autoxidation of green fluorescent protein. *Proc. Natl. Acad. Sci. USA* **91,** 12,501–12,504.
5. Ausubel, F. M., Brent, R., Kingston, R. E., et al. (1994) *Current Protocols in Molecular Biology*, Wiley, New York.
6. Chalfie, M. and Kain, S. (eds.) (1998) *Green Fluorescent Protein: Properties, Applications, and Protocols*, Wiley-Liss, New York.
7. Margolin, W. (2000) Green fluorescent protein as a reporter for macromolecular localization in bacterial cells. *Methods* **20,** 62–72.
8. Cormack, B. P. and Struhl, K. (1993) Regional codon randomization: defining a TATA-binding protein surface required for RNA polymerase III transcription. *Science* **262,** 244–248.
9. Doyle, T. and Botstein, D. (1996) Movement of yeast cortical actin cytoskeleton visualized in vivo. *Proc. Natl. Acad. Sci. USA* **93,** 3886–3891.
10. Webb, C. D., Decatur, A., Teleman, A., et al. (1995) Use of green fluorescent protein for visualization of cell-specific gene expression and subcellular protein localization during sporulation in *Bacillus subtilis. J. Bacteriol.* **177,** 5906–5911.
11. Cormack, B. P., Valvidia, R. H., and Falkow, S. (1996) FACS-optimized mutants of the green fluorescent protein (GFP). *Gene* **173,** 33–38.
12. Delagrave, S., Hawtin, R. E., Silva, C. M., et al. (1995) Red-shifted excitation mutants of the green fluorescent protein. *BioTechnology* **13,** 151–154.
13. Heim, R., Cubitt, A.B., and Tsien, R.Y. (1995) Improved green fluorescence. *Nature* **373,** 663–664.

14. Heim, R. and Tsien, R.Y. (1996) Engineering green fluorescent protein for improved brightness, longer wavelengths and fluorescence resonance energy transfer. *Curr. Biol.* **6**, 178–182.
15. Li, X., Zhao, X., Fang, Y., et al. (1998) Generation of destabilized enhanced green fluorescent protein as a transcription reporter. *J. Biol. Chem.* **273**, 34,970–34,975.
16. Haseloff, J. and Amos, B. (1995) GFP in plants. *Trends Genet.* **11**, 328–329.
17. Hawley, T. S., Telford, W. G., Ramezani, A., et al. (2001) Four-color cytometric detection of retrovirally expressed red, yellow, green, and cyan fluorescent proteins. *BioTechniques* **30**, 1028–1034.
18. Chamberlain, C. and Hahn, K. M. (2000) Watching proteins in the wild: fluorescence methods to study protein dynamics in living cells. *Traffic* **1**, 755–762.
19. Roessel, P. and Brand, A. H. (2002) Imaging into the future: visualizing gene expression and protein interactions with fluorescent proteins. *Nature Cell Biol.* **4**, E15–E20.
20. Reits, E. A. J. and Neefjes, J. J. (2001) From fixed to FRAP: Measuring protein mobility and activity in living cells. *Nature Cell Biol.* **3**, E145–E147.
21. Yang, T. T., Kain, S. R., Kitts, P., et al. (1996) Dual color microscopic imagery of cells expressing the green fluorescent protein and a red-shifted variant. *Gene* **173**, 19–23.
22. Patterson, G. H., Knobel, S. M., Sharif, W. D., et al. (1997) Use of the green fluorescent protein and its mutants in quantitative fluorescence microscopy. *Biophys. J.* **73**, 2782–2790.
23. Cheng, L., Fu, J., Tsukamoto, A., et al. (1996) Use of green fluorescent protein variants to monitor gene transfer and expression in mammalian cells. *Nat. Biotechnol.* **14**, 606–609.
24. Green, G., Kain, S. R., and Angres, B. (2000) Dual color detection of cyan and yellow derivatives of green fluorescent protein using conventional fluorescence microscopy and 35-mm photography. *Methods Enzymol.* **327**, 89–94.
25. Sullivan, K. F. and Kay, S. A. (1998) *Methods in Cell Biology: Green Fluorescent Proteins*, Academic, San Diego, CA.
26. Tombolini, R. and Jansson, J. K. (1998) Monitoring of GFP tagged bacterial cells. *Methods Mol. Biol.* **102**, 285–298.
27. Feilmeier, B. J., Isemiger, G., Schroeder, D., et al. (2000) Green fluorescent protein functions as a reporter for protein localization in *Escherichia coli*. *J. Bacteriol.* **182**, 4068–4076.
28. Morin, J. G. and Hastings, J. W. (1971) Energy transfer in a bioluminescent system. *J. Cell Physiol.* **77**, 313–317.
29. Wang, S. and Hazelrigg, T. (1994) Implications for bed mRNA localization from spatial distribution of exu protein in *Drosophila* oogenesis. *Nature* **369**, 400–403.
30. Ward, W. W., Cody, C. W., Hart, R. C., et al. (1980) Spectrophotometric identity of the energy-transfer chromophores on *Renilla* and *Aequorea* green-fluorescent proteins. *Photochem. Photobiol.* **31**, 611–615.
31. Chalfie, M., Tu, Y., Euskirchen, G., et al. (1994) Green fluorescent protein as a marker for gene expression. *Science* **263**, 802–805.
32. Ropp, J. D., Donahue, C. J., Wolfgang-Kimball, D., et al. (1995) *Aequorea* green fluorescent protein analysis by flow cytometry. *Cytometry* **21**, 309–317.
33. Valvidia, R. H., Hromockyj, A. E., Monack, D., et al. (1996) Applications for green fluorescent protein (GFP) in the study of host–pathogen interactions. *Gene* **173**, 47–52.

Index

A

Affinity purification,
recombinant proteins, 270,
271, 283, 285
Agarose gel electrophoresis,
buffers, 139, 178, 179
colony screening, 170,
172–173
DNA standards, 179–180
extraction of DNA, 137–139,
159
low-melt, 138
plasmid analysis, 77, 78,
175–177
resolution, 179, 180
size fractionation of genomic
DNA, 158
staining,
ethidium bromide, 179,
180–181
GelStar, 180
methylene blue, 180
Alkaline lysis plasmid isolation,
76, 92–93
principle, 75
yield, 75
Alkaline phosphatase,
dephosphorylation of DNA,
123–124, 129–130
promoter, expression vectors,
261
reporter gene assay, 294
substrate, 293
troubleshooting, 294–295
α-Complementation. *See* Blue-
white selection

Ampicillin resistance, 23
Antibiotic resistance. *See*
Plasmids, traits encoded by;
Selectable markers
Aph, 23
Arcing, 57, 58
Arabinose operon, 263–264
expression vectors, 264
Autoradiography, 188, 191

B

Bacterial artificial chromosome
(BAC), 99
copy number, 22
culture, 100, 101
DNA isolation, 100–101
high-throughput, 93–94
commercial kits, 101,
242–243
RNA contamination, 99, 102
yield, 101
E. coli hosts, 240–242
electroporation, 242–244
insert size, 21
propagation, 243–244
sequencing, 197–198
storage, 242
transposon mutagenesis, 234,
239–240, 241–242
Bacteriophage lambda. *See* λ
phage
Bacteriophage. *See also* Infection
extraction of phage particles,
104–105, 164
titering, 104, 163
*Bal*31 nuclease, 225

BIACORE, 248
β-Galactosidase, 277. *See also*
  Blue-white selection
  reporter gene, 290, 291
    assay, 294
    substrates, 292
    troubleshooting, 294–295
β-Glucuronidase, 290
β-Lactamase, 23
Blue-white screening,
  principle, 38–39, 170
  host strains, 39
  preparing agar plates, 172, 173–174
Blunt-ending DNA termini,
  overhang fill-in, 122, 123
    Klenow DNA polymerase, 128
    T4 DNA polymerase, 128–129
  overhang polishing, 123
    Klenow DNA polymerase, 219
    T4 DNA polymerase, 129, 220
Boiling lysis plasmid isolation, 80
  host strains, 79
  outline, 79
5-Bromo-4-chloro-3-indoxyl-β-D-galactopyranoside. *See* X-Gal

C

Calf intestinal alkaline phosphatase, 123, 129
Canine microsomal membranes, 250
CAT. *See* Chloramphenicol acetyltransferase
Catabolite activator protein, 278, 280
CcdAB, 11
Chi site, 38
Chloramphenicol acetyltransferase,
  reporter gene, 290
  chloramphenicol resistance, 24
Chromosomal DNA. *See* Genomic DNA
CIAP. *See* Calf intestinal alkaline phosphatase
Cleavage of proteins, 101
Cloning, 24. *See also* DNA library construction; PCR, cloning products
  blunt-ending termini,
    overhang fill-in, 122, 123, 128–129
    overhang polishing, 123, 129
  compatible ends, 122
  dephosphorylation, 123, 129–130
  ligation, 124–125, 130, 133–134
  partial fill-in, 154–155, 162
  phosphorylation, 123, 130
  restriction site introduction, 122–123
    linkers, 127–128
    PCR products, 128, 142
  software, 107–120
Cloning vectors,
  copy number, 22
  incompatibility, 22
  insert size, 19–21
  multiple cloning sites, 24–25
  selectable markers, 22–24
Codon usage, 41, 44

ColE1, 4
  copy number, 2, 22, 41
  control, 7
  incompatibility, 22
  replication, 6–7
Colony screening, 169–170
  blue-white selection, 170, 172
  direct electrophoresis, 170, 172–173
  hybridization, 184–194
  PCR, 170–171, 173
  restriction analysis, 175–181
Competent cells. *See also* Electrocompetent cells
  commercial, 52
  preparation, 50–52
  transformation, 52
  storage, 52
Conjugation,
  filter-mating, 63
  frequency, 64
  liquid-mating, 63
  mechanism, 61–62
  natural, 12, 61
  plasmids, 61–63
  transposon delivery, 234–236
  transposons, 61
Coomassie blue stain, 285
Copy number, 3, 22
  control, 6–7
  expression vectors, 261
  reporter gene vectors, 290
Cosmids, 19, 67
  cos sites, 67
  DNA isolation, 100–101
    high-throughput, 93–94
    RNA contamination, 99, 102
    yield, 101
  infection of *E. coli*, 68–69, 71
  insert size, 20
  library construction, 67, 69, 70–71
  packaging, 68, 71
Crosslinking DNA, to nitrocellulose, 186, 190
CspA, 265

D

Dam, 35–36
Dcm, 35–36
Dephosphorylation of DNA, 123, 129–130
DNA,
  molar conversion, 132
  spectrophotometric quantification, 77
  non-radioactive-labeling, 187
  radioactive-labeling, 186–187, 190
DnaA, 5
DNA gyrase, 3, 11
DnaJ/K, 44
DNA library construction,
  cosmids, 67–70
    ligation, 70–71
    packaging, 71
    host strains, 70, 72
    insert preparation, 157–162
  λ vectors, 153–155
    amplification, 165
    ligation, 162
    packaging, 162
    storage, 162
    validation, 163–165, 167
DNA library screening,
  hybridization, 184, 187–188, 191
  membrane preparation, 184–186, 189–190
  probes, 186–187, 190
  immunodetection, 183–184

phenotypic, 183
DNA methylation, 35–37
DNA polymerases,
   Klenow,
      blunt-ending termini,123, 128
      primer extension, 186–187, 190
   T4,
      blunt-ending termini, 123, 128–129
   thermostable. *See* PCR
DNA precipitation,
   ethanol, 76, 198
   isopropanol, 80, 127, 131
DNA purification. *See also* Plasmid isolation
   from agarose gels, 137–139
   Geneclean, 217–218
   phenol : chloroform extraction, 76, 127
   Sephadex beads, 138, 199
DNase I, 225
DNA sequencing,
   automated, 196
   dideoxy method, 195
   fluorescent, 196, 197–198
   polyacrylamide gels, 199–200
   radioactive, 195
   troubleshooting, 200–201
*Dpn*I, 212, 217
dut ung hosts, 39

E

E. coli S-30 extract, 249–250, 251
E. coli. *See also* Conjugation; Electroporation; Infection; Transformation
   auxotrophy, 32
   glycerol archive, 92, 95
   growth of, 51
   growth rate, 27
   hosts for BACs, 242
   hosts for cloning, 32–39
   hosts for expression, 39, 41–44
   hosts for infection, 72
   hosts for library construction, 70
   hosts for mutagenesis, 39, 40
   hosts for plasmid isolation, 75
   hosts for protein expression, 282
   K-12, 27, 44
   recombination, 37–38
   restriction, 35–37
   suppliers, 32
*Eco*K, 36
Electrocompetent cells,
   commercial, 56
   preparation, 57
   storage, 57
Electrophoresis. *See* Agarose gel electrophoresis; Polyacrylamide gel electrophoresis
Electroporation, 57
   apparatus, 56
   BACs, 242
   efficiency, 55–56, 58, 242
   mechanism, 55
elk, 44
Endonuclease A, 75, 78
Enterokinase, 271
Escherichia coli. *See E. coli*
Ethanol precipitation. *See* DNA precipitation
Ethidium bromide,
   DNA estimation, 77
   safety, 180
   staining agarose gels, 179, 180–181
   stock solution, 178

Index

Exonuclease III,
  activity, 225–226, 229–231
  digestion of DNA, 228
Expression vectors. *See also*
    Protein expression in *E. coli*
  copy number, 261
  eukaryotic genes, 258, 272
  fusion tags, 266–270
  promoters, 261–265, 280–281
  transcription vector, 257–258
  translation vector, 258

F

Factor Xa, 271
Fertility factor. *See* F plasmid
F plasmid, 2
  active partition, 21
  conjugative transfer, 11–12, 61, 62
  nomenclature, 32, 35
  origin of transfer, 62
  toxin-antitoxin, 11
  vectors. *See* Bacterial artificial chromosome
Flow cytometry,
  autofluorescence, 301
  GFP, 301–302
  live/dead staining, 303
Fluorescence microscopy, GFP, 300
  filter sets, 298, 302
  light source, 302
  quantification, 300
Fluorometry, GFP, 300–301
  filter sets, 302
  photobleaching, 298, 301
  quantification, 301
  sensitivity, 302–303
FRAP. *See* Green fluorescent protein, applications

FRET. *See* Green fluorescent protein, applications
FRETWORKS assay, 270

G

gam, 38
Gel extraction, 137
  buffer exchange, 138
  low-melt agarose, 138, 159
  spin protocol, 138–139
GelStar, 180
Geneclean, 217–218
Gene Construction Kit, 108–120
Genomic DNA,
  isolation, 157–158
  ligation into λ vector arms, 162
  partial digestion, 158, 160
Genotype, 27–31
  nomenclature, 31–32
GFP. *See* Green fluorescent protein
Glutathione-S-transferase, 271, 283
gor, 44
GroES/L, 44, 271
Green fluorescent protein (GFP),
  applications, 297
    fluorescence recovery after photobleaching (FRAP), 298
    fluorescence resonance energy transfer (FRET), 298
  assays,
    flow cytometry, 301–302
    fluorescence microscopy, 300
    fluorometry, 300–301
    troubleshooting, 298, 301–302
    UV illumination, 299
  derivatives, 298
  spectral properties, 297
  thermosensitivity, 298
GUS. *See* β-Glucuronidase

H

Helper phage, 103
High frequency recombinant (Hfr), 35
  chromosome mapping, 62
  conjugation, 61
Histidine tag, 270, 283, 285
Host strains. See E. coli
hsdRMS, 36
Hybridization, 184, 187–188, 191
  troubleshooting, 192–194
  washing conditions, 188, 191, 192

I

Imidazole, 285, 286
Immunodetection, library screening, 183–184
Immunoprecipitation, recombinant proteins, 270
Inclusion bodies, 266
Incompatibility, 22
Infection,
  cosmids, 71
  efficiency, 71–72
  host strains, 70
  natural, 11
  preparation of E. coli, 71
  principle, 68–69
  recombinant λ phage, 163
Insoluble proteins, 266, 282–283
  chaperones, 44, 271
  media, 285
  tags enhancing solubility, 270, 271
  temperature, 286
Integrating plasmids, 290–291
Intein, 271
Inverse PCR. See Site-directed mutagenesis
IPTG,
  for blue-white selection, 39
  induction protein expression, 261, 265, 278, 281, 284
  preparing agar plates, 172, 173–174
Isopropanol precipitation. See DNA precipitation
Isopropyl-β-D-thiogalactoside. See IPTG
Iterons, 5–6

K

Kanamycin resistance, 23
  reporter gene, 290
Klenow DNA polymerase,
  blunt-ending DNA termini, 123, 128, 219
  partial fill-in, 154–155, 162
  primer extension, 186–187, 190
Kunkel mutagenesis, 39

L

Labeling,
  DNA,
    non-radioactive, 187
    radioactive, 186–187, 190
    random-primed, 186–187, 190
  proteins, 248, 253
  RNA, 248, 252
Lac operon, 36–37, 277–278. See also Expression vectors; Protein expression in E. coli
  α-complementation. See Blue-white selection
  lac promoter, 41, 262, 278
  LacI, 262, 277–278
  LacI^q, 280, 281, 282
  lacUV5 promoter, 262, 280
  LacY, 41, 262, 277, 293

LacZ, 277. *See also* β-
  galactosidase
*lacZ*ΔM15, 39
lamB, 68
λ phage, 151. *See also* Bacteriophage
  DNA purification, 164
  commercial kits, 167
  E. coli hosts, 21, 38
  infection of *E. coli*, 163
  lifecycle, 21, 153
  packaging extract, 68, 162
  promoter, 265
  recombination, 38
  replication, 21
  vectors. *See also* DNA library
    construction
    insert size, 21
    lambdagen-12, 153–154
Large insert clones. *See* Bacterial
  artificial chromosome;
  Cosmid
Lateral gene transfer. *See*
  Conjugation
Library construction. *See* DNA
  library construction
Ligation, 124–125,
  blunt-end, 130, 134
  cosmids, 70–71
  linkers, 127–128
  self-ligation, 123
  sticky-end, 130, 134
Linkers, 123, 127–128
Lipopolysaccharide,
  contamination of plasmid
    DNA, 83, 85
lon, 41
LPS. *See* Lipopolysaccharide
Luciferase, 290
LuxAB, 290
Lysogeny, 21, 153

Lysozyme,
  in plasmid isolation, 77–78,
    80, 100
  stock solution, 81

M

M13 vectors. *See also* Phagemids
  mutagenesis, 209
  propagation, 94
  replication, 7
  single-strand DNA rescue, 35
    high-throughput, 94
Maltose binding protein, 265, 271, 283
Marker-exchange mutagenesis, 62
Mating assay. *See* Conjugation
Mcr, 36–37
Megaprimer mutagenesis. *See*
  Site-directed mutagenesis
Methylation of DNA, 35–37
Methylene blue, 180
Mobilization. *See* Conjugation
Mrr, 36–37
Multiple cloning sites, 24–25
Mutagenesis,
  marker-exchange, 62
  nested deletions, 225–231
  site-directed,
    inverse PCR, 210–223
    megaprimer, 203–207
  transposon, 233–244
mutD, 39

N

Nickel affinity purification, 270
  Ni-NTA resin, 283, 285
Nitrocellulose membranes,
  colony growth, 184–186,
    189–190
  crosslinking DNA, 186, 190
Nonsense mutations, 32

O

Oligonucleotide annealing, 149
Origins of replication, 5–7, 22
oriT, 12, 62

P

P1, partition, 9
p15A,
  compatibility, 22
  copy number, 22
pACYC, 20
  antibiotic resistance, 23, 24
  compatibility, 22
  copy number, 22
Packaging in vitro,
  cosmids, 71
  phagemids, 103
  λ extracts, 68
    storage, 72
  recombinant λ phage, 162
Partition, 7, 9–10
  F plasmid, 21
  P1, 9
pBAD, 258, 264, 270
pBluescript, 20, 103, 282
  phage promoters, 25
pBR322, copy number, 22
PCR, 141
  cloning products,
    blunt-end ligation, 142
    restriction endonuclease-directed, 122–123, 128, 142
    TA cloning, 142–151
  colony screening, 170–171, 173
  thermostable DNA polymerases,
    fidelity, 206, 221
    strand-displacement activity, 212
    terminal transferase activity, 144–145, 150
  inverse. *See* Site-directed mutagenesis
  mutagenesis. *See* Site-directed mutagenesis
pET, 258, 261, 265, 267, 271
Pfu DNA polymerase,
  fidelity, 206, 221
  strand-displacement activity, 212
Phage. *See* Bacteriophage; λ phage
Phagemids, 25, 103
  helper phage, 103
  isolation of single-stranded DNA, 104–105
  packaging, 103
Phenol : chloroform extraction, 76, 127
Phenotype, 27, 31
PhoA. *See* Alkaline phosphatase
Phosphorylation of DNA, 123, 130
Pilus, 12, 62
pir, 234, 290–291
Plasmid maps, software, 107–120
Plasmids,
  conjugative, 61–63
  cryptic, 3
  definition, 1
  dimerization, 9
  dissemination, 3, 11–12
  evolution, 12–13
  incompatibility, 22
  integrating, 290–291
  linear, 3
  natural, 1–5
  partition, 7, 9–10, 21
  recombination, 9, 11
  replication, 5–8, 21–22

segregation, 7, 9–11
temperature-sensitive, 236
topology, 3
traits encoded by, 3–5
  antibiotic resistance, 1, 3
  virulence genes, 4, 13
Plasmid isolation. *See also*
  Bacterial artificial
  chromosome, isolation;
  Cosmid, isolation
  alkaline lysis, 75–76, 84–85,
    100–101
  boiling lysis, 79–80
  E. coli hosts, 75
  high-throughput, 92–93
    equipment, 89–91, 95
    yield, 83
  lipopolysaccharide
    contamination, 83, 85
  low-copy-number, 101
  RNA contamination, 78, 99,
    102
  silica oxide-based, 84–85
    chaotropic agents, 83–84
    large-scale, 86
pLysS/E, 41, 265, 280, 282
pMAL, 267, 271
Polishing. *See* Blunt-ending DNA
  termini
Polyacrylamide gel
  electrophoresis,
  for DNA sequencing,
    preparation of gels, 199–200
    resolution, 195
  proteins, 284–285
    staining, 285
  RNA, 252
Polymerase chain reaction. *See*
  PCR
Promoters,

for recombinant protein
  expression, 261–265,
  280–281
for RNA synthesis, 25, 247
Propidium iodide staining, 303
Proteinase K, 101
Protein cleavage, 271
Protein expression in *E. coli*, 39,
  41–44, 284. *See also*
  Expression vectors
  optimization, 281–282
  troubleshooting, 285–286. *See
  also* Toxic proteins;
  Insoluble proteins
Protein expression in vitro, 253
  applications, 249
  commercial kits, 249–251
  coupled system, 251
  labeling, 248
  post-translational processing,
    250
  preparative scale, 251
  template preparation,
    DNA, 253–254
    RNA. *See* Transcription in
      vitro
Protein purification, 267–270,
  283–285
pSC101, 261
  copy number, 22
ptsM, 68
pUC, 20, 24
  copy number, 22

Q

Quikchange mutagenesis,
  220–221
  efficiency, 223
  primer design, 220, 221–222
  principle, 212–214

R

R6K gamma origin, 234, 290–291
  compatibility, 22
Rabbit reticulocyte lysate, 250, 253
Random-prime labeling, 186–187, 190
Recombination,
  in *E. coli*,
    recA, 37
    recBCD, 37–38
  λ vectors, 38
  plasmid resolution, 9, 11
red, 38
Replication, 12
  ColE1, 6–7, 22
  handcuffing, 6
  iterons, 5–6
  rolling-circle, 7, 8, 21
Replicon, 5–7, 22
Reporter genes, 289
  alkaline phosphatase, 290, 293–294
  β-galactosidase, 290, 291–293, 294
  β-glucuronidase, 290
  chloramphenicol acetyltransferase, 290
  green fluorescent protein, 290, 297–303
  kanamycin resistance, 290
  luciferase, 290
Reporter gene vectors,
  choosing, 291
  copy number, 290
  fusions, 291, 297
  integrative, 290–291
  subcellular localization, 293
  transcription terminators, 291
Restriction endonucleases,
  methylation sensitivity, 179
  Type II, 122
  Type IIS, 210, 222
  units, 180
Restriction digestion, , 178
  buffers, 179–180
  compatible ends, 122
  complete, 126
  double, 126–127
  hemimethylated DNA, 212, 217
  inhibitors, 77–78
  partial, 121, 127
    genomic DNA, 158
  plasmid analysis, 175–177
  software, 114–115, 118
  troubleshooting, 131
Restriction-modification systems, 35–37
Ribonuclease A (RNaseA), 76, 78, 100
RNA synthesis. *See* Transcription in vitro
rne, 41
Rolling-circle replication, 7, 8
RP4, conjugation, 61, 63
rpoH, 41
rrnB, 291
RSF1010, conjugation, 61

S

S1 nuclease,
  activity, 226
  digestion of DNA, 228–229
S-30 extract, 249–250, 251
SAP. *See* Shrimp alkaline phosphatase
Screening colonies. *See* Colony screening; DNA library screening

Secretion of recombinant
  proteins, 267
Segregation, 7, 9–11
Selectable Markers, 22
  antibiotic resistance, 22–24
Sephadex, 138, 199
Sequencing DNA. *See* DNA
  sequencing
Shrimp alkaline phosphatase,
  123, 129–130, 133
Signal sequence, 267
Silica oxide, 83–84
Single-stranded DNA, 103
  isolation, 104–105
  high-throughput, 94
Site-directed mutagenesis,
  inverse PCR, 216–221
    efficiency, 223
    primer design, 216, 218,
      220, 221–222
    principle, 209–215
    troubleshooting, 221–223
  megaprimer, 205–206
    efficiency, 204, 207
    primer design, 205
    principle, 203–204
    troubleshooting, 206–207
    single-strand DNA template,
      209
SP6 RNA polymerase, 25, 247
Spectrophotometry,
  bacterial density, 51
  DNA analysis, 77
Subcellular localization,
  recombinant proteins,
    cytoplasm, 266
    extracellular milieu, 267
    periplasm, 266–267
    reporter genes, 293, 297
Suppressor mutations, 32, 35

T

T4 bacteriophage,
  DNA ligase, 124–125, 128,
    130
  DNA polymerase, 123,
    128–129, 220
  polynucleotide kinase,
    123–124, 130, 133
T7 bacteriophage,
  lysozyme, 41, 265, 282
  promoter, 25, 265, 281
  RNA polymerase, 41, 247,
    265, 281
  T7*lac* promoter, 281, 282
TA cloning, 142–151, 258
tac promoter, 263, 280
Talon resin, 283
Taq DNA polymerase,
  fidelity, 221
  terminal transferase activity,
    144–145, 150
Temperature-sensitive promoters,
  265
Tetracycline resistance, 24
Thrombin, 271
Tn7,
  in vitro mutagenesis, 236–239,
    241
  structure, 236–237
Tn5, 236
  transposome mutagenesis, 239,
    241–242
Toxic proteins, 265, 282
  plasmid copy number, 261
  in vitro synthesis, 249
Toxin-antitoxin systems, 11
Tra, 12, 35, 62
Transconjugants, 61
Transcription in vitro, 252
  applications, 248

labeling, 248
promoters, 247, 251
RNA polymerases, 247–248, 252
secondary structure, 247–248
transcript purification, 252–253
Transcription terminators, 291
Transduction. *See* Infection
Transformation, 52. *See also* Electroporation
competent cells, 50–52
efficiency, 50–51, 53
natural, 11, 49
mechanism, 49
Translation. *See* Protein expression
Transposome, 239. *See also* Transposon, mutagenesis
Transposons,
antibiotic resistance, 3
class I, 233
class II, 233
conjugative, 61
mutagenesis, 241–242
principle, 233–239
troubleshooting, 242–244
Transprimer, 236. *See also* Transposon, mutagenesis
trc promoter, 263, 280
trp promoter, 264
trxB, 44, 271
T vectors, 142
commercial, 143
custom,

Taq-directed, 144–145, 150
*Xcm*I-directed, 144, 149–150
ligation, 150, 151

U

UP elements, 266

V

Vectors. *See* Bacterial artificial chromosome; Cloning vectors; Cosmids; Expression vectors; M13 vectors; Phagemids; Reporter gene vectors
Vent DNA polymerase, fidelity, 221

W

Weiss units, 133
Western blot, 270

X

*Xcm*I, 144–146
XerC/D, 11
X-Gal,
for blue-white selection, 39
preparing agar plates, 172, 173–174
reporter gene assay, 292

Y

Yeast artificial chromosome (YAC), 99